Quantal Response Equilibrium

Quantal Response Equilibrium

A Stochastic Theory of Games

JACOB K. GOEREE

CHARLES A. HOLT

THOMAS R. PALFREY

PRINCETON UNIVERSITY PRESS
Princeton and Oxford

Copyright © 2016 by Princeton University Press
Published by Princeton University Press, 41 William Street,
Princeton, New Jersey 08540
In the United Kingdom: Princeton University Press, 6 Oxford Street,
Woodstock, Oxfordshire OX20 1TW

press.princeton.edu

ISBN: 978-0-691-12423-0

Library of Congress Control Number: 2015958001

British Library Cataloging-in-Publication Data is available

This book has been composed in Times Roman

Printed on acid-free paper. ∞

Printed in the United States of America
Typeset by Nova Techset Pvt Ltd, Bangalore, India

10   9   8   7   6   5   4   3   2   1

Contents

Preface

This book project began a decade ago, when the three of us decided that the accumulated research on quantal response equilibrium, scattered around various journals in economics and political science, would benefit from an organized presentation in one volume. As it turns out, this accumulation has continued unabated over the last ten years and continues to build. On the one hand it is a relief to know that a research agenda the three of us have spent a big chunk of our careers working on is still alive and kicking, but on the other hand it means that as the book evolved there was always a temptation to include new material that did not exist when we embarked on the project. At this point, with some helpful nudging from our incredibly patient publishing editor, we have reached an imaginary finish line and hope that readers benefit from the final product.

While the book project may have started only ten years ago, the ideas presented here have been developing in our minds for more than double that time, and are foreshadowed by research by econometricians and game theorists going back fifty years or more. The original paper on quantal response equilibrium has its roots in two landmark papers: one in econometrics (McFadden, 1976), and the other in game theory (Harsanyi, 1973). In fact, McFadden's terminology of "quantal choice" provides the etymological basis for the equilibrium concept that is the subject of this book. These and other connections with precursors of quantal response equilibrium are discussed and explained in the book.

The material presented here would not have existed without the many excellent colleagues who have collaborated with us. Above all is Richard McKelvey, to whose memory we dedicate the book. He was a scholar's scholar, endowed with a remarkable blend of creativity and modesty. Richard was not only a founder of quantal response equilibrium theory, but also played a key role in the early development of methods for computing and estimating quantal response equilibrium models (Palfrey, 2007).

Much of the material in this book draws on collaborative work on quantal response equilibrium and closely related subject matter with many of our coauthors. The long list includes Lisa Anderson, Simon Anderson, Marco Battaglini, AJ Bostian, Jordi Brandts, Christoph Brunner, Colin Camerer, Monica Capra, Juan Carrillo, David Clark, Pete Coughlan, Mahmoud El-Gamal, Mark Fey, Rosario Gomez, Serena Guarnaschelli, Helios Herrera, Susan Laury,

David Levine, Philippos Louis, Massimo Morelli, Rebecca Morton, Timothy Nordstrom, Salvatore Nunnari, Kirill Pogorelskiy, Jeffrey Prisbrey, William Reed, Brian Rogers, Howard Rosenthal, Katri Sieberg, Angela Smith, Roberto Weber, and Jingjing Zhang. Many graduate and undergraduate students, too many to list here, provided invaluable assistance in computational and numerical aspects, as well as helping collect experimental data. Without the published work generated by these collaborations, this book would not exist, and our CVs would be much shorter. We are truly lucky to have had so many excellent coauthors and research assistants. A heartfelt thanks to each and every one of you.

More generally, we have benefited from interactions with many colleagues in the profession, especially in the experimental economics community, even if we weren't fortunate enough to have had formal collaborations on the topic of quantal response equilibrium. In this regard we express our gratitude to (without implicating) Olivier Armantier, Isabelle Brocas, Alessandra Casella, Tim Cason, Syngjoo Choi, Jim Cox, Vince Crawford, André de Palma, John Duffy, Ernst Fehr, Daniel Friedman, James Friedman, Drew Fudenberg, Douglas Gale, Jens Grosser, Glenn Harrison, John Kagel, Shachar Kariv, John Ledyard, Daniel Levin, Daniel McFadden, Rosemarie Nagel, Charles Plott, Alvin Roth, Andy Schotter, Arthur Schram, Curtis Signorino, Vernon Smith, Dale Stahl, Jacques Thisse, and Ted Turocy.

Of these, we owe a special thanks to Tim Cason, Jens Grosser, Kirill Pogorelskiy, Andrew Shotter, and Ted Turocy for taking time to read all or parts of earlier versions of the book. Their insightful comments helped enormously as we tried to polish those early drafts into something resembling a well-written book. We surely did not succeed in every respect, and any shortcomings are ours and ours alone. Special thanks also go to Philippos Louis, Xavier del Pozo, and Jingjing Zhang for reading the manuscript and preparing the book's figures and references.

Because one of our goals in writing the book is to bring together and coherently organize many scattered results about quantal response equilibrium that have been published in journal articles, we naturally have drawn on material from some of our previously published articles. Specifically, it reuses selected parts of the following articles, with minor changes of notation and exposition where needed: chapter 2: Anderson, Goeree, and Holt (2001, 2002); Goeree, Holt, and Palfrey (2005); McKelvey and Palfrey (1995, 1996); chapter 3: McKelvey and Palfrey (1998); chapter 4: Goeree and Holt (2004); Rogers, Palfrey, and Camerer (2009); chapter 5: Anderson, Goeree, and Holt (2004); Goeree and Holt (2004); chapter 7: McKelvey and Palfrey (1998); Capra et al. (1999); Goeree et al. (2007); Goeree, Holt, and Palfrey (2008); Carrillo and Palfrey (2009); chapter 8: Guarnaschelli, McKelvey, and Palfrey (2000); Goeree and Holt (2005b); Levine and Palfrey (2007); Battaglini and Palfrey (2012); Sieberg et al. (2013); chapter 9: Anderson, Goeree, and Holt (1998a); Capra et al. (2002); Goeree, Holt, and Palfrey (2002); Camerer, Nunnari, and Palfrey (2014). Sections of chapters that rely heavily on excerpted material are noted where appropriate. We are grateful to the publishers and professional organizations who own the copyrights for allowing us to reuse published material, and to Catriona Reid for obtaining the permissions.

We are grateful for the patience and encouragement of the editors and staff at Princeton University Press, which has a long history of promoting game theory research that dates back to the publication of von Neumann and Morgenstern's (1944) *Theory of Games and Economic Behavior*. They enthusiastically agreed to publish the book ten years ago, and then waited a long time for final delivery. In particular, we would like to thank Seth Ditchik, Peter Dougherty, Tim Sullivan, and Samantha Nader for their support and editorial advice. Finally, we will always remember the enthusiastic, contagious energy of the late Jack Repcheck, who was so instrumental in publishing related work on experimental economics and game theory while at Princeton, and later at Norton Press.

The research embodied in this book could not have been accomplished without generous financial support from many different organizations. Much of it was supported through grants each of us has received from the Economics and Political Science programs of the Australian Research Council and the US National Science Foundation over the last two decades,[1] and Goeree received significant funding from the European Research Council (ERC Advanced Investigator Grant, ESEI-249433), Holt from the Bankard Fund and Quantitative Collaborative at the University of Virginia, and Palfrey from the Gordon and Betty Moore Foundation at Caltech. Our home universities provided us with both financial and infrastructure support, as well as a nurturing scientific environment. For that we thank the California Institute of Technology, Princeton University, the University of Technology Sydney, and the University of Virginia. Thomas Palfrey is deeply grateful to the Russell Sage Foundation for providing financial support that allowed him to spend the 2014–15 academic year in residence there as a Visiting Scholar. It is an extraordinary place to visit for many reasons, and an especially wonderful environment for working on a book project. Thank you to all the staff, administration, and fellow 2014–15 Visiting Scholars at RSF.

[1] The current grant numbers are ARC-DP150104491 for Goeree, SES-1459918 for Holt, and SES-1426560 for Palfrey.

Quantal Response Equilibrium

1
Introduction and Background

1.1 ERRARE HUMANUM EST

That humans do not always optimize perfectly is not the least bit controversial, even among economists and game theorists. In games of skill, we often see experts making errors. Chess grandmasters blunder against weaker opponents. Downhill skiers fall. People get busy and forget to return important phone calls. Some of these imperfections may be rationalizable by incomplete information. The batter whiffs because of a sudden gust of wind; the skier falls after hitting an invisible patch of ice. But some of these errors are harder to explain, and even for these simple ones it is often difficult or impossible to identify explicitly the source. It turns out that this mundane observation, captured in a nutshell as a Latin proverb, has important implications for the theory of games, which brings us to the subject of this book: quantal response equilibrium (QRE).

The theory of noncooperative games, as embodied by the Nash equilibrium, is based on three central tenets, which combined, rule out errors in decisions and beliefs. First, choice behavior by players in the game depends on *expectations* about the choice behavior of other players in the game. Second, individual choice behavior is *optimal* given these expectations. Third, expectations about the behavior of other players are *correct*, in a probabilistic sense. When these three elements are in place, an internally consistent model of behavior emerges, in the form of an equilibrium distribution of action profiles, that is stable.

Quantal response equilibrium relaxes the second of these assumptions, that is, that individual choice behavior is always optimal, by allowing for players to make mistakes. In particular, the expected payoffs calculated from players' expectations may be subject to "noise" elements, as indicated by the "+ Error" component in the top part of figure 1.1, which may lead players to choose suboptimal decisions.

In turn, this has significant implications about what constitutes equilibrium play in the game. If we retain the other central tenets of game theory, then players in a game, who form correct expectations about how others behave, must take account of the fact that other players don't always perfectly optimize. But this implies that errors have *equilibrium effects*: one player's expectations about the

FIGURE 1.1. Quantal response equilibrium as a fixed point in beliefs and error-prone actions.

possibility of errors by other players alters the strategic calculus in ways that fundamentally change behavior in the game.

Examples abound. One might be advised to drive more cautiously late on New Year's Eve because of the presence of error-prone drunk drivers. Expert chess players may lay traps in hopes that their opponents will fall into them. While such moves are disadvantageous and can even lead to certain defeat when the opponent sees them coming, they are advantageous if the probability that the opponent will not respond optimally is high enough. In sports (and other cases), players even take into account the possibility of their own errors as well in adjusting their behavior. Tennis players may let up on their first serve because of the risk of also missing their second serve. A quarterback may take a sack rather than risking an errant throw. Some batters in baseball strike out often because they are waiting for the pitcher to make a mistake and throw a pitch that can be slugged out of the park.

There are countless other examples where players' awareness of their own or others' imperfect optimization alters strategic incentives. But the story does not stop there. In game theory, there are never-ending levels of strategic calculation. If one's tennis opponent is likely to ease up on the first serve, one should take a more aggressive position in returning the serve by moving in a few steps, but this tempts the server to serve harder, which increases the probability of error. A quantal response equilibrium takes account of all of these—potentially infinite—levels of strategic play. It recognizes that players make errors, that players understand and take account of the fact that errors are made, that they understand that other players recognize the possibility of error, and so forth ad infinitum. A quantal response equilibrium is a stable point, where everyone optimizes as best as they can subject to their behavioral limitations (in the sense that they may make errors), in full recognition that everyone else optimizes as best as they can subject to their behavioral limitations.

Quantal response equilibrium theory was developed largely in response to data from laboratory experiments that regularly show systematic and sometimes large departures from the basic predictions of standard game theory. This is true even in the simplest imaginable games, such as two-person bimatrix zero-sum games with a unique mixed-strategy equilibrium (O'Neill, 1987; Brown and Rosenthal, 1990; Ochs, 1995). When games have multiple equilibria, further restrictions are typically imposed on the data, such as those implied by subgame perfection, and the news is even worse (Güth, Schmittberger, and Schwarze, 1982; Thaler, 1988). The evidence is overwhelming that, as a predictive model of behavior, the standard theory of equilibrium in games needs an overhaul, and fortunately game theorists and economists have been working hard to improve the theory.[1] The need for doing this is made even more urgent since traditional game theory is regularly applied as a predictive tool in policy. For example, mechanism design theory applies game theory to develop and improve incentive structures for firms and regulatory agencies, to design better auction and bargaining mechanisms, and improve institutions for the provision of public goods. Game theory is also used to make policy prescriptions for government intervention in imperfectly competitive industries; and in fact this is one of the most significant innovations in the field of industrial organization and applied microeconomics in the last few decades. More recently, game-theoretic predictions have been applied to a variety of issues that arise in the study of international trade.[2] QRE is squarely placed in this broad effort to develop a better (and more useful) theory of games.

One possible response might be to abandon all three of the central tenets of Nash equilibrium and explain these deviations in terms of psychological propensities. The subject of this book, QRE, represents a more incremental approach. While judgment biases, and other sources of deviations from rational behavior, surely play some role in these departures, there is still a need for models that account for the equilibrium interaction of these effects. Initial attempts at this sought to explain systematic deviations from Nash equilibrium by presuming the existence of specific types of players who do not seek to maximize their expected payoffs in the experimental games. While this approach, and related applications of psychological theories of bias, have had success in explaining some of the departures from Nash equilibrium, they are open to the criticism of being post hoc, in the sense that the "invention" of types is specific to the game under consideration. What would be more desirable is a *general* model that does not need to be tailored to each game and each data set.

---

[1] See Camerer (2003).

[2] In fact, if one goes back further, to the 1950s when the first surge of research in game theory began, the bulk of the research was largely motivated (and funded) by applications to arms races and other problems of international conflict and national security.

This book is all about one such class of general models, QRE, and its applications to economics, political science, and pure game theory. By viewing choice behavior as including an inherently random component, QRE is a statistical generalization of Nash equilibrium. Its *statistical* foundation is closely related to quantal or discrete choice models that have been widely employed by statisticians in biometrics, psychology, and the social sciences, to study a wide range of scientific phenomena, including dosage response, survey response, and individual choice behavior. But it goes a step further by studying *interactive choice behavior*, and is based on the notion of rational expectations, a concept at the heart of modern economics and game theory. It is a *generalization* of the standard model of Nash equilibrium in the sense that Nash equilibrium is formally nested in QRE as a limiting case in the absence of decision error.

Thus, from a methodological perspective, QRE combines the stochastic choice approach, now commonly used in discrete choice econometrics, with the Nash equilibrium theory of games, the standard paradigm for studying strategic interaction in economics and political science. Strategies in a game are chosen based on their relative expected utility, but the choices are not necessarily best responses, which brings a flavor of limited rationality to the study of games, and does so in a general way. But, while closely related to ideas of limited rationality, it has its foundations in the theory of games of incomplete information. One of the central theoretical results of the next chapter is that there is a direct correspondence between quantal response equilibria of a game and the Bayesian equilibria of an expanded game with incomplete information.

The hybridization of a standard econometric approach with game theory opens up the possibility of using standard statistical models for quantal choice in a game-theoretic setting. With quantal response equilibrium, best-response functions are probabilistic (at least from the point of view of an outside observer) rather than deterministic. While better strategies are more likely to be played than worse strategies, there is no guarantee that best responses are played with certainty. Importantly, this imperfect response behavior is understood by the players. Like Nash equilibrium, this leads to a complicated equilibrium interplay of strategizing by the players, where they have to think about how other players are playing, realizing that those other players are thinking about them, too. The fixed point of such a process is a quantal response equilibrium. In the simplified perspective provided in figure 1.1, the fixed point requires that the probability distributions representing players' expectations (left-hand side) match the distributions of players' actual decisions (right-hand side), that is, the vector of equilibrium distributions gets mapped into itself. This equilibrium analysis can provide a limiting point of a learning process in which expectations evolve with observed distributions of others' decisions. In this book, we also discuss models of stochastic learning and introspection that have proved to be useful in dynamic or novel settings where players' expectations may not be aligned with the actual distributions of others' decisions.

The first paper on QRE, published two decades years ago in *Games and Economic Behavior*, proposed a structural version of QRE based on the additive random-utility model of choice that provides the choice-theoretic foundation for discrete choice econometrics (McKelvey and Palfrey, 1995). The basic idea is that there are many different possible "types" of players, but these types are known only to the agents themselves, and cannot be observed directly by the econometrician or, in the case of interactive games, by the other players. The statistical variation in behavior that is always observed in experiments can thus be interpreted as the result of some underlying distribution of these types. Hence observed or anticipated behavior will appear to be stochastic, even if all agents know their own type and always adopt pure strategies. This way of viewing QRE owes significant debt to Harsanyi's (1967, 1973) landmark contributions to our understanding of games of incomplete information, which today is called Bayesian Nash equilibrium; in fact, as originally defined by McKelvey and Palfrey (1995, 1998), QRE corresponds to a Bayesian equilibrium relative to some distribution of additive random-utility disturbances to the game.

These connections help explain why QRE has become an important tool for the statistical analysis of data from experimental games. It serves as a formal structural econometric framework to estimate behavioral parameters using both laboratory and field data, and can also lead to valuable insights into theoretical questions such as equilibrium selection and computation of equilibrium. There is now an extensive literature that either uses QRE in the analysis of data, or is focused exclusively on studying the theoretical properties of QRE in specific classes of games. The time is ripe for a book that pulls this work together into a single volume. The aim of the book is to lay out for an informed reader the broad array of theoretical and experimental results based on QRE. It contains some genuinely new material, offering new directions and open issues that have not been explored in depth yet, as well as delving into a range of peripheral issues, tangencies, and more detailed data analysis than the typical journal-article format allows for. There is also a "how-to" aspect to the book, as it provides details and sample programs to show readers how to compute QRE and how to take it to the data.

1.2 WHY A STATISTICAL VERSION OF NASH EQUILIBRIUM?

The simple answer is that it is a better descriptor of real human behavior, which is inevitably error prone. But there are deeper methodological reasons. Laboratory

TABLE 1.1. A dominance solvable game (Lieberman, 1960).

	a_1	a_2	a_3
a_1	0	15	−2
a_2	−15	0	−1
a_3	2	1	0

studies of games, ranging from pedantically simple to highly complex, all share two features. First, as pointed out above, systematic contradictions to the most basic predictions of standard game theory are commonplace. But a more severe problem for formally testing or rejecting theories is that, even where classical game theory predicts reasonably well "on average," there is still a huge amount of variation in behavior that is inconsistent with the theory. This second problem is particularly troublesome for games with a pure-strategy equilibrium, because standard theoretical models generally don't allow for any variation in the data.[3] Point predictions about behavior are simply too sharp and easy to reject, even if they can approximate observed behavior reasonably well on average. When games have multiple equilibria, one typically needs to place further restrictions on the data, such as the restrictions implied by subgame perfection and other refinements or selection criteria, and the performance of standard theory is even worse (e.g., in ultimatum and centipede games). However, these further restrictions generally share the same weaknesses (or are more severe) with respect to point predictions.

These two problems—systematic contradictions to predicted behavior and the overly restrictive nature of point predictions—are nicely illustrated by the following symmetric two-player zero-sum game that is strictly dominance solvable (and conceptually much easier than many children's games such as tic-tac-toe). Each player has three possible action choices. The payoff matrix is shown in table 1.1, with the entries being Row's payoffs (and the negative of Column's payoffs). Action a_2 is strictly dominated by a_3. Once these dominated actions are removed, then a_1 is strictly dominated by a_3. Thus the unique prediction from any standard equilibrium notion, or rationalizability, is for both players to choose a_3. Yet, in the laboratory, one observes all three strategies being played by both the Row and Column players. Indeed, both Row and Column very often choose a_1.[4]

In order to model behavior in games like this in a rigorous way for statistical analysis, one needs a model that admits the statistical possibility for any strategy

---

[3] A similar problem arises with respect to market environments in the laboratory, where the theoretical models being studied (competitive equilibrium) are typically inconsistent with price dispersion. One imagines it may be possible to adapt the quantal response equilibrium approach to develop a statistical theory of market equilibrium. There were some early attempts to apply this approach to demand theory and utility theory (e.g., Block and Marschak, 1960).

[4] A laboratory experiment based on this game was conducted by Lieberman (1960).

to be used. Indeed, without such a statistical model, for any game with a unique pure-strategy equilibrium, a single observation of a player using a different strategy (as is invariably the case in the laboratory) leads to immediate rejection of Nash equilibrium at any level of significance. This is sometimes referred to as the zero-likelihood problem, since the theoretical model assigns zero-likelihood to some data sets.[5] Call a model that assigns positive probability to all data sets a *statistical model*. Without a statistical model, standard maximum-likelihood methods are virtually useless for analyzing most data sets generated by laboratory experiments, due to unspecified variation outside the model.

Aside from statistical problems, the zero-likelihood problem also causes headaches for game theory itself. The classical problem of specifying beliefs "off the equilibrium path" in extensive-form games has never been resolved satisfactorily. Bayes' rule does not apply if the denominator is zero. And once Bayes' rule is lost, one is in a theoretical wilderness without a compass, left with ad hoc assumptions about subsequent behavior. In signaling games, for instance, what one ends up with is a morass of "belief-based refinements" that led theorists down a dead end in search of the holy grail of equilibrium concepts.[6] In repeated games, we are left with subtle and questionable models of "renegotiation" when a player happens to wander off the equilibrium path.

In contrast, a statistical approach to game theory has no wilderness, no off-path behavior, because no move is ever completely unexpected and hence no player is ever completely surprised. The problem, of course, is how to formulate such a statistical theory, the subject of this book.

1.3 SIX GUIDING PRINCIPLES FOR STATISTICAL GAME THEORY

Quantal response equilibrium theory is based on six guiding principles. The first principle, *completeness* (or *interiority*), requires the model to avoid the zero-likelihood problem discussed above. That is, all possible actions occur with positive probability. The second principle, *continuity*, requires that choice probabilities vary continuously in expected payoffs. The third, *responsiveness*, requires that the probability of choosing an action rises with its expected payoff. The fourth principle, *monotonicity*, requires that one action is more likely to be chosen than another if its expected payoff is higher. The fifth, *equilibrium*, requires rational expectations. That is, each player's choice probabilities correspond to quantal

---

[5] In the example of table 1.1, the Nash equilibrium assigns zero likelihood to nearly all data sets.

[6] Brandts and Holt (1993) report a series of laboratory experiments that provide strong evidence that equilibrium refinements are not useful in terms of explaining which Nash equilibrium is most commonly observed. See chapter 7.

response behavior relative to the expected payoffs in a game, evaluated at the actual choice probabilities of others. The sixth principle, *generality*, requires the model to be specified in sufficient generality to apply to arbitrary games. A general model is not ad hoc; it does not need to be specially tailored for each game and/or each data set. Variations are possible, of course, and the book discusses several that have been developed that follow only a subset of these principles.

1.4 A ROAD MAP

The remaining chapters of this book explore the theoretical properties of QRE, its implications for the analysis of experimental and field data, and applications to a variety of games of particular interest to political scientists, economists, and game theorists. We discuss how QRE can organize and explain many observed behavioral anomalies, how QRE can be computed efficiently, and how QRE can be used as a structural econometric model to fit data and to measure parameters from decision-theoretic and behavioral models.

The book is organized into two parts: "Part I: Theory and Methods" (chapters 2–6), and "Part II: Applications" (chapters 7–9), although part I also includes a number of simple examples to illustrate the basic principles and for a clearer presentation of the theory.

Part I begins by laying out the general theory of quantal response equilibrium for normal-form and extensive-form games, in chapters 2 and 3 respectively. Proofs of some of the main propositions are included. Chapter 4 explores issues related to individual heterogeneity with respect to player error rates, or what we refer to as *skill*. That chapter also describes some extensions of QRE that relax the assumption that player expectations about the choice behavior of other players are correct. For example, in games that are played only once, players are not able to learn from others' prior decisions, and expectations must be based on introspection. Chapter 4 develops the implications of *noisy introspection* embedded in a model of iterated thinking. Chapter 5 explores questions related to learning and dynamics. Part of that chapter lays out the theory for how to apply QRE to repeated and dynamic stochastic games with an infinite horizon and stationary structure. Chapter 6 provides some user-friendly examples to illustrate how QRE can be computed numerically in games, and how QRE can be used as a structural model for estimation. Sample computer programs in MATLAB for specific examples are provided, which the reader can use as a template to adapt to specific applications.

Part II analyzes in detail many different applications of QRE to specific classes of games, in order to give a sense of the wide range of problems that QRE is useful for, and to illustrate many of the ideas that were developed in chapters 2–6. The applications in chapter 7 include a variety of simple games that illustrate the effects of QRE in several canonical settings that are of widespread interest to game theorists and social scientists: social dilemmas (the traveler's dilemma), adverse selection (the compromise game), signaling (the poker game), and social learning. Chapter 8 examines the effect of QRE in a range of games of particular interest to political scientists, including free riding, voter turnout, information aggregation in committees and juries, crisis bargaining in international relations, and dynamic bargaining in legislatures. Many of the sharply nonintuitive Nash predictions for voting and political bargaining games are reversed or modified by adding sensible amounts of noise in an equilibrium analysis. Chapter 9 is oriented toward economic applications, including games with continuous strategy spaces, such as auction, pricing, and bargaining games. Most of the applications in these three chapters also include a summary of experimental results that have used QRE in the analysis of the data.

The final chapter 10 speculates about directions in which future developments in QRE theory and experimental work could contribute new insights about strategic behavior.

2
Quantal Response Equilibrium in Normal-Form Games

As explained in the opening chapter, the main idea behind quantal response equilibrium (QRE) is to replace the knife-edge best-response functions that underlie the Nash equilibrium with "smoothed" best responses that are continuous and increasing in expected payoffs. In other words, QRE models players as "better responders" rather than best responders: they are more likely to choose strategies with higher expected payoffs, but they do not always choose the strategy with the highest expected payoff. In addition, players realize that others are better responders as well, that is, they anticipate others' noisy behavior. Because of this awareness, noise may have feedback effects resulting in systematic (and sometimes quite large) deviations from Nash equilibrium predictions. This chapter considers the basic properties of QRE for normal-form games, allowing for both discrete and continuous choice sets.

There are (at least) two ways to model QRE, which are neither isomorphic nor mutually exclusive. The *structural approach* specifies an underlying stochastic process in the form of privately observed payoff disturbances, and provides a rational foundation for stochastic choice functions as best responses given the unobserved payoff disturbances. To an outsider (or other player of the game) who cannot observe the disturbances, choice behavior of a player will appear as payoff-responsive stochastic choice. QRE based on this approach is called *structural QRE*. The *reduced-form approach* specifies some particular functional form of imperfect best responses with the property that better strategies are used more often than worse strategies. QRE based on this approach is called reduced-form or *regular QRE*.[1]

The idea of smoothed best responses originated in work by mathematical psychologists to explain why observed decisions appear to be more random as

[1] Our terminology of reduced-form versus structural quantal response functions follows the econometrics literature where the reduced-form approach typically involves the specification of a demand function, while in the structural approach this demand function is derived from optimal consumer choice behavior subject to individual heterogeneity.

the strength of a stimulus becomes weaker (e.g., Thurstone, 1927; Luce, 1959; Luce and Suppes, 1965). For example, a subject in an experiment may be asked which of two lights is brighter or which of two sounds is louder. Subjects seldom make mistakes when one signal is a lot stronger than another, but the probability that the subject recognizes which signal is stronger falls to nearly one-half as the difference in the two intensities goes to zero. The new dimension in applying the model to the analysis of games of strategy is that the choice probabilities of one player affect the expected payoffs of the other players' choice alternatives, so that the choice probabilities themselves are endogenously determined. That is, random or noisy individual behavior has equilibrium effects.

There are a number of ways to think about players adopting strategies that are not necessarily best responses. Two immediately come to mind. First, one could interpret this as a departure from rational decision making. In this sense, the model falls into the class of "bounded rationality" models. This is the basic motivation behind the large literature in mathematical psychology called "probabilistic choice." The idea here is that behavior is subject to decision errors (i.e., nonoptimal choice) by individuals, but these decision errors are payoff sensitive: big mistakes (that involve a large loss relative to the opti-mal choice) are made less often than small ones. This is the reduced-form approach.

An alternative interpretation of smoothed best responses follows results in Harsanyi's (1973) work on Bayesian equilibrium for games of incomplete information. The basic idea there is to think of a complete information game as an approximation to a more complicated game of incomplete information, in which the actual payoffs of the players are private information. More specifically, a player's actual payoff is equal to the complete-information game payoff augmented by an additive payoff disturbance known only to the player, that is, the player's privately known *type*. This generates a game of incomplete information with player types determined by the individual-specific payoff disturbances.

This second interpretation is related to the additive random-utility models (ARUM) developed by structural econometricians (McFadden, 1976), which is the reason it is called the structural approach. In that literature, models of probabilistic choice are applied to cross-sectional data containing the decisions of many individuals, and the stochastic elements are interpreted as *interpersonal* variation, or heterogeneity, in preferences. The idea is to interpret the random disturbances as latent variables that cannot be observed by an outsider (say an econometrician or another player in the game). However, each player is assumed to observe the sum of this latent variable and the true expected payoff, and choose optimally with respect to this sum.

The structural econometricians' interpretation thus contrasts with that in the mathematical psychology literature, where the stochastic elements represent *intrapersonal* variations in utility levels, or perception errors, which may cause a subject to choose differently when faced with the same stimuli. One way to reconcile the two approaches is to think of the perturbed payoffs as simply being each player's estimate of the expected payoff of a particular strategy, with the latent variable representing the estimation error. It is important to note that the different interpretations result in the same mathematical model, and hence, in this sense, there is no need to favor one or the other. Most experimental economists, for instance, are agnostic about the precise interpretation of the noise. In the laboratory, observed noise may be ascribed to either distractions, perception biases, miscalculations, or possible behavioral, psychological, or emotional factors too numerous to list here. However, if the data from an experiment consists of a panel with a sequence of observations for each person, then some of the variability may be due to individual-specific heterogeneity that stays the same for a specific person from period to period and some may be due to decision error that is random across periods for a specific person. In this case, it can be useful to model both the random perturbations that determine individual propensities and the random perturbations that cause the same person to make errors (Goeree, Holt, and Smith, 2015).

The remainder of the chapter is divided into two parts. First, because it is more intuitive, it is simpler to start with the reduced-form approach to QRE, based on the direct specification of "regular" quantal or smoothed best-response functions that are required to satisfy four intuitive axioms of stochastic choice. Regular QRE was introduced first in McKelvey and Palfrey (1996) as *R-response equilibrium*, a reduced-form model of the structural QRE analysis in McKelvey and Palfrey (1995). It was not fully developed until a decade later, where it was given the name it now goes by (Goeree, Holt, and Palfrey, 2005). The analysis of regular QRE in this chapter is based on the latter paper. A simple asymmetric matching pennies game is used to illustrate these ideas and to show that QRE imposes strong restrictions on the data, even without parametric assumptions on the quantal response functions. Particular attention is given to the logit QRE, since it is the most commonly used approach taken when QRE is applied to experimental or other data. The discussion includes the topological and limiting properties of logit QRE, and connections with refinement concepts. QRE is also related to several other equilibrium models of imperfectly rational behavior in games, including a game-theoretic equilibrium version of Luce's (1959) model of individual choice, Rosenthal's (1989) linear response model, and Van Damme's (1987) control cost model, and these connections are explained in the chapter.

Second, the chapter formally develops the structural approach to QRE, which was the original formulation in McKelvey and Palfrey (1995) built on the additive random-utility framework commonly employed in the econometrics literature. This section starts with a much simpler structural approach to studying equilibrium in games where players can make errors. In that alternative approach, errors are not payoff sensitive. While this means that such models are often much easier to work with analytically (the reason it is used to introduce the section), the simplification comes at some cost in terms of motivation, plausibility, and explanatory power. Models of this type were used in early statistical analysis of experimental data, now replaced by QRE analysis.

Structural QRE based on additive random utility, in its most general form, can generate surprising and nonintuitive results unless reasonable restrictions are imposed on the structure of the additive payoff disturbances. Indeed, a section of this chapter explains the need to impose further conditions on the distribution of payoff disturbances, since otherwise structural QRE can rationalize any observed data in a game (Haile, Hortaçsu, and Kosenok, 2008). In particular, these implausible and nonintuitive properties are derived from specifications that violate standard assumptions (independent and identically distributed errors) that are used in all empirical applications of QRE. For example, if the error associated with one action is larger than that for another with high probability, but with small probability much smaller, then to preserve a zero mean, obviously the action with the higher expected payoff might not be chosen more often. A specific example is provided in the section on the empirical content of QRE later in this chapter. Finally, regular QRE is extended to continuous games to include applications of interest to economists, for example, Bertrand and Cournot pricing games, auctions, etc.

2.1 THE REDUCED-FORM APPROACH: REGULAR QRE

The traditional best-response functions underlying Nash equilibrium are generally multivalued and upper hemicontinuous, but not continuous. The departure considered here is to let players' response functions be smooth, that is, continuously differentiable in expected payoffs. A quantal response function for player i, denoted R_i, maps the vector of expected payoffs for i's alternative actions into a probability distribution of choices over these strategies. Players no longer choose the best strategy with probability 1, but to ensure some degree of (stochastic) rationality that players are "better" responders, that is, strategies with higher expected payoffs are chosen more frequently. This is captured by the intuitive notions of responsiveness and monotonicity defined below.

Let $\Gamma = [N, \{A_i\}_{i=1}^n, \{u_i\}_{i=1}^n]$ be a finite normal-form game, where

1. $N = \{1, \ldots, n\}$ is the set of **players**;
2. $A_i = \{a_{i1}, \ldots, a_{iJ_i}\}$ is player i's set of J_i **actions** (or **pure strategies**) and $A = A_1 \times \cdots \times A_n$ is the set of **action profiles**;
3. $u_i : A \to \Re$ is player i's **payoff function**.

Furthermore, let Σ_i be the set of probability distributions over A_i. An element $\sigma_i \in \Sigma_i$ is a **mixed strategy** for player i, which is a mapping from A_i to $[0, 1]$, where $\sigma_i(a_i)$ is the probability that player i chooses pure strategy a_i. Denote by $\Sigma = \prod_{i \in N} \Sigma_i$ the set of mixed-strategy profiles. For each $j = 1, \ldots, J_i$ and any $\sigma \in \Sigma$, player i's expected payoff for choosing action a_{ij} is $U_{ij}(\sigma) = \sum_{a_{-i} \in A_{-i}} p(a_{-i}) u_i(a_{ij}, a_{-i})$, where $p(a_{-i}) = \prod_{k \neq i} \sigma_k(a_k)$. Since player i has J_i possible actions, the vector of expected payoffs for these actions, denoted by $U_i = (U_{i1}, \ldots, U_{iJ_i})$, is an element of $\Re^{J_i}$.

Definition 2.1: *$R_i : \Re^{J_i} \to \Sigma_i$ is a **regular quantal response function** if it satisfies the following four axioms.*

(R1) ***Interiority:*** *$R_{ij}(U_i) > 0$ for all $j = 1, \ldots, J_i$ and for all $U_i \in \Re^{J_i}$.*

(R2) ***Continuity:*** *$R_{ij}(U_i)$ is a continuous and differentiable function for all $j = 1, \ldots, J_i$ and for all $U_i \in \Re^{J_i}$.*

(R3) ***Responsiveness:*** *$\partial R_{ij}(U_i)/\partial U_{ij} > 0$ for all $j = 1, \ldots, J_i$ and for all $U_i \in \Re^{J_i}$.*

(R4) ***Monotonicity:*** *$U_{ij} > U_{ik} \Rightarrow R_{ij}(U_i) > R_{ik}(U_i)$ for all $j, k = 1, \ldots, J_i$.*

These axioms are economically and intuitively compelling. *Interiority* ensures that the model has full domain, that is, it is logically consistent with all possible data sets. This is important for empirical application of the QRE framework. *Continuity* is a technical restriction, which ensures that R_i is nonempty and single valued. Furthermore, continuity seems natural from a behavioral standpoint, since arbitrarily small changes in expected payoffs should not lead to jumps in choice probabilities. *Responsiveness* requires that if the expected payoff of an action increases, ceteris paribus, the choice probability must also increase. *Monotonicity* is a stochastic form of rational choice that involves binary comparisons of actions: an action with higher expected payoff is chosen more frequently than an action with a lower expected payoff. Without monotonicity, players are not necessarily "better responders," suggesting some form of *anti*rationality, which is not the intent of the approach. One key implication, which follows directly from continuity and monotonicity, is that strategies with the same expected payoff are chosen with equal probability: $U_{ij} = U_{ik} \Rightarrow R_{ij}(U_i) = R_{ik}(U_i)$ for all $j, k = 1, \ldots, J_i$.

Recall that Σ is the set of mixed strategies for all players, so that an element σ of this set specifies a mixed strategy for each player. Since players' "beliefs" can be represented by probability distributions over other players' decisions, one can think of σ in two ways: as representing a common set of beliefs, and as stochastic responses to the expected payoffs determined by those beliefs. Loosely speaking, a quantal response equilibrium is a set of belief distributions for which the stochastic responses are consistent with the beliefs, that is, it is a set of mixed strategies that gets mapped into itself via the response functions. More formally, define $R(U) = (R_1(U_1), \ldots, R_n(U_n))$ to be *regular* if each R_i satisfies axioms (R1)–(R4). Since $R(U) \in \Sigma$ and $U = U(\sigma)$ is defined for any $\sigma \in \Sigma$, then $R \circ U$ defines a mapping from Σ into itself.

> **Definition 2.2:** *Let R be regular. A **regular quantal response equilibrium** of the normal-form game Γ is a mixed-strategy profile σ^* such that $\sigma^* = R(U(\sigma^*))$.*

Since regularity of R includes continuity, $R \circ U$ is a continuous mapping. Existence of a regular QRE therefore follows directly from Brouwer's fixed-point theorem.

> **Proposition 2.1** (Existence): *There exists a regular quantal response equilibrium σ^* of Γ for any regular R.*

2.2 TWO SIMPLE REGULAR QRE MODELS

This section illustrates the calculations underlying quantal response equilibrium with two simple regular QRE models, which differ in their assumptions about the R-function that maps the vector of expected payoffs into a vector of choice probabilities, and with respect to the restricted domains for which they are well defined.[2] The first approach defines a simple parametric class of quantal response functions as a game-theoretic adaptation of the pioneering axiomatic formulation of stochastic choice developed by Luce (1959), who assumed that choice probabilities are simply proportional to expected payoffs. The second class

---

[2] The first model is well defined only for games with nonnegative payoff functions. The second approach applies only to games with two pure strategies for each player. These two models are introduced mainly for illustration and because they are used in some applications later in the book.

TABLE 2.1. An asymmetric matching pennies game.

	H	T
H	$X, 0$	$0, 1$
T	$0, 1$	$1, 0$

of quantal response functions assumes that choice probabilities are proportional to a cumulative distribution function of the payoff difference between the two actions, where the distribution has a density function that is symmetric around zero. This second approach has, as special cases, two of the most commonly used functions for estimation of binary-choice models in econometrics: logit and probit. This leads naturally into the following section of the chapter, which develops in detail a number of important properties of logit QRE, the most widely used QRE model for application and analysis of data.

Example 2.1 (Asymmetric matching pennies game)*:* To illustrate, these two regular QRE models are applied to the simple matching pennies game in table 2.1. For $X > 0$, this game has a unique mixed-strategy Nash equilibrium in which the Row player mixes uniformly between heads (H) and tails (T) and the Column player selects heads with probability $\frac{1}{X+1}$ and tails with complementary probability. So when Row's payoff X of the (H, H) outcome is increased, the Nash equilibrium prediction is that Column plays T more frequently while Row's behavior remains unchanged. As shown in the table, this nonintuitive invariance of Row's Nash equilibrium strategy with respect to own payoffs is not a property of the QRE models. As will be shown later in the chapter, for *any* regular quantal response function, the equilibrium probability that Row chooses H is strictly increasing in X. The asymmetric matching pennies example focuses the discussion in this section and is revisited at several points later in the book.

2.2.1 The Luce Model and a Power-Law Generalization

To explain seemingly noisy data from individual decision-making experiments, the mathematical psychologist Duncan Luce (1959) posited that choice frequencies are proportional to intensities. For instance, consider the case where subjects have to choose which of two lights is brighter. If the true intensities of the lights are l_1 and l_2 respectively, the Luce rule implies that option 1 is chosen with probability $l_1/(l_1 + l_2)$.

This simple rule illustrates some of the main features of the quantal response approach. First, when the two intensities are equal, both options have an equal chance of being selected. Second, when one intensity is higher, the corresponding option is chosen more frequently. However, in this case, the option with the higher intensity is not necessarily chosen with probability 1. For this reason, the Luce rule may be interpreted as a "boundedly rational" choice rule.

The Luce model can be generalized for interactive strategic games by replacing the exogenous intensities by the expected payoffs of the game. To illustrate, consider the two-person matching pennies game in table 2.1 for $X > 0$. Let U_{11} and U_{12} denote player 1's expected payoffs of playing H and T respectively, with similar definitions for player 2. Similarly, let σ_{11} and σ_{12} denote player 1's probability of choosing H and T respectively, with corresponding definitions for player 2. The Luce rule then becomes

$$\sigma_{ij} = \frac{U_{ij}}{U_{ij} + U_{ik}}, \quad i, j, k = 1, 2, \quad k \neq j. \tag{2.1}$$

Note that (2.1) is not an explicit solution, since the payoffs on the right-hand side are themselves functions of the choice probabilities σ_{ij}. For the above game,

$$U_{11} = X\sigma_{21}, \qquad U_{12} = \sigma_{22},$$
$$U_{21} = \sigma_{12}, \qquad U_{22} = \sigma_{11}. \tag{2.2}$$

To compute the Luce quantal response equilibrium, the probabilities used to calculate the expected payoffs in (2.2) must match the probabilities that follow from (2.1) as determined by the expected payoffs. That is, the equations given by (2.2) and (2.1) must hold simultaneously. For the game in table 2.1 the Luce quantal response equilibrium is easily computed as

$$\sigma_{11}^* = \frac{\sqrt{X}}{1 + \sqrt{X}}, \qquad \sigma_{21}^* = \frac{1}{1 + \sqrt{X}}, \tag{2.3}$$

with $\sigma_{12}^* = 1 - \sigma_{11}^*$ and $\sigma_{22}^* = 1 - \sigma_{21}^*$. The Luce quantal response equilibrium thus predicts that as player 1's payoff (X) of the (H, H) outcome rises, player 1 is more likely to choose H. This prediction contrasts sharply with that of the unique Nash equilibrium in that there is an "own-payoff" effect, that is, player 1's probability of choosing H rises when X increases.

The Luce rule can be generalized by taking monotone transformations of the "intensities," or, in the case of game theory, the expected utilities associated with each strategy. One obvious transformation is to use the power function, where the intensities are each raised to the same specific positive exponent. The QRE conditions for the Luce power rule for a two-person game with parameter $\lambda \geq 0$ are given by

$$\sigma_{ij} = \frac{\left(U_{ij}\right)^{\lambda}}{\left(U_{ij}\right)^{\lambda} + \left(U_{ik}\right)^{\lambda}}, \quad i, j, k = 1, 2, \quad k \neq j, \tag{2.4}$$

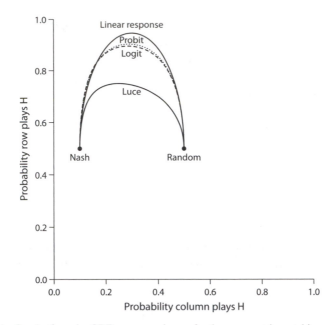

FIGURE 2.1. Graph of regular QRE correspondences for the asymmetric matching pennies game in table 2.1 with $X = 9$. The lines trace equilibria for values of the precision parameter that range from $\lambda = 0$ to $\lambda = \infty$.

where λ can be interpreted as a *responsiveness* or *precision* parameter. When $\lambda = 0$, strategy is unresponsive to expected payoffs, and both strategies are used with equal probability. As λ increases, the strategy with higher expected payoff is played more often, and when $\lambda \to \infty$, players almost always use best replies. Thus $\lambda \in [0, \infty)$ parameterizes an entire *family* of "power" Luce QRE as the solutions to (2.4).

For the game in table 2.1, the power Luce QRE can be computed as

$$\sigma_{11}^* = \frac{X^{\lambda/(\lambda^2+1)}}{1 + X^{\lambda/(\lambda^2+1)}}, \qquad \sigma_{21}^* = \frac{1}{1 + X^{\lambda^2/(\lambda^2+1)}}, \qquad (2.5)$$

with $\sigma_{12}^* = 1 - \sigma_{11}^*$ and $\sigma_{22}^* = 1 - \sigma_{21}^*$. Note that (2.5) reduces to the Luce QRE in (2.3) when $\lambda = 1$. The lowest curve in figure 2.1 graphs the equilibrium probabilities of the power Luce QRE for the case of $X = 9$, for $\lambda \in [0, \infty)$. With $\lambda = 0$ (pure noise), the curve starts at the center point with equal probabilities for each decision. As the noise is reduced, Column's probability of H declines, and Row's probability of choosing H rises at first and then falls back to 0.5 as $\lambda \to \infty$, and converges to the mixed-strategy Nash equilibrium.

2.2.2 QRE Based on Distribution Functions

The QRE derived from the (power) Luce model is not defined for all games, because the expected utilities (intensities) have to be nonnegative numbers. Furthermore, such quantal response functions are not invariant to adding a constant to all payoffs, so one cannot avoid the negative payoff problem by simply converting games with negative to positive payoffs by adding a constant to all payoffs. The resulting Luce equilibrium will depend on exactly which constant was added.[3] For practical applications, it is often more useful to work with smooth differentiable response functions that apply to *all games* with real-valued payoff functions. There are many such quantal response functions.

For the case of 2×2 games, let $F : \Re \to [0, 1]$ denote a strictly increasing continuously differentiable distribution function with $F(x) = 1 - F(-x)$ for all $x \in \Re$. Notice that $F(x)$ is a cumulative distribution function for a probability density function that is symmetric around 0. Then for 2×2 games one can represent a family of quantal response functions using F, which takes the form

$$\sigma_{i1} = F(\lambda(U_{i1} - U_{i2})),$$

with $\sigma_{i2} = 1 - \sigma_{i1}$. This distribution-based approach was used by Goeree and Holt (2005b) to derive general properties of QRE in "participation games" with two decisions, for example, vote or not, enter or not, etc. These games are discussed in chapter 8.

As in the power law generalization of the Luce model of choice, the distribution function model contains a nonnegative precision parameter, λ, which can be thought of as representing the degree of rationality of the players. Indeed, as λ approaches infinity, the linear response rule approaches rational behavior: the option with the higher expected payoff is chosen with probability 1. At the other extreme, when $\lambda = 0$, behavior is essentially random and choice probabilities are uniform. The single degree of freedom introduced by λ is an essential feature of most applications of quantal response models presented in this book, and allows for more than simple point predictions.

One commonly used choice for $F(\cdot)$ in scientific applications is the distribution function, $\Phi(\cdot)$, of a standard normal random variable (i.e., with mean 0 and standard deviation 1). For the game in table 2.1, the fixed-point equations for the *probit QRE* are

$$\sigma_{11}^* = \Phi\left(\lambda((X + 1)\sigma_{21}^* - 1)\right), \qquad \sigma_{21}^* = \Phi\left(\lambda(1 - 2\sigma_{11}^*)\right), \qquad (2.6)$$

---

[3] Except for the limitation to positive payoffs, the Luce and power Luce QRE will generally satisfy the interiority, continuity, responsiveness, and monotonicity axioms of regular QRE.

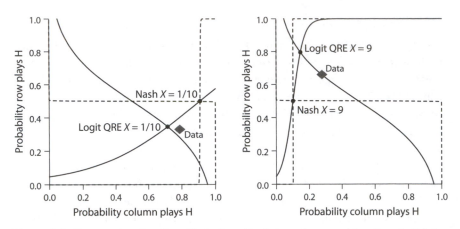

FIGURE 2.2. Best-response functions (dashed) and logit quantal response functions (solid) for the asymmetric matching pennies game in table 2.1 with $X = \frac{1}{10}$ and $X = 9$.

with $\sigma_{12}^* = 1 - \sigma_{11}^*$ and $\sigma_{22}^* = 1 - \sigma_{21}^*$. There are no analytic solutions to these equations, but they can be easily solved numerically. The dotted line in figure 2.1 shows the family of probit QRE predictions as a function of the precision parameter, λ, for the matching pennies game with $X = 9$. Note that also the probit QRE predicts that both players will overplay H relative to mixed-strategy Nash equilibrium.

Another common choice is the logistic distribution $F(x) = \frac{1}{1+\exp(-x)}$. For the asymmetric pennies example, the *logit QRE* conditions can be expressed as ratios of exponentials of expected payoffs:

$$\sigma_{11}^* = \frac{e^{\lambda X \sigma_{21}^*}}{e^{\lambda X \sigma_{21}^*} + e^{\lambda(1-\sigma_{21}^*)}}, \qquad \sigma_{21}^* = \frac{e^{\lambda(1-\sigma_{11}^*)}}{e^{\lambda(1-\sigma_{11}^*)} + e^{\lambda \sigma_{11}^*}}, \qquad (2.7)$$

with $\sigma_{12}^* = 1 - \sigma_{11}^*$ and $\sigma_{22}^* = 1 - \sigma_{21}^*$. Again, while there are no analytic solutions to these equations, they can be easily solved numerically. The family of logit equilibria when $X = 9$ is shown as the dashed curve in figure 2.1 and shares the same qualitative properties as the other QRE models.

The intuition behind the tendency for Row to overplay H relative to Nash in this game when $X > 1$, and to underplay H when $X < 1$, is due to the fact that QRE models smooth the sharp edges of best-response functions. Without any noise, the best-response function for Column is a perfectly flat line at the indifference point for Column (where Row plays H with probability 0.5). See the horizontal dashes in the left and right panels of figure 2.2. Therefore, shifts in Row's best-response function caused by changes in Row's payoff, X, result in intersection points at the same 0.5 probability of Row playing H as indicated by the labels "Nash X=1/10" and "Nash X=9". It is analogous to the case of a flat demand function, for which shifts in supply do not affect the price. But when noise smooths off the sharp corners, Column's quantal response function will be decreasing; see the solid decreasing lines in the left and right panels of figure 2.2. (Note that Column's best response and Column's quantal response are identical across the two panels because Column's payoffs are unaffected by changes in X.) Now shifts in Row's response function caused by own-payoff changes do result in changes in the choice probabilities for both players, as indicated by the labels "Logit QRE X=1/10" and "Logit QRE X=9".

Several researchers have conducted experiments based on the asymmetric matching pennies game, by varying the value of X and observing how choice behavior changes.[4] These experiments are easy to run and are a useful demonstration for game theory classes. As an illustration, figure 2.2 presents choice averages (labeled "Data") from three sessions conducted with either 10 or 12 subjects in each session. The subjects first played the game with Row's payoff parameter $X = \frac{1}{10}$. Players were grouped into pairs, and played the game 25 times, with random re-pairing between plays. The observed frequencies with which Row and Column chose H in this game were 0.33 and 0.8 respectively, as indicated by the diamond in the left panel. This is close to the intersection of logit quantal responses, which were computed for a value of λ *not* estimated from the data but chosen to highlight the smooth characteristics of quantal responses. At $\lambda = 3$, the logit QRE prediction is that Row and Column choose H with probabilities 0.34 and 0.72 respectively. The same subjects played an additional 25 rounds (with random re-pairing) for the game with $X = 9$. In the experiment, Row's propensity to choose H exhibits a strong own-payoff effect, as shown by the diamond located at approximately (0.3, 0.65) in the right panel of figure 2.2. Note that logit QRE predicts an even stronger "own-payoff" effect and overshoots the probability with which Row plays H. Of course, the logit QRE predictions depend on the value of λ and a better fit would be obtained by estimating a single λ using data from both the $X = \frac{1}{10}$ and $X = 9$ games. Chapter 6 describes how standard maximum-likelihood techniques can be applied to estimate λ, as well as other parameters of interest (e.g., risk-aversion parameters, fairness parameters, etc.) from experimental game data.

---

[4] Ochs (1995) reports the first such experiment. See also McKelvey, Palfrey, and Weber (2000), Goeree and Holt (2001), and Goeree, Holt and Palfrey (2003).

2.3 PROPERTIES OF LOGIT QRE

The logit choice rule is widely applied by econometricians and experimental game theorists in the analysis of data. This section derives some important general properties of logit QRE, which will be used throughout the book. A logit equilibrium for a general normal-form game Γ is a solution to the following system of equations:

$$\sigma_{ij} = \frac{e^{\lambda U_{ij}}}{\sum_{k=1}^{J_i} e^{\lambda U_{ik}}}, \quad i \in I, \quad j = 1, \ldots, J_i. \tag{2.8}$$

One property of the logit model is that the ratio of choice probabilities depends on the expected payoffs for those decisions but is independent of the payoffs for the other decisions that are "irrelevant" to the comparison:

$$\frac{\sigma_{ij}}{\sigma_{ik}} = \frac{e^{\lambda U_{ij}}}{e^{\lambda U_{ik}}}, \quad i \in I, \quad a_j, a_k \in A_i,$$

which is known as the "independence of irrelevant alternatives" (IIA) property. A second property is that the addition of a constant to all of player i's expected payoffs will not affect logit choice probabilities. As Luce (1959) noted, the reverse logic also applies; that is, any reduced-form quantal response function with these properties (independence of irrelevant alternatives and invariance with respect to adding a constant to all payoffs) must have the logit form. These properties have a certain appeal, but they can be controversial in some situations.

The next result is that any limit point of a sequence of logit equilibria as λ approaches infinity is a Nash equilibrium of the underlying game.

Proposition 2.2 (Convergence of logit equilibria to Nash)*: Let $\{\lambda_t\}_{t=1}^{\infty}$ be a sequence such that $\lim_{t \to \infty} \lambda_t = \infty$, and let σ_t^* denote the logit equilibrium for λ_t. Then any accumulation point of $\{\sigma_t^*\}_{t=1}^{\infty}$ is a Nash equilibrium.*

Proof: If σ^* is an accumulation point of $\{\sigma_t^*\}_{t=1}^{\infty}$, then one can extract a subsequence such that σ^* is a limit point of this subsequence. Suppose, in contradiction, that σ^* is not a Nash equilibrium. Then there is some player i, and some pair of actions a_{ij} and a_{ik}, with $\sigma^*(a_{ik}) > 0$ and $U_i(a_{ij}, \sigma_{-i}^*) > U_i(a_{ik}, \sigma_{-i}^*)$. Since the expected payoff functions are continuous, it follows that for sufficiently small ϵ there exists a T such that for $t \geq T$, $U_i(a_{ij}, \sigma_{-i}^*) > U_i(a_{ik}, \sigma_{-i}^*) + \epsilon$. This implies that as $t \to \infty$ (and hence $\lambda_t \to \infty$), the ratio $\sigma_t^*(a_{ik})/\sigma_t^*(a_{ij}) = e^{\lambda_t(U_i(a_{ik}, \sigma_{-i}^*) - U_i(a_{ij}, \sigma_{-i}^*))}$ tends to 0. Hence, $\sigma_t^*(a_{ik})$ tends to 0, which contradicts the assumption that $\sigma^*(a_{ik}) > 0$. ∎

In the spirit of Harsanyi (1973), call a Nash equilibrium, σ^*, **approachable** if there exists a sequence of λ's approaching infinity and a corresponding sequence of logit equilibria that converge to σ^*.

Proposition 2.3 (Properties of logit equilibrium correspondence): *The logit equilibrium correspondence,*

$$\sigma^* : \lambda \mapsto \left\{ \sigma : \sigma_{ij} = \frac{e^{\lambda U_{ij}(\sigma)}}{\sum_{k=1}^{Ji} e^{\lambda U_{ik}(\sigma)}} \quad \forall i, j \right\},$$

has the following properties:

1. $\sigma^*(\lambda)$ *is odd for almost all* λ.
2. $\sigma^*(\lambda)$ *is upper hemicontinuous.*
3. *For almost all games, the graph of* $\sigma^*(\lambda)$ *contains a unique branch which connects the centroid to exactly one Nash equilibrium. All other Nash equilibria are connected as pairs to each other.*

Proof: See McKelvey and Palfrey (1995, Appendix) for a detailed proof. Properties (1) and (2) are standard properties of equilibrium correspondences that follow from standard arguments. Property (3) is more subtle, but has the following intuition. First, for small enough values of λ, the logit response function is a contraction mapping, so there is a unique logit equilibrium close to the "centroid" of the game at $\lambda = 0$, that is, the point where all strategies are adopted with equal probability. One can trace out this *principal branch of the logit correspondence* for progressively higher values of λ. For generic games, the principal branch is a one-dimensional manifold and hence can be followed for all values of λ ranging from 0 to ∞. The result follows.[5] ∎

The limit point of the principal branch as λ approaches infinity is called the *logit solution* of the game. Thus the logit equilibrium correspondence can be used to define one weak refinement (approachability) and one strong refinement (the logit solution). The result is not limited to the logit specification, but applies more generally to families of smooth quantal response functions, parameterized by a precision parameter λ ranging from 0 to ∞. These ideas are illustrated with a few simple examples below, using the logit correspondence.

---

[5] The basic idea and the mathematical tools needed to prove this are reminiscent of the tracing procedure of Harsanyi (1975) and Harsanyi and Selten (1988). The selections that are made by the two approaches are different.

TABLE 2.2. Game with unique perfect equilibrium (D, D).

	U	M	D
U	1, 1	0, 0	1, 1
M	0, 0	0, 0	0, X
D	1, 1	X, 0	1, 1

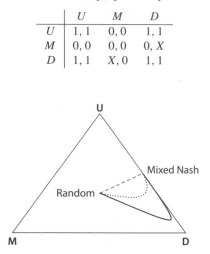

FIGURE 2.3. Logit QRE does not select the perfect equilibrium (D, D).

2.3.1 Examples to Illustrate Logit-QRE-Based Refinements

Example 2.2 (Non-trembling-hand perfectness)**:** One may be tempted to conjecture that limit points as λ grows will not only be Nash equilibria, but will also be trembling-hand perfect. To see that this is not true, consider the game in table 2.2 for $X > 0$. This game has a unique perfect equilibrium (D, D), and the Nash equilibria consist of all mixtures of U and D for both Row and Column. The limit of logit equilibria selects $\sigma_1^* = \sigma_2^* = (\frac{1}{2}, 0, \frac{1}{2})$ as the unique limit point, which differs from the (D, D) perfect equilibrium. Along the sequence, for finite λ, $\sigma_{13}^* > \sigma_{11}^* \gg \sigma_{12}^*$ and $\sigma_{23}^* > \sigma_{21}^* \gg \sigma_{22}^*$, but as λ becomes large, σ_{12}^* and σ_{22}^* converge to 0. So the middle strategies are eliminated in the limit.[6]

While the limit solution does not depend on the magnitude of X, the logit equilibrium for intermediate values of λ is quite sensitive to X. Figure 2.3 illustrates the logit QRE graph for both Row and Column (since the game is symmetric) as a function of λ for the cases $X = 0$ (dashes), $X = 5$ (dots), and $X = 50$ (solid). One might consider the fact that the limiting logit equilibrium is not always perfect to be a drawback. Alternatively, it could be viewed as a consequence of the "independence of irrelevant alternatives" property of the

---

[6] In the game without the middle strategies it is obvious that the unique limit point of logit equilibrium is $(\frac{1}{2}, \frac{1}{2})$.

TABLE 2.3. An asymmetric game of chicken.

	T	S
T	0, 0	6, 1
S	1, 14	2, 2

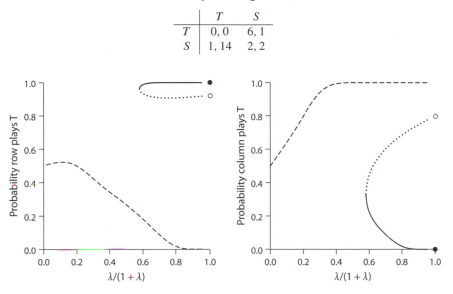

FIGURE 2.4. Logit equilibrium correspondence for the asymmetric chicken game in table 2.3.

limiting logit QRE: strategies that have near-zero probability in the logit QRE do not affect the play of the rest of the game.

Example 2.3 (Asymmetric game of chicken): An asymmetric version of the game of chicken is shown in table 2.3. If both players play "soft" (S) then each gets an intermediate payoff. If they both play "tough" (T) then they both get low payoffs. If they choose different strategies, then the tough player receives a large payoff, and the soft player receives a low payoff. In this game there are three Nash equilibria, two pure-strategy equilibria where one player is tough and the other is soft, and a third equilibrium where both players mix.[7] Usual refinements provide no help in selecting among equilibria.

Figure 2.4 displays the logit equilibrium correspondence for this game. As one can see, the weak refinement of approachability by a sequence of logit equilibria does not provide any help, as all three Nash equilibria are approachable. The mixed-strategy Nash equilibrium corresponds to the open circles on the far right of each panel in figure 2.4 while the filled circles

---

[7] In the unique mixed-strategy Nash equilibrium, Row plays tough with probability $\frac{12}{13}$ and Column plays tough with probability $\frac{4}{5}$. Note that Row "has to" play tough with a higher probability in order to keep Column indifferent, since Column has a high payoff of 14 in the lower-left part of table 2.3.

correspond to the pure-strategy Nash equilibrium where Row plays tough and Column plays soft. The principal branch of the logit QRE correspondence converges to the pure-strategy equilibrium where Row plays soft and Column plays tough, the logit solution of this game. This is the equilibrium favoring the player (Column) who benefits the most from playing tough. In fact, this would seem to be a general property of the class of asymmetric chicken games where the only asymmetry is in the payoff to the tough player when one player is tough and the other is soft.[8]

Example 2.4 (A 3×3 ultimatum game)*:* The third example is a very simple version of a bargaining game, known as the "ultimatum game" (Güth, Schmittberger, and Schwarze, 1982; Thaler, 1988). It has been the subject of a great deal of experimental work by both economists and psychologists, largely because the results strongly contradict the predictions of subgame-perfect Nash equilibrium. In this game, there are two players, who are given the task of dividing a fixed sum of money according to the following simple rules. Player 1, the proposer, offers a split of the money, and then player 2, the responder, either accepts or rejects the proposed division. If the offer is rejected, the game ends and neither player receives anything. Assume for the purposes of analysis that there are a finite number of feasible splits (say, in 1 dollar increments).[9] Then subgame-perfect Nash equilibrium implies that the responder will receive, in equilibrium, either zero or the lowest possible positive amount.

A simplified version of the ultimatum game is easy to study in the context of logit QRE. In this simplified version of the game, the pie consists of $2J$ dollars and there are J permissible splits of the pie, ranging from the fifty-fifty split (where each player receives J dollars) and the most favorable split for the proposer, subject to the constraint that the responder receives a positive amount (i.e., the proposer receives $2J - 1$ dollars and the responder receives 1 dollar). The responder writes down a minimum acceptable offer (some number between 1 and J). If the proposer offers at least this minimum level, then the outcome is exactly the proposed split. If not, then both players receive 0. The normal form of this game, with $J = 3$, is given in table 2.4. For example, in the upper-left cell, when the proposer offers 3 and the responder demands at least 3, then this is implemented and both earn 3. But in the

---

[8] This conjecture lacks a formal proof, but extensive computations have not produced a counterexample. A similar property appears to hold for asymmetric battle-of-the-sexes games.

[9] In experiments, there are a finite number of choices available for the proposer to offer the responder.

TABLE 2.4. A 3×3 ultimatum game. The numbers labeling rows correspond to the proposer's offer and the numbers labeling columns correspond to the responder's minimum demand.

	3	2	1
3	3, 3	3, 3	3, 3
2	0, 0	4, 2	4, 2
1	0, 0	0, 0	5, 1

lower-left cell, the proposer only offers 1 and the responder demands 3, so both earn 0. Before considering the properties of logit QRE for this game, it is important to note that the analysis of experimental data would typically also incorporate a fairness consideration that can play an important role in bargaining games.[10]

It is well known that this game has multiple Nash equilibria (in fact a continuum of Nash equilibria), only one of which is trembling-hand perfect. In the perfect equilibrium, the proposer offers 1 and the responder's minimum demand is 1 (i.e., the responder accepts any offer of 1, 2, or 3). Figure 2.5 shows the logit solution (unique continuous selection from the logit equilibrium correspondence) as a function of λ. This shows clearly that the logit solution to this game is *not* the unique perfect equilibrium, but rather results in the (4, 2) split. The responder does not adopt a pure strategy, but rather mixes evenly between the choices of demanding 1 (accept anything) and demanding 2 (reject only the worst offer). In addition to the limiting properties of this logit equilibrium mapping, it is also instructive to look at the solution for low and intermediate values of λ. At low values of λ both players are nearly random, so the proposer makes each offer with probability $\frac{1}{3}$. The probability of using the strategy of making the greediest offer (only offer 1) declines monotonically in λ. For higher values of λ, the modal offer becomes 2, that is, a (4, 2) split. For the responder, the strategy of accepting only the equal split (by demanding 3) disappears quickly.

Besides looking at the unique equilibrium implied by a continuous selection from the logit equilibrium correspondence, or logit solution of the game, it is also possible that for large values of λ there could be other logit equilibria. These "high λ" logit equilibria would be separate components of the logit correspondence that are not connected to the centroid of the strategy space ($\lambda = 0$), but may have limit points approaching Nash equilibria as λ goes to infinity. In this sense they are limiting logit equilibria, and the limit points are "approachable" Nash equilibria, a refinement of the set of Nash

---

[10] See Goeree and Holt (2000) for a QRE analysis of the Fehr–Schmidt (1999) model of inequity aversion applied to two-person bargaining games with alternating offers.

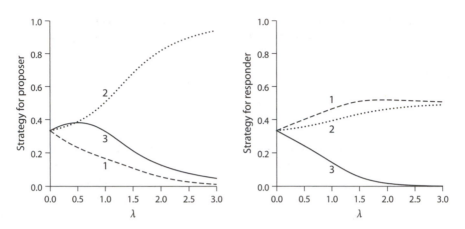

FIGURE 2.5. Logit solution graph for the 3×3 ultimatum game. Probabilities of proposer offers of 1, 2, or 3 are shown in the left panel and probabilities of responder minimum demands of 1, 2, or 3 are shown in the right panel.

equilibria defined earlier. As it turns out, there are indeed many limiting logit equilibria, even though there is a unique logit solution. The main result is summarized in the following proposition.

Proposition 2.4 (Logit solution for the discrete ultimatum game)*: If $J > 2$, then*

1. *for each pure-strategy offer by the proposer less than J, there exists a limiting logit equilibrium where that offer is made with probability 1 and is accepted by the responder with probability 1;*
2. *in every limiting logit equilibrium, the proposer uses a pure strategy and the responder mixes uniformly over all pure strategies that allow the offer to be accepted.*

In other words, all possible splits are supportable as limiting logit equilibria. This is in stark contrast to perfect equilibrium, which eliminates all the equilibria where the receiver gets more than 1 dollar, the smallest positive share.[11]

---

[11] This proposition follows from a more general result obtained by Yi (2005). His simplified ultimatum game allows the proposer to make any offer from 0 to $2J$, rather than requiring the offer to range from 1 to J. Interestingly, this does not increase the set of limiting logit equilibria. That is, there are no limiting logit equilibria except those where the proposer offers some share from 1 to $J - 1$.

Example 2.5 (A coordination game)*:* Consider two students who work together on a project and who simultaneously have to decide whether to exert "low" effort, $L = 1$, or "high" effort, $H = 2$. The grade or payoff from the project to both is determined by the lowest of the efforts exerted: if both students work hard the payoff is 2 for each, but if one or both students slack off, the payoff is only 1. Moreover, every unit of effort costs $c < 1$. The normal-form representation of this game (see table 2.5) represents a coordination game in the following sense. If Row chooses L, Column wants to choose L (and vice versa) and if Row chooses H, Column wants to choose H (and vice versa). These are the two pure-strategy Nash equilibria of the game. In addition, there exists a mixed-strategy Nash equilibrium of the game in table 2.5 where both players choose L with probability $1 - c$ and H with probability c. The mixed-strategy Nash equilibrium has nonintuitive comparative statics properties in that the probability of exerting high effort rises when the cost of effort goes up.[12] Coordination games have been widely studied and applied because of the possibility that rational players may "get stuck" in the low-payoff equilibrium. Indeed, it is sometimes claimed that minimum-effort games will, in fact, yield the minimum-effort outcome, but as shown below, cost matters.

The theoretical construct most commonly used to "select" among the equilibria in 2×2 coordination games is Harsanyi and Selten's (1988) notion of risk dominance. One appealing feature of risk dominance is its sensitivity to cost that determines the losses associated with deviations from best responses to others' decisions. When both players choose low effort, the cost of a unilateral deviation to high effort is just the cost of the extra effort, c, which will be referred to as the "deviation loss." Similarly, the deviation loss at the (H, H) equilibrium is $1 - c$, since a unilateral reduction in effort reduces the minimum by 1 but saves the marginal effort cost c. In this case, risk dominance selects the equilibrium that has the highest deviation loss (in an asymmetric game, the equilibrium with the highest product of deviation losses for the two players would be selected). The deviation loss from the low-effort equilibrium is greater than that from the high-effort equilibrium if $c > 1 - c$, or equivalently, if $c > \frac{1}{2}$, in which case the low-effort equilibrium is risk dominant. Risk dominance, therefore, has the desirable property that it selects the low-effort outcome if the cost of effort is sufficiently high.

An alternative selection criterion is provided by the logit solution, that is, the unique approachable Nash equilibrium that belongs to the principal

---

[12] Moreover, the mixed equilibrium in coordination games is generally thought to be unstable (Anderson, Goeree, and Holt, 2001).

TABLE 2.5. A 2 × 2 coordination game.

	L	H
L	$1-c, 1-c$	$1-c, 1-2c$
H	$1-2c, 1-c$	$2-2c, 2-2c$

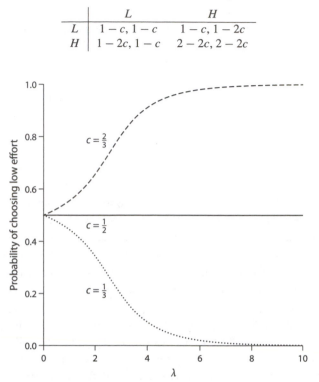

FIGURE 2.6. Logit solution for three versions of the coordination game. The graph shows the logit equilibrium probability of choosing L for Row and Column as a function of λ.

branch of the logit equilibrium correspondence. Figure 2.6 shows the principal branch as a function of λ for three values of the effort cost: $c = \frac{1}{3}$ (dots), $c = \frac{1}{2}$ (solid), and $c = \frac{2}{3}$ (long dashes). For each case, the line shows Row's and Column's equilibrium probability of choosing low effort as a function of the precision λ. Note that the logit solution reacts to the effort cost in the same intuitive way as the risk-dominant solution: when costs are high, the low-effort equilibrium is selected but when costs are low, the high-effort equilibrium results.[13] These issues come up again in example 2.9 in the context of coordination games with continuous strategy spaces.

The properties of the logit correspondence described above, including the notions of approachability and the logit solution, can be defined more

---

[13] For theoretical conditions that guarantee the logit solution will coincide with the risk-dominant solution, see Turocy (2005).

generally by parameterizing any family of quantal response functions by a precision parameter λ, where $\lambda = 0$ corresponds to completely random choice and $\lambda = \infty$ corresponds to best-response behavior. This was illustrated earlier in this chapter for the simple binary-choice case applied to 2×2 games. More generally, given any regular quantal response function, R, one simply defines $R(U; \lambda) = R(\lambda U)$. A detailed analysis of QRE correspondences and selection criteria using this general functional form for QRE correspondences parameterized by a precision parameter λ is carried out in Zhang (2013) and Zhang and Hofbauer (2013).

2.4 ALTERNATIVE MODELS RELATED TO REGULAR QRE

2.4.1 Rosenthal's Linear Response Model

Rosenthal's (1989) T-equilibrium was one of the first models developed to represent *equilibrium* boundedly rational behavior in games where players are better-responders. Like the distribution function approach of section 2.2.2, it applies only to games with binary action spaces. That model specifies that the probability that an individual adopts a particular strategy is a linearly increasing function of the difference between the expected payoff for the strategy and the expected payoff for the other strategy. If the two strategies yield the same expected payoff (given the choice probabilities of the other players) then each is used with probability $\frac{1}{2}$. More generally,

$$\sigma_{i1} = \min\left\{\max\{0, \tfrac{1}{2} + \lambda(U_{i1}(\sigma) - U_{i2}(\sigma))\}, 1\right\} \tag{2.9}$$

and $\sigma_{i2} = 1 - \sigma_{i1}$. The "min" and "max" functions in (2.9) prevent the choice probability from going above 1 or below 0. The linear response model does not satisfy interiority and (strict) responsiveness. More importantly, the linear response model is limited to 2×2 games, for which it provides an analytically tractable model.

As in the distribution function model of regular quantal response equilibrium, the linear response model in (2.9) contains a nonnegative precision parameter, λ, which can be thought of as representing the degree of rationality of the players. In fact, the linear response model is a special case of the distribution function approach, where the distribution is uniform on an interval $[-A, A]$, and λ is inversely proportional to A.

For the matching pennies game in table 2.1 the linear response equilibrium can be worked out as

$$\sigma_{11}^* = \frac{1}{2} + \frac{(X-1)\lambda}{2 + 4(X+1)\lambda^2}, \qquad \sigma_{21}^* = \frac{1}{2} - \frac{(X-1)\lambda^2}{1 + 2(X+1)\lambda^2}, \tag{2.10}$$

with $\sigma_{12}^* = 1 - \sigma_{11}^*$ and $\sigma_{22}^* = 1 - \sigma_{21}^*$. Note that behavior is random ($\sigma_{11}^* = \frac{1}{2}$) when $X = 1$ or $\lambda = 0$. More generally, for $X > 1$ or $\lambda > 0$, the linear response model predicts an own-payoff effect. Finally, when λ tends to ∞, the linear response equilibrium limits to the Nash equilibrium where Row selects H with probability $\frac{1}{2}$ and Column selects H with probability $\frac{1}{X+1}$. For $X = 9$, the equilibrium probabilities are graphed in figure 2.1 as functions of the precision parameter λ.

2.4.2 Van Damme's Control Cost Model

In Van Damme's (1987) control cost model, players cannot choose with perfect discrimination, due to inherent implementation difficulties. Players realize that an option may be erroneously selected, and that the extent of such errors can be controlled at a cost. The selection probability for an option derives from an optimization problem that incorporates both the profitability of the best option and the costs of improving control. Assume that the lowest control cost is associated with no control at all (i.e., random choice across options), and better control (greater precision) is more costly. To capture these ideas formally, suppose that there exists a control cost function $C(\sigma_{ij})$ that can be associated with the ability of player i to play strategy j with probability σ_{ij}. Following Van Damme (1987), assume that the control cost function is twice differentiable and strictly convex.

The problem that decision maker i faces is to maximize the total payoff:

$$\max_{\sigma_i} \sum_{j=1}^{J_i} \left\{ \sigma_{ij} U_{ij} - \frac{C(\sigma_{ij})}{\lambda} \right\} \tag{2.11}$$

for $\lambda > 0$, subject to the constraint that the probabilities σ_{ij} are nonnegative and sum to 1. The solution to this maximization problem will be interior if one assumes $C'(0) = -\infty$ but one allows for corner solutions (pure strategies) as well.[14] The parameter λ is introduced into the formulation in order to provide a way to weight the importance of control costs. As $\lambda \to \infty$, the problem reverts to the standard optimization problem, so the option with the highest expected payoff is chosen with probability 1.

To illustrate the nature of control cost equilibria, and their connection with QRE, consider the case with only two possible strategies, so that the above control

---

[14] Van Damme (1987) assumes there is an infinite cost associated with placing a zero probability on any given decision so that pure strategies cannot occur. He uses this assumption to ensure that the decision maker will not try to control decisions so tightly as to reduce any probability to near-zero levels. Thus all choice distributions will be completely mixed, and Van Damme considers equilibria in sequences of games in which the weight attached to the control cost function goes to 0, as a way of analyzing equilibrium selection issues.

cost problem can be written as

$$\max_{0 \le \sigma_{i1} \le 1} \left\{ \sigma_{i1} U_{i1} + (1 - \sigma_{i1}) U_{i2} - \frac{\gamma(\sigma_{i1})}{\lambda} \right\},$$

where $\gamma(\sigma) \equiv C(\sigma) + C(1 - \sigma)$. Note that γ is strictly convex since C is, and that $\gamma(\sigma)$ is minimized at $\sigma = \frac{1}{2}$, that is, control costs are lowest when decisions are random. Since γ is strictly convex, its derivative γ' is increasing and invertible. So the solution to the control cost problem can be written as

$$\sigma_{i1} = \gamma'^{(-1)} \left(\lambda(U_{i1} - U_{i2}) \right).$$

In other words, with only two possible strategies, Van Damme's control cost model is isomorphic to a distribution-based QRE with cumulative distribution function $F(\cdot) = \gamma'^{(-1)}(\cdot)$. Conversely, with only two options, for any (distribution-based) quantal response function $F(\cdot)$ one can construct a control cost function, $C(\cdot)$, as follows. Integrating $F(\cdot) = \gamma'^{(-1)}(\cdot)$ yields

$$\gamma(\sigma) = \gamma\left(\frac{1}{2}\right) + \int_0^{F^{(-1)}(\sigma)} t \, dF(t),$$

from which one can, in principle, construct the control cost function $C(\cdot)$.

For example, it is readily verified that Rosenthal's linear response model is equivalent to a control cost model with quadratic costs $C(\sigma) = \frac{1}{4}(\sigma - \frac{1}{2})^2$ since then $\gamma'^{(-1)}(x) = x + \frac{1}{2}$. Likewise, the probit QRE follows by assuming

$$C(\sigma) = \frac{1 - \exp(-\frac{1}{2}\Phi^{(-1)}(\sigma)^2)}{2\sqrt{2\pi}}, \tag{2.12}$$

which yields $\gamma'^{(-1)}(x) = \Phi(x)$, that is, the standard normal distribution function. Finally, the logit QRE can be derived from a control cost model with entropic costs,

$$C(\sigma) = \sigma \log(\sigma) - \frac{1}{2} \log\left(\frac{1}{2}\right), \tag{2.13}$$

for which $\gamma'^{(-1)}(x) = e^x/(1 + e^x)$. Figure 2.7 shows the relation between the control cost functions and the associated quantal response models for the case when there are only two possible strategies.

More generally, the solution to (2.11) is given by

$$\sigma_{ij} = C'^{(-1)} \left(\lambda(U_{ij} - \underline{U}) \right), \quad i \in I, \quad j \in A_i, \tag{2.14}$$

where $\underline{U}$ is chosen such that $\sum_{j=1}^{J_i} C'^{(-1)}(\lambda(U_{ij} - \underline{U})) = 1$. Since $C(\cdot)$ is convex, the left-hand side is strictly increasing so there is a unique solution for $\underline{U}$. For instance, when control costs are entropic, $C(\sigma) = \sigma \log(\sigma)$, then $\sigma_{ij} = e^{\lambda(U_{ij} - \underline{U} - 1)}$ for $j = 1, \ldots, J_i$, and the condition that choice probabilities sum to 1

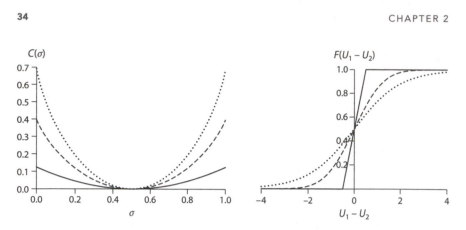

FIGURE 2.7. The left panel shows three different control cost functions and the right panel shows the implied quantal response models: the solid lines correspond to Rosenthal's linear model for quadratic costs, the dashed lines to the logit model for entropic costs, and the dotted lines to the probit model for the cost function in (2.12).

yields $e^{\lambda(-\underline{U}-1)} = \sum_{j=1}^{J_i} e^{\lambda(U_{ij})}$. Hence, choice probabilities take the logit form

$$\sigma_{ij} = \frac{e^{\lambda U_{ij}}}{\sum_{k=1}^{J_i} e^{\lambda U_{ik}}}, \quad i \in I, \quad j \in A_i.$$

It is straightforward to verify that for general convex cost functions, the choice probabilities (2.14) satisfy the regularity assumptions (R1)–(R4) of regular QRE; see definition 2.1. However, the control cost model is not as general as regular QRE since it implies some other restrictions on the choice probabilities. For instance, it predicts that choice probabilities are "translation invariant," that is, they are unaffected when a constant is added to all expected payoffs. Moreover, an increase in the expected payoff of one option raises the probability with which that option is chosen (which is the responsiveness property (R3) in definition 2.1) and lowers the choice probabilities of all other options. This "substitutability" property is not necessarily a feature of regular QRE, nor is translation invariance. Both properties are shared by structural QRE models, as shown in section 2.6.4 below.

OTHER RELATED APPROACHES

Chen, Friedman, and Thisse (1997) observed that the original formulation of McKelvey and Palfrey (1995) could be reinterpreted as a reduced-form model of bounded rationality. They represent choice using the power Luce model and characterize some stability properties of regular QRE within that parametric framework. In particular, they show that the class of quantal response functions satisfies a Lipschitz condition if the error rates are small enough, and hence the equilibrium is globally stable for high error rates. This stability property is similar to the result established in McKelvey and Palfrey (1995), that logit quantal response functions are contraction mappings for sufficiently low values of λ. Schmidt (1992) investigates games of reputation building in which players choose an optimal strategy with probability ε and a suboptimal strategy with probability $1 - \varepsilon$, solves for the Bayesian Nash equilibrium of the resulting game, when this is common knowledge among the players, and applies this to the analysis of multiperiod chain-store-paradox games. Zauner (1999) models behavior in the centipede game by introducing Gaussian payoff disturbances to the agent form of the game and then computes the resulting Bayesian equilibrium. This is essentially the same as looking at a quantal response equilibrium with "probit" response curves, but using the extensive form instead of the normal form of the game.[15] Stahl and Wilson (1995) apply a version of the logit equilibrium model to analyze data from a series of experimental two-person games. Beja (1992) proposes a model of imperfectly rational play in games, where players are limited in their ability to exactly implement targeted strategy choices. Ma and Manove (1993) apply a similar idea to study bargaining games when the players' cannot perfectly implement their bargaining strategies. Friedman and Mezzetti (2005) develop a model where the source of error is in noisy beliefs rather than actions or payoff disturbances.

2.5 STRUCTURAL APPROACH TO QRE

The main goal of this section is to explain structural QRE as it applies to normal-form games. But before doing so, it is instructive to go through a simpler generalization of Nash equilibrium, partly because it was a stepping stone to the development of structural QRE and partly because of its resemblance to trembling-hand perfect equilibrium, but mainly because it is a very simple QRE model to explain and can be solved analytically. This simpler version unfortunately does not satisfy all six "principles" outlined in the previous chapter. It satisfies completeness (interiority), generality, and equilibrium, but fails to satisfy continuity, (strict) responsiveness, and (strict) monotonicity.

2.5.1 An Equilibrium Model of Action Trembles

In a series of papers written prior to McKelvey and Palfrey (1995), *errors in actions* were introduced in a primitive way: the standard game theory model was extended by allowing players to imperfectly implement their equilibrium

---

[15] QRE for extensive-form games is developed in chapter 3 and centipede games are analyzed as the leading example.

strategy.[16] In particular, each player was assumed to play their equilibrium strategy with probability $1 - \epsilon$ and to randomize uniformly across all available actions (i.e., make errors) with probability ϵ.[17] However, in a game-theoretic model, it is not enough just to tack an error term onto the strategy choices and leave it at that, because errors by one player affect the payoffs to all the other players, and vice versa. Therefore, as in QRE, the occurrence of errors in action is assumed to be common knowledge among the players, implying equilibrium effects of the error terms because players are responding to the noisy play of the other players. Call this an ϵ-*perturbed Nash equilibrium*.

Let $\epsilon \in [0, 1]$ be the *tremble parameter*. Given the tremble parameter, ϵ, when player i adopts strategy σ_i, the ϵ-*perturbed strategy* of player i is equal to $\tilde{\sigma}_i^\epsilon = (1 - \epsilon)\sigma_i + \epsilon\sigma^o$ where σ^o is a uniform distribution over the elements of A_i.

Definition 2.3: $\sigma^* \in \Sigma$ *is an ϵ-perturbed Nash equilibrium if σ_i^* is a best response to $\tilde{\sigma}^{*\epsilon}$ for all i.*

Note that an ϵ-perturbed Nash equilibrium is defined in terms of the strategy that i is adopting, σ_i, not the perturbed strategy that i actually implements, $\tilde{\sigma}_i^\epsilon$. Behaviorally, however, it is the perturbed strategies that govern the probability distribution over observed actions. Note that for any finite normal-form game and for any ϵ, an ϵ-perturbed Nash equilibrium exists by standard arguments. Furthermore (and useful for comparison with QRE), for any finite game there is an upper hemicontinuous correspondence of ϵ-perturbed Nash equilibria, as a function of ϵ.[18]

To explain how ϵ-perturbed Nash equilibrium is computed, and to compare it to QRE, the asymmetric matching pennies game in table 2.1 and the asymmetric game of chicken in table 2.3 are revisited. As it turns out, ϵ-perturbed Nash equilibria and QREs share important properties: (1) players tremble, that is, they do not play optimal strategies with probability 1; (2) players are aware that other players make errors and take this into account when calculating their equilibrium

---

[16] See Beja (1992), El-Gamal and Palfrey (1995, 1996), Ma and Manove (1993), McKelvey and Palfrey (1992), Schmidt (1992), and El-Gamal, McKelvey, and Palfrey (1993a, 1993b, 1994).

[17] Another way was to introduce unobservable player types. In the original experimental study of the centipede game McKelvey and Palfrey (1992) incorporate both unobservable player types and errors in actions in their statistical analysis of the data. That game is analyzed in detail in chapter 3.

[18] It is important to realize that ϵ-perturbed Nash equilibrium is not the same as ϵ-Nash equilibrium. In an ϵ-Nash equilibrium, the players choose a strategy with a payoff within ϵ of the payoff from a best reply. On the other hand, an ϵ-perturbed Nash equilibrium satisfies the conditions of ϵ-perfect Nash equilibrium, but with an additional restriction imposed that all nonoptimal actions are played with the same probability. Also, the interest in ϵ-perturbed Nash equilibrium is not just limited to its properties when ϵ is close to 0.

expected payoffs; and (3) implemented strategies are interior (if $\epsilon > 0$). However, there are some important differences. First, strategies are not payoff responsive: all inferior actions are played with equal probability, and if the payoff to an inferior strategy increases slightly, the probability it will be played does not increase. Second, as indicated in the example that follows, response functions are neither single valued nor continuous; rather they are upper hemicontinuous just like standard best reply correspondence. In a sense, ϵ-perturbed Nash equilibrium goes "halfway" between Nash equilibrium and QRE.

Example 2.6 (Example 2.3 continued: asymmetric game of chicken)*:*
Figure 2.8 shows the *perturbed actions* of the ϵ-perturbed Nash equilibrium for the asymmetric game of chicken in table 2.3. For ease of comparison with later figures, $1 - \epsilon$ is displayed on the horizontal axis, so the origin corresponds to the value of ϵ at which play is completely random ($\epsilon = 1$). At this point, the strict best response for both players is T, and the unique ϵ-perturbed Nash equilibrium is equal to $\sigma_R^* = \sigma_C^* = (1, 0)$, where the subscripts refer to Row and Column, respectively. When $\epsilon = 1$ (maximum noise), the perturbed strategy of each player is $\widetilde{\sigma}_R^* = \widetilde{\sigma}_C^* = (\frac{1}{2}, \frac{1}{2})$, a uniform mixture between T and S. As ϵ decreases, T continues to be a strict best reply to $\widetilde{\sigma}_R^* = \widetilde{\sigma}_C^* = (1 - \frac{\epsilon}{2}, \frac{\epsilon}{2})$ for both players, and the unique ϵ-perturbed Nash equilibrium continues to be a noisy mix of the tough best response and pure noise until $\epsilon = \frac{2}{5}$. At this point Row is exactly indifferent between T and S and Row's best responses consist of all mixtures between T and S while Column's best response continues to be T. So at $\epsilon = \frac{2}{5}$ there is a continuum of ϵ-perturbed Nash equilibria. For $\epsilon > \frac{2}{5}$, there is an ϵ-perturbed Nash equilibrium equal to $\sigma_R^* = (0, 1)$, $\sigma_C^* = (1, 0)$, which (if there were no perturbations) is the pure-strategy equilibrium where Row plays S and Column plays T. The perturbed strategies generated by this component of the equilibrium correspondence are graphed as the dashed lines in figure 2.8.

However, there is a second component to the ϵ-perturbed Nash equilibrium correspondence that exists for values of $\epsilon \leq \frac{2}{13}$. As is typically the case with equilibrium correspondences, two new ϵ-perturbed Nash equilibria arise at this point. One of them is the perturbed version of the other pure Nash equilibrium where Row plays tough, $\sigma_R^* = (1, 0)$, and Column plays soft, $\sigma_C^* = (0, 1)$. The latter requires $\epsilon \leq \frac{2}{13}$. At $\epsilon = \frac{2}{13}$, Column is indifferent between T and S while Row's best response continues to be T. The perturbed strategies for this component of the equilibrium correspondence are shown by the solid line in figure 2.8. The second new ϵ-perturbed Nash equilibrium has players mixing in such a way that their perturbed strategies implement the mixed-strategy Nash equilibrium: $\widetilde{\sigma}_R^*(T) = \frac{12}{13}$ and $\widetilde{\sigma}_C^*(T) = \frac{4}{5}$; see the dotted

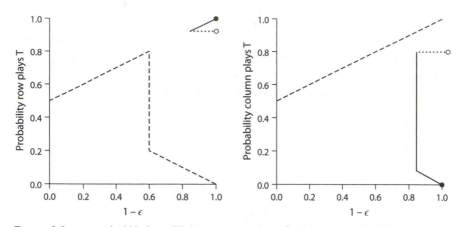

FIGURE 2.8. ϵ-perturbed Nash equilibrium correspondence for the asymmetric chicken game in table 2.3.

line in figure 2.8. Since $\widetilde{\sigma}_i = (1 - \epsilon)\sigma_i + \epsilon\sigma^o$ and σ_i is at most 1, this equilibrium also requires $\epsilon \le \frac{2}{13}$.

The equilibrium correspondence shown in figure 2.8 has three key properties.[19] First, if there is enough noise in the implementation of strategies (ϵ large on the left-hand side of the figure), the equilibrium is unique, with both players "trying to play" tough. Second, the limit points of the correspondence as ϵ goes to 0 (on the right-hand side of the figure) are Nash equilibria. In this example there are three limit points, one for each of the Nash equilibria of the game. The mixed-strategy Nash equilibrium corresponds to the open circles on the far right of each panel in figure 2.8 while the filled circles correspond to the pure-strategy Nash equilibrium where Row plays tough and Column plays soft. Finally, the equilibrium in which Column plays tough and Row plays soft is the limit point of the unique connected component of the graph when $\epsilon \to 0$. These three properties are also true for the logit equilibrium correspondence (see figure 2.4), which is basically a smoothed-out version of the ϵ-perturbed Nash equilibrium correspondence in figure 2.8. In particular, the logit solution also picks out the pure-strategy equilibrium favoring the player who benefits the most from being tough.

Example 2.7 (Example 2.1 continued: asymmetric matching pennies game): The equilibrium effects of trembles in actions are also nicely illustrated by the simple asymmetric matching pennies game in table 2.1. Consider, for

---

[19] A natural conjecture is that these three properties of the ϵ-perturbed Nash equilibrium correspondence hold generically for finite games.

instance, the case $X > 1$. For high values of ϵ, Row's best response is H while Column's best response is T. So even though Row prefers to match Column in a standard analysis of the game, Row chooses not to match when the possibility that Column trembles is sufficiently high. Indeed, it is readily verified that for $\epsilon > \frac{2}{X+1}$, the unique ϵ-perturbed Nash equilibrium is given by $\sigma_R^* = (1, 0)$ and $\sigma_C^* = (0, 1)$, yielding perturbed strategies $\widetilde{\sigma}_R^* = (1 - \frac{\epsilon}{2}, \frac{\epsilon}{2})$ and $\widetilde{\sigma}_C^* = (\frac{\epsilon}{2}, 1 - \frac{\epsilon}{2})$. When $\epsilon < \frac{2}{X+1}$ the ϵ- perturbed Nash equilibrium is also unique and has the players mixing in such a way that their perturbed strategies implement the mixed-strategy Nash equilibrium: $\widetilde{\sigma}_R^* = (\frac{1}{2}, \frac{1}{2})$ and $\widetilde{\sigma}_C^* = (\frac{1}{X+1}, \frac{X}{X+1})$. Finally, when $\epsilon = \frac{2}{X+1}$ there is a continuum of ϵ-perturbed Nash equilibria where Column's best response is T and Row's best responses consist of all mixtures in which H is chosen with probability at least $\frac{1}{2}$ to ensure Column prefers T.

Compared to the QRE correspondences in figure 2.1, the equilibrium correspondence for the ϵ-perturbed Nash equilibrium is less smooth, in particular, it is upper hemicontinuous but not continuous. One reason for its "knife edge" characteristics is that action trembles occur irrespective of their payoff consequences. The next section concerns the structural QRE model of payoff disturbances, which, under standard assumptions on the distribution of the disturbances, implies the regular QRE property that the probability of a mistake is sensitive to its cost.

2.5.2 Structural QRE: An Equilibrium Model of Payoff Trembles

So far, noisy behavior has been interpreted as suboptimal, boundedly rational decision making. An alternative is to assume that observed behavior is, in fact, consistent with perfect rationality but that some aspects of the players' payoffs are unobservable. Indeed, statisticians realized early on that mixed strategies could be transformed into pure strategies (*purified*) by introducing private information that may cause one player's action to be random from another player's point of view.[20] Harsanyi (1973) demonstrated that this purification process yields a refinement of the Nash equilibrium when the payoff disturbances become arbitrarily small.[21]

This alternative approach is related to the additive random-utility models (ARUM) developed by "structural" econometricians (McFadden, 1976). In this

---

[20] The first discoveries of this sort emerged in tandem with the theory of statistical decision functions (Dvoretsky, Wald, and Wolfowitz, 1951), and these ideas resurfaced in a game-theoretic context via the work of Harsanyi (1973). These purification results involved extending a game of complete information to one of incomplete information by introducing additive payoff disturbances. Of course, it was known before that one could purify mixing by adding payoff-irrelevant information, such as the private observation of a randomizing device. The innovation here was that the private information actually could change the payoff of the game, and players were using *strictly* optimal strategies.

[21] In this landmark paper, three key roles played by stochastic choice in game theory (mixed strategies, equilibrium refinements, and private information) were tied together.

literature, models of probabilistic choice are applied to cross-sectional data containing the decisions of many individuals, and the stochastic elements are interpreted as interpersonal variation, or heterogeneity, in preferences. The idea is to interpret the random disturbances as latent variables which cannot be observed by an outsider (say an econometrician or another player in the game). However, each player is assumed to observe the sum of this latent variable and the true expected payoff, and choose optimally accordingly.

This section describes the structural approach to QRE for normal-form games, which was the original definition of QRE in McKelvey and Palfrey (1995). Assume that for each pure strategy a_{ij}, there is an additional privately observed payoff disturbance, $\varepsilon_{ij} \in \Re$, and denote i's disturbed expected payoff by

$$\widehat{U}_{ij}(\sigma) = U_{ij}(\sigma) + \varepsilon_{ij}.$$

For each i, player i's profile of payoff disturbances, $\varepsilon_i = (\varepsilon_{i1}, \ldots, \varepsilon_{iJ_i})$, has a joint distribution with several properties. First, one assumes the distribution of ε_i can be represented by a strictly positive continuously differentiable density function $f_i(\varepsilon_i)$ and the marginal densities are assumed to exist for each ε_{ij}.[22] Furthermore, the disturbances are *independent across players* (not necessarily across strategies) and *unbiased*, in the sense that $E(\varepsilon_i) = 0$ for all i. Call $f = (f_1, \ldots, f_n)$ **admissible** if f_i satisfies the above properties for all i. One interpretation of this formulation of payoff disturbances is that each player estimates the expected payoff of each strategy, with the estimates subject to unbiased errors.

The assumed choice behavior is that each player chooses strategy a_{ij} when $\widehat{U}_{ij}(\sigma) \geq \widehat{U}_{ik}(\sigma)$ for all $k = 1, \ldots, J_i$. Given this choice behavior, $U = (U_1, \ldots, U_n)$ and $f = (f_1, \ldots, f_n)$ together induce a distribution over the actual choices by each player. To be more specific, for any U, define $B_{ij}(U)$ to be the set of realizations of ε_i such that strategy a_{ij} has the highest disturbed expected payoff, $\widehat{U}$. Then

$$R_{ij}(U) = \int_{B_{ij}(U)} f(\varepsilon)d\varepsilon \qquad (2.15)$$

is the induced choice probability that player i selects strategy j. The R_{ij}, which map U_i into Σ_i, are called player i's **(structural) quantal response functions**.

---

[22] The original definition of QRE in McKelvey and Palfrey (1995) was more general and did not require a strictly positive density (full support). Full support guarantees interiority. The ϵ-perturbed Nash equilibrium is an example of a more general definition of structural QRE if full support is not required and mass points are allowed. The ϵ-perturbed Nash equilibrium for the asymmetric game of chicken corresponds to a structural QRE where the error distribution is equal to 0 with probability $1 - \epsilon$ and is uniformly distributed over the region $[-10, -9] \cup [9, 10]$ with probability ϵ.

Perhaps the best-known example of a structural quantal response function is based on the assumption that the additive payoff shocks follow an extreme value ("double-exponential") distribution, in which case the choice probabilities have the familiar logit form. Suppose, for instance, that there are only two possible actions, 1 and 2. Then action 1 is chosen if and only if $U_{i1} + \varepsilon_{i1} > U_{i2} + \varepsilon_{i2}$, where both shocks are distributed according to $G(\varepsilon) = \exp(-\exp(-\lambda\varepsilon))$ with density $g(\varepsilon) = \lambda\exp(-\lambda\varepsilon - \exp(-\lambda\varepsilon))$. For a given value of ε_{i1}, action 1 is chosen with probability $G(U_{i1} - U_{i2} + \varepsilon_{i1})$, which can be integrated using the density $g(\varepsilon_{i1})$ to derive the logit quantal response function

$$R_{11} = \int_{-\infty}^{\infty} \exp\Big(-\exp(-\lambda(U_{i1} - U_{i2} + \varepsilon_{i1}))\Big)\lambda\exp\Big(-\lambda\varepsilon_{i1} - \exp(-\lambda\varepsilon_{i1})\Big)d\varepsilon_{i1}$$

$$= \int_0^{\infty} \exp\Big(-t(1 + \exp(-\lambda(U_{i1} - U_{i2})))\Big)dt$$

$$= \frac{e^{\lambda U_{i1}}}{e^{\lambda U_{i1}} + e^{\lambda U_{i2}}},$$

where in going from the first to the second line there is a transformation of variables $\epsilon_{i1} = \log(\frac{1}{t})/\lambda$.

Note that the power Luce formulation in (2.4) can be obtained by applying the logit rule to $\log(U_{ij})$, rather than U_{ij}, assuming that expected payoffs are positive. This implies that the power Luce quantal response functions can be obtained in the structural formulation by adding extreme value disturbances ε_{ij} to $\log(U_{ij})$, or equivalently, by *multiplying* the U_{ij} by disturbances ν_{ij} that are distributed according to $G(\nu) = \exp(-\nu^{-\lambda})$ for $\nu \in \Re_+$. Note that multiplicative errors do not diminish as payoffs are scaled up, as would be the case with additive errors. This interpretation of the Luce quantal response function as resulting from multiplicative errors helps clarify why scaling up all payoffs will not reduce noise in that formulation.

More generally, the structural response functions in (2.15) can be combined to define a mapping $R(U) \in \Sigma$. Since $U = U(\sigma)$ is defined for any $\sigma \in \Sigma$, $R \circ U(\sigma) = R(U(\sigma))$ defines a mapping from Σ into itself.

Definition 2.4: *Let $f(\varepsilon)$ be admissible. A (**structural**) **quantal response equilibrium** of the normal-form game Γ is a mixed-strategy profile σ^* such that $\sigma^* = R(U(\sigma^*))$, with R defined in (2.15).*

Existence of a structural QRE again follows from Brouwer's fixed-point theorem since R, and hence $R \circ U$, is continuous.

TABLE 2.6. Illustration of structural QRE disturbances for a 3×3 game.

	a_{21}	a_{22}	a_{23}
a_{11}	$u_2(a_{11}, a_{21}) + \varepsilon_{21}$ $u_1(a_{11}, a_{21}) + \varepsilon_{11}$	$u_2(a_{11}, a_{22}) + \varepsilon_{22}$ $u_1(a_{11}, a_{22}) + \varepsilon_{11}$	$u_2(a_{11}, a_{23}) + \varepsilon_{23}$ $u_1(a_{11}, a_{23}) + \varepsilon_{11}$
a_{12}	$u_2(a_{12}, a_{21}) + \varepsilon_{21}$ $u_1(a_{12}, a_{21}) + \varepsilon_{12}$	$u_2(a_{12}, a_{22}) + \varepsilon_{22}$ $u_1(a_{12}, a_{22}) + \varepsilon_{12}$	$u_2(a_{12}, a_{23}) + \varepsilon_{23}$ $u_1(a_{12}, a_{23}) + \varepsilon_{12}$
a_{13}	$u_2(a_{13}, a_{21}) + \varepsilon_{21}$ $u_1(a_{13}, a_{21}) + \varepsilon_{13}$	$u_2(a_{13}, a_{22}) + \varepsilon_{22}$ $u_1(a_{13}, a_{22}) + \varepsilon_{13}$	$u_2(a_{13}, a_{23}) + \varepsilon_{23}$ $u_1(a_{13}, a_{23}) + \varepsilon_{13}$

Proposition 2.5 (Existence): *There exists a (structural) quantal response equilibrium of the normal-form game* Γ *for any admissible* $f(\varepsilon)$.

Before moving on to the next section, it is useful to elaborate briefly on the connection with Bayesian–Nash equilibria of perturbed games à la Harsanyi. Table 2.6 illustrates the structural approach to QRE for a 3×3 game.

This differs only slightly from the Harsanyi (1973) setup which assumes a separate disturbance $\varepsilon_i(a)$, for i's payoff to each action profile, a, while the QRE approach assumes that this disturbance for i is the same for payoffs of all strategy profiles in which i uses the same strategy. That is, one assumes $\varepsilon_i(a_i, a_{-i}) = \varepsilon_i(a_i, a'_{-i})$ for all i and for all $a_{-i}, a'_{-i} \in A_{-i}$. This violates Harsanyi's condition 16 (1973, p. 5) that requires the existence of a density function for $\varepsilon(a)$. In spite of this, it is easy to see that the main results in Harsanyi (1973) are still true under the weaker assumption that for each i a density function exists for $\varepsilon_i = (\varepsilon_{i1}, \ldots, \varepsilon_{iJ_i})$. This weaker assumption is met in the structural QRE model.

Therefore, structural QRE inherits the following properties of Bayesian equilibrium in Harsanyi's disturbed game approach:

1. Best replies are "essentially unique" pure strategies.
2. Every equilibrium is "essentially strong" and in "essentially pure" strategies.
3. There exists an equilibrium.

2.6 ON THE EMPIRICAL CONTENT OF QRE

As seen in section 2.2, alternative specifications for regular quantal response functions yield numerically different equilibrium predictions, but are similar qualitatively. One natural question that arises is, can any observed data point be rationalized by *some* regular QRE? If this is the case, then additional restrictions would be required for regular QRE as defined to be useful empirically.

If this is not the case, then the next question is whether all QRE models based on quantal response functions satisfying axioms (R1)–(R4) (i.e., interiority, continuity, responsiveness, and monotonicity) share similar qualitative properties. The analysis in this section shows that axioms (R1)–(R4) collectively have very strong empirical implications, even without *any* parametric assumptions on the response functions.[23]

This section also addresses the question of whether any observed data point can be rationalized by *some* structural QRE. Haile et al. (2008) show that without restrictions on the payoff disturbances the answer is yes, meaning that structural QRE in its most general formulation is not empirically falsifiable. Further restrictions or assumptions would be needed to give structural QRE empirical content. It is shown, however, that with identically and independently distributed payoff disturbances, structural QRE is also regular and the nonfalsifiability problem disappears. This is reassuring, as all applications of structural QRE to data from experimental games use the i.i.d. assumption, either explicitly or implicitly.

2.6.1 Empirical Restrictions of Regular QRE

There are (at least) three alternative ways to evaluate the empirical restrictions of regular QRE: in terms of possible outcomes, and in terms of nonparametric and parametric comparative statics predictions.

Consider again the asymmetric matching pennies game in table 2.1 for $X \geq 1$. It is readily verified that Row's expected payoff for choosing T is lower (respectively, higher) than for choosing B when $\sigma_{21} < \frac{1}{X+1}$ (respectively, $\sigma_{21} > \frac{1}{X+1}$). Monotonicity therefore implies that, if (σ_1^*, σ_2^*) define a QRE, they must satisfy the inequalities

$$\sigma_{11}^* \leq \tfrac{1}{2} \quad \text{if } \sigma_{21}^* \leq \frac{1}{X+1},$$

$$\sigma_{11}^* \geq \tfrac{1}{2} \quad \text{if } \sigma_{21}^* \geq \frac{1}{X+1}.$$

Likewise, Column's expected payoff for choosing L is higher (respectively, lower) than for choosing R when $\sigma_{11} < \tfrac{1}{2}$ (respectively, $\sigma_{11} > \tfrac{1}{2}$). Thus, (σ_1^*, σ_2^*) must satisfy

$$\sigma_{21}^* \geq \tfrac{1}{2} \quad \text{if } \sigma_{11}^* \leq \tfrac{1}{2},$$

$$\sigma_{21}^* \leq \tfrac{1}{2} \quad \text{if } \sigma_{11}^* \geq \tfrac{1}{2}.$$

---

[23] This section includes material from Goeree, Holt and Palfrey (2005).

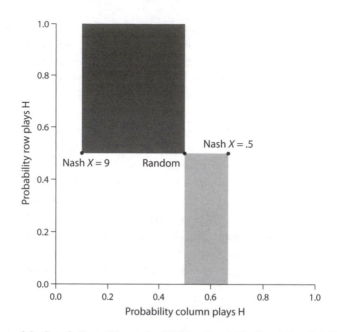

FIGURE 2.9. Set of all possible regular QRE for $X = 9$ (dark) and $X = 0.5$ (light).

The region defined by these inequalities characterizes the set of possible QREs, for any choice of the quantal response functions. For the specific case of $X = 9$, this region is given by the dark shaded area in figure 2.9: only outcomes in this set are consistent with some QREs.[24] Note that the equilibrium correspondences of the different QRE models considered in figure 2.1 all define one-dimensional curves that belong to this set.[25]

Conversely, any point in the dark shaded area can be obtained as a QRE outcome of the game in table 2.1. To see this, let the quantal response model be given by

$$\sigma_{11}^* = F\big(\lambda_1((X+1)\sigma_{21}^* - 1)\big), \qquad \sigma_{21}^* = F\big(\lambda_2(1 - 2\sigma_{11}^*)\big),$$

for some $F(\cdot)$ and allow the precision parameter to vary across players. The fixed-point equations can be inverted to yield

$$\lambda_1 = \frac{F^{(-1)}(\sigma_{11}^*)}{(X+1)\sigma_{21}^* - 1}, \qquad \lambda_2 = \frac{F^{(-1)}(\sigma_{21}^*)}{1 - 2\sigma_{11}^*},$$

[24] Of course, QRE is a statistical theory so data from a finite sample could fall outside the dark shaded area. However, as the sample size grows the probability that this occurs tends to 0.

[25] Also the ϵ-perturbed Nash equilibrium correspondence (see example 2.7) falls in this area although part of its nonsmooth curve belongs to the edge of the set.

with $F^{(-1)}(p) < 0$ when $p < \frac{1}{2}$ and $F^{(-1)}(p) > 0$ when $p > \frac{1}{2}$. Note that λ_1 and λ_2 so defined are positive for all $\sigma_{11}^* \in (\frac{1}{2}, 1)$ and $\sigma_{12}^* \in (\frac{1}{X+1}, \frac{1}{2})$. A concrete example is obtained by considering the power Luce model with individual-specific precision parameters λ_1 and λ_2. The power Luce choice equilibrium probabilities are

$$\sigma_{11}^* = \frac{X^{\lambda_1/(\lambda_1\lambda_2+1)}}{1 + X^{\lambda_1/(\lambda_1\lambda_2+1)}}, \qquad \sigma_{21}^* = \frac{1}{1 + X^{\lambda_1\lambda_2/(\lambda_1\lambda_2+1)}}, \qquad (2.16)$$

and any (p, q) in the dark shaded area of figure 2.9, that is, $p \geq \frac{1}{2}$ and $\frac{1}{X+1} \leq q \leq \frac{1}{2}$, can be obtained by using the following precision parameters:

$$\lambda_1 = \frac{\log(1/(1-p)-1)}{\log(X) + \log(1/(1-q)-1)}, \qquad \lambda_2 = \frac{\log(1/q-1)}{\log(1/(1-p)-1)}.$$

An alternative way to evaluate the empirical restrictions of QRE is to consider its nonparametric comparative statics properties that arise by varying X in table 2.1. The set of possible QRE for $X = 0.5$, for instance, is given by the light shaded area in figure 2.9, with the only point of overlap being the midpoint $(0.5, 0.5)$. This sharp prediction does not depend on any assumptions about the underlying quantal response functions, and holds whether or not the response functions R_{ij} are assumed to be identical across the two treatments.

Finally, consider parametric comparative statics properties and ask whether the intuitive own-payoff effect shown for logit QRE in figure 2.2 is true generally: holding constant the response functions as X is varied, does Row's probability of choosing H rise with X and does Column's probability of choosing H fall in any regular QRE?

Proposition 2.6 (Own-payoff effects in asymmetric matching pennies game)*: In any regular QRE of the asymmetric matching pennies game of table 2.1, Row's probability of choosing H is strictly increasing in X and Column's probability of choosing H is strictly decreasing in X.*

Proof: The equilibrium probabilities satisfy

$$\sigma_{11}^* = R_{11}\left((X+1)\sigma_{21}^* - 1, 1 - 2\sigma_{21}^*\right), \qquad (2.17a)$$

$$\sigma_{21}^* = R_{21}\left(1 - 2\sigma_{11}^*, 2\sigma_{11}^* - 1\right). \qquad (2.17b)$$

Responsiveness implies that R_{ij} is strictly increasing in U_{ij}, and since $R_{i2}(U_{i1}, U_{i2}) = 1 - R_{i1}(U_{i1}, U_{i2})$, R_{i1} is strictly increasing (respectively, decreasing) in its first (respectively, second) argument. It follows from (2.17a) that σ_{11}^* is a strictly increasing function of σ_{21}^*, while (2.17b) implies that σ_{21}^* is a strictly decreasing function of σ_{11}^*. Hence, the solution to the fixed-point conditions (2.17) is unique. Moreover, (2.17b) implies that either σ_{11}^* rises with X and σ_{21}^* falls, σ_{11}^* falls with X and σ_{21}^* rises, or both σ_{11}^* and σ_{21}^* remain constant as X rises. The latter two cases are impossible, however, since (2.17a) implies that σ_{11}^* rises with X if σ_{21}^* remains constant or rises with X. Hence, an increase in X results in a strict increase of σ_{11}^* and a strict decrease of σ_{21}^*. ■

This comparative static "own-payoff" effect contrasts with the Nash prediction of no change in Row's choice probabilities (since they are determined by the requirement that Column is indifferent). Own-payoff effects have been consistently reported in published experiments on asymmetric matching pennies games, and do not go away with experience. In all these cases, the QRE predictions track the qualitative features of the data, and the observed choice frequencies generally fall in the QRE region, as illustrated in figure 2.9.

The main point of this section is to illustrate that even without parametric assumptions about the quantal response functions R_{ij}, regular QRE provides clear testable predictions about observable choice frequencies in games (figure 2.9), and it provides unambiguous testable comparative statics predictions as well (proposition 2.6). Furthermore, while there is no a priori reason to commit to a particular set of response functions, in many games the predictions of different parametric families of QRE will share similar properties (figure 2.1). Hence, the choice for the quantal response functions R_{ij} is often guided by practical considerations, that is, computational or technical issues that can arise, for example, when the model is applied to estimate players' precision parameters (and possibly other behavioral parameters) from observed data.

2.6.2 Empirical Restrictions of Structural QRE

Even though the most general versions of structural QRE do not restrict the nature of the shock distributions (beyond admissibility), such restrictions are needed to obtain useful empirical predictions. Indeed, Haile, Hortaçsu, and Kosenok (2008) showed quite generally that any pattern of choice probabilities can be rationalized with an additive random-utility model if one relaxes the i.i.d. assumption *and* if one has the freedom to customize the joint distribution of payoff disturbances differently for each data set.

In order to assess the methodological implications of the Haile, Hortaçsu, and Kosenok (2008) result, consider two simple examples to illustrate how, in the absence of restrictions on the joint distribution of disturbances, a standard, single decision-maker, two-alternative, probabilistic choice model can be rigged so that the probability of choosing the low-payoff outcome is arbitrarily close to 1. Suppose expected payoffs are $U_{11} = 1$ and $U_{12} = 2$. The probability of choosing option 1 is given by

$$R_{11} = \text{Prob}[1 + \varepsilon_{11} > 2 + \varepsilon_{12}],$$

and the probability of choosing option 2 is $R_{12} = 1 - R_{11}$. The epsilon shocks are not required to be independently and identically distributed, but each will have mean 0.[26] The next result is that for any $p \in (0, 1)$ one can find an admissible $f(\varepsilon_{11}, \varepsilon_{12})$ such that $R_{11} = p$, if the ε_{1j} are either (i) independently but not identically distributed or (ii) identically but not independently distributed.

(i) *Independently but not identically distributed errors.* The idea is to let ε_{11} exceed ε_{12} by more than 1 with probability p. To ensure that ε_{11} has mean 0, it has to be sufficiently negative with the complementary probability, $1 - p$. For example, let $\varepsilon_{12} = 0$ with probability 1, and let $\varepsilon_{11} = 2$ with probability p and $\varepsilon_{11} = -\frac{2p}{1-p}$ with probability $1 - p$. Then

$$R_{11} = \text{Prob}[1 + \varepsilon_{11} > 2] = p.$$

(ii) *Identically but not independently distributed errors.* The idea is to let ε_{11} be perfectly correlated with ε_{12}, such that ε_{11} exceeds ε_{12} by more than 1 for a large set of ε_{12} values. For example, let ξ be a random variable, uniformly distributed on $[0, \frac{2}{1-p}]$. Furthermore, let $\varepsilon_{12} = \xi$ and $\varepsilon_{11} = (\xi + 2)\text{mod}(\frac{2}{1-p})$. By construction, the marginal distributions of ε_{11} and ε_{12} are both uniform on $[0, \frac{2}{1-p}]$. Furthermore, ε_{11} equals $\varepsilon_{12} + 2$ except when this shift puts ε_{11} above the upper bound of the support of the distribution of ε_{12}, in which case it is inserted in the lower part of the support. Hence, $\varepsilon_{11} = \varepsilon_{12} + 2 > \varepsilon_{12}$ when ε_{12} is between 0 and $\frac{2}{1-p} - 2$ and $\varepsilon_{11} = \varepsilon_{12} + 2 - \frac{2}{1-p} < \varepsilon_{12}$ otherwise, so

$$R_{11} = \frac{2/(1-p) - 2}{2/(1-p)} = p.$$

To summarize, *without further restrictions on the error distributions, probabilistic choice models can be constructed to predict any observed behavior.*[27] In particular, when the i.i.d. assumption is dropped, the choice probability of the less attractive option can be made arbitrarily close to 1. This raises important questions about fitting the structural QRE to data, because structural QRE can

---

[26] Note that without the mean-zero restriction required by admissibility, it would be trivial to find a distribution of disturbances such that $\text{Prob}[1 + \varepsilon_{11} > 2 + \varepsilon_{12}] = p$ for any $p \in (0, 1)$. Simply let ε_{11} be uniform over $[0, 1]$ and let ε_{12} be degenerate at $\varepsilon_{12} = -p$.

[27] While the examples above concern a one-person game with only two possible strategies, the logic of the construction can be extended in a straightforward way to allow for an arbitrary number of players and strategies (Haile, Hortaçsu, and Kosenok, 2008).

fit the aggregate frequencies of any data set perfectly if one allows arbitrary error structures. However, this does not imply that probabilistic choice models, or structural QRE in particular, are without empirical content. It simply means that economically sensible assumptions about the epsilon shocks are needed before one can reach meaningful conclusions. *It is worth noting that all empirical implementations of structural QRE models maintain the i.i.d. assumption.*

2.6.3 Regularity with Structural QRE

In the definition of structural QRE, axioms (R1)–(R3) are satisfied by any choice probabilities generated by admissible payoff disturbances satisfying a full-support condition.[28] But in general, it is the monotonicity axiom (R4) that can be violated for some disturbance distributions (such as the ones in the examples of the previous section). However, as an axiom about economic choice behavior, monotonicity seems quite reasonable. So, a natural question to ask is, what admissible distributions of payoff disturbances generate regular quantal response functions?

A sufficient condition for monotonicity is i.i.d., but this restriction can obviously be relaxed. For example, a joint normal distribution with identical means, variances, and covariances will generate monotone responses. In McKelvey and Palfrey (1994), a condition called "label independence" is identified to ensure regularity. Label independence means that choice probabilities depend on expected payoffs only and not on the labels of the strategies. For example, suppose there are three options with expected payoffs $U_{11} = 1$, $U_{12} = 2$, and $U_{13} = 3$, and corresponding choice probabilities $R_{11} = \frac{1}{6}$, $R_{12} = \frac{1}{3}$, and $R_{13} = \frac{1}{2}$. Label independence implies that if expected payoffs changed to $U_{11} = 2$, $U_{12} = 3$, and $U_{13} = 1$, the choice probabilities would become $R_{11} = \frac{1}{3}$, $R_{12} = \frac{1}{2}$, and $R_{13} = \frac{1}{6}$. In other words, when the payoffs are permuted, so are the choice probabilities.

The next proposition shows that label independence of the choice probabilities can be ensured by requiring the additive payoff disturbances to be interchangeable random variables, which includes i.i.d. as a special case (Karlin, 1966). Let Ψ_{J_i} denote the set of all possible permutations of J_i objects. The density of payoff disturbances satisfies **interchangeability** if, for all $\psi \in \Psi_{J_i}$, $f(\varepsilon_1, \ldots, \varepsilon_{J_i}) = f(\varepsilon_{\psi(1)}, \ldots, \varepsilon_{\psi(J_i)})$.

> **Proposition 2.7** (Monotonicity for structural QRE): *The quantal response function defined in (2.15) is regular if $f(\varepsilon)$ satisfies admissibility and interchangeability.*

---

[28] Without full support, the inequalities in conditions (R1) and (R3) hold only weakly. For example, interiority and *strict* responsiveness are violated with uniformly distributed disturbances.

Proof: Admissibility of $f(\varepsilon)$ guarantees axioms (R1)–(R3). To prove monotonicity, it is first established that interchangeability implies label independence:

$$
\begin{aligned}
R_{ij}(U_{i\psi(1)}, \ldots, U_{i\psi(J_i)}) &= \text{Prob}\left[U_{i\psi(j)} + \varepsilon_{ij} \geq U_{i\psi(k)} + \varepsilon_{ik}, \; k = 1, \ldots, J_i\right] \\
&= \text{Prob}\left[U_{i\psi(j)} + \varepsilon_{i\psi(j)} \geq U_{i\psi(k)} + \varepsilon_{i\psi(k)}, \right. \\
&\qquad\qquad \left. k = 1, \ldots, J_i\right] \\
&= R_{i\psi(j)}(U_{i1}, \ldots, U_{iJ_i}),
\end{aligned}
$$

where interchangeability of the ε_i is used in going from the first to the second line. Label independence implies that $R_{ij}(U) = R_{ik}(U)$ if $U_{ij} = U_{ik}$, as can be seen by restricting ψ to be a pairwise permutation of j and k. Proposition 2.8 below shows that R_{ij} rises with U_{ij} and falls with all the U_{ik} for $k \neq j$. Likewise, R_{ik} falls with U_{ij} when $j \neq k$. Together with $R_{ij} = R_{ik}$ when $U_{ij} = U_{ik}$, this implies that $R_{ij} > R_{ik}$ when $U_{ij} > U_{ik}$, because as U_{ij} rises from U_{ik}, R_{ij} strictly increases and R_{ik} strictly decreases. ∎

Interchangeability is not a necessary condition in the sense that the quantal response function defined in (2.15) may be regular without it. Consider, for instance, the case of two options and let ε_{11} and ε_{12} be independent normal random variables, both with mean 0 but with different variances σ_{11}^2 and σ_{12}^2 respectively, so that the difference $\varepsilon_{12} - \varepsilon_{11}$ is normally distributed with mean 0 and variance $\sigma^2 = \sigma_{11}^2 + \sigma_{12}^2$. The quantal response functions are given by $R_{i1} = \Phi((U_{i1} - U_{i2})/\sigma)$ and $R_{i2} = \Phi((U_{i2} - U_{i1})/\sigma)$, with $\Phi(\cdot)$ the distribution function of a standard normal variable. In this example, monotonicity holds even though ε_{11} and ε_{12} are not interchangeable random variables.

2.6.4 Structural versus Reduced-Form QRE

Thus, regularity is a property of structural QRE if the additive disturbances are interchangeable random variables. The resulting label independence of the choice probabilities is an appealing property, which implies that choice probabilities depend on expected payoffs but not on the precise labeling of the strategies. However, as is well known, the *additive* structural approach also imposes other conditions on the choice probabilities.

Proposition 2.8 (Restrictions on structural QRE implied by additivity)*: The structural quantal response functions defined in (2.15) satisfy*

(i) ***translation invariance****: $R_{ij}(U_i + ce_{J_i}) = R_{ij}(U_i)$ for all $c \in \Re$ and $U_i \in \Re^{J_i}$, where $e_{J_i} = (1, \ldots, 1)$;*

(ii) **symmetry:** $\partial R_{ij}/\partial U_{ik} = \partial R_{ik}/\partial U_{ij}$ for all $j, k = 1, \ldots, J_i$ and $U_i \in \Re^{J_i}$;

(iii) **strong substitutability:** $(-1)^\ell \partial^\ell R_{ij}/\partial U_{ik_1} \cdots \partial U_{ik_\ell} \geq 0$ for all $1 \leq \ell \leq J_i - 1$, $k_1 \neq \cdots \neq k_\ell \neq j$ and $U_i \in \Re^{J_i}$.

Proof: Translation invariance and strong substitutability follow more or less directly from the definition of the structural quantal response functions:

$$R_{ij} = \int_{-\infty}^{\infty} \int_{-\infty}^{(U_{ij}-U_{i1})+\varepsilon_{ij}} \cdots \int_{-\infty}^{(U_{ij}-U_{iJ_i})+\varepsilon_{ij}} f(\varepsilon_{i1}, \ldots, \varepsilon_{ij}, \ldots, \varepsilon_{iJ_i}) d\varepsilon_i,$$

where the leftmost integral is over the entire range of possible ε_{ij} values while the other integrals restrict the ε_{ik} with $k \neq j$ such that action j is (stochastically) optimal. Since the expression for R_{ij} involves only payoff differences, adding a constant to all payoffs changes nothing in the limits of integration, or elsewhere. Differentiating with respect to $\partial U_{ik_1} \cdots \partial U_{ik_\ell}$ and multiplying by $(-1)^\ell$ removes ℓ integrals and leaves us with $J_i - \ell$ integrals of a (positive) density function. The next step is to prove symmetry:

$$\frac{\partial R_{ij}}{\partial U_{ik}} = - \int_{-\infty}^{\infty} \int_{-\infty}^{(U_{ij}-U_{i1})+\varepsilon_{ij}} \cdots \int_{-\infty}^{(U_{ij}-U_{iJ_i})+\varepsilon_{ij}} f(\varepsilon_{i1}, \ldots, \varepsilon_{ij}, \ldots, (U_{ij} - U_{ik})$$

$$+ \varepsilon_{ij}, \ldots, \varepsilon_{iJ_i}) d\varepsilon_i,$$

where the integral with respect to ε_{ik} is removed. Now transform $t = (U_{ij} - U_{ik}) + \varepsilon_{ij}$, and rewrite:

$$\frac{\partial R_{ij}}{\partial U_{ik}} = - \int_{-\infty}^{\infty} \int_{-\infty}^{(U_{ik}-U_{i1})+t} \cdots \int_{-\infty}^{(U_{ik}-U_{iJ_i})+t} f(\varepsilon_{i1}, \ldots, (U_{ik} - U_{ij})$$

$$+ t, \ldots, t, \ldots, \varepsilon_{iJ_i}) d\varepsilon_i$$

$$= \frac{\partial R_{ik}}{\partial U_{ij}},$$

where the last step follows from the definition of R_{ik}. ∎

These conditions are easy to interpret. Translation invariance implies that if a constant is added to all payoffs, choice probabilities do not change. Symmetry says that the effect of an increase in strategy k's payoff on the probability of choosing strategy j is the same as the effect of an increase in strategy j's payoff on the probability of choosing strategy k. Finally, strong substitutability implies (among other things) that if the payoff of strategy k rises, the probability of choosing any of the other strategies $j \neq k$ falls.

While they are easy to interpret, these conditions do *not* translate into sensible empirical restrictions on possible QRE outcomes, since they are not related to monotonicity. Moreover, they result from a modeling assumption (of *additive* payoff disturbances) but are *not* derived from economic principles. As a consequence, they may lead to implausible or empirically false restrictions in certain contexts.

For example, translation invariance is not plausible in settings where the magnitudes of perception errors or preference shocks depend on the magnitudes of expected payoffs. For instance, a 25 cent error is unlikely for a decision involving pennies, but such an error would be common in decisions involving hundreds of dollars. Power-law choice rules, like the Luce rule, have been shown to provide better fits to laboratory choice data when the experiment involves large changes in payoff scale across treatments, for example, from several dollars to several hundred dollars (Holt and Laury, 2002). In particular, they found that the variation in choice data did not diminish with increases in the payoff scale, so the multiplicative nature of the shock specification provided a better fit. The power-law choice model was also used by Goeree, Holt, and Palfrey (2002) to analyze data from different first-price auction treatments.

Luce's power-law rule can be generalized to allow for negative payoffs. For instance, let $g : \Re \to \Re_+$ be a strictly positive and strictly increasing function, and define

$$R_{ij} = \frac{g(\lambda_i U_{ij})}{\sum_{k=1}^{J_i} g(\lambda_i U_{ik})}, \quad i = 1, \ldots, n, \quad j = 1, \ldots, J_i. \quad (2.18)$$

For general $g(\cdot)$ functions, these choice probabilities do not necessarily satisfy translation invariance, nor symmetry, and hence cannot be derived via the structural approach with additive random shocks. Note that the choice probabilities in (2.18) do satisfy strong substitutability. However, this condition may not be desirable also because there is no a priori reason why the probability of choosing strategy 1, say, should fall when the payoff of strategy 3 rises.[29] Responsiveness implies that the probability of choosing a certain strategy should go up if its payoff rises, but there is no reason that the probabilities with which the other strategies are chosen should all go down.

---

[29] Indeed, in the structural empirical industrial organization (IO) literature, symmetry and strong substitutability are often seen as weaknesses of the logit model because of the implied substitution patterns. One solution in this literature is to define demand over product attributes (rather than products) and to allow for interaction effects between consumer demographics and product attributes. This solution is not applicable in the context of the abstract (matrix) games studied here.

To summarize, regular QREs derived from structural QREs are only a subset of all regular QREs. The reduced-form approach admits additional quantal response functions that satisfy conditions (R1)–(R4) and are not derived via the structural approach, as the next proposition illustrates.

Proposition 2.9: *The quantal response functions*

$$R_{11}(U_{11}, U_{12}, U_{13}) = \frac{e^{U_{11}}/(1 + U_{12}^2 + U_{13}^2)}{\frac{e^{U_{11}}}{1+U_{12}^2+U_{13}^2} + \frac{e^{U_{12}}}{1+U_{11}^2+U_{13}^2} + \frac{e^{U_{13}}}{1+U_{11}^2+U_{12}^2}},$$

$$R_{12}(U_{11}, U_{12}, U_{13}) = \frac{e^{U_{12}}/(1 + U_{11}^2 + U_{13}^2)}{\frac{e^{U_{11}}}{1+U_{12}^2+U_{13}^2} + \frac{e^{U_{12}}}{1+U_{11}^2+U_{13}^2} + \frac{e^{U_{13}}}{1+U_{11}^2+U_{12}^2}},$$

$$R_{13}(U_{11}, U_{12}, U_{13}) = \frac{e^{U_{13}}/(1 + U_{11}^2 + U_{12}^2)}{\frac{e^{U_{11}}}{1+U_{12}^2+U_{13}^2} + \frac{e^{U_{12}}}{1+U_{11}^2+U_{13}^2} + \frac{e^{U_{13}}}{1+U_{11}^2+U_{12}^2}}$$

are regular, but violate translation invariance, symmetry, and strong substitutability, and, hence, cannot be generated by the additive structural approach.

Proof: Interiority and continuity are obvious. To show responsiveness, consider R_{11}:

$$R_{11}(U_{11}, U_{12}, U_{13}) = \left[1 + \frac{e^{U_{12}}}{e^{U_{11}}}\left(\frac{1 + U_{12}^2 + U_{13}^2}{1 + U_{11}^2 + U_{13}^2}\right)\right.$$
$$\left. + \frac{e^{U_{13}}}{e^{U_{11}}}\left(\frac{1 + U_{12}^2 + U_{13}^2}{1 + U_{11}^2 + U_{12}^2}\right)\right]^{-1}.$$

Define $f(x, y) = e^x(1 + x^2 + y^2)$ and note that its derivative is $\partial f/\partial x > 0$, so f is a strictly increasing function of x. Hence, R_{11} is strictly increasing in U_{11}. The proofs of responsiveness for R_{12} and R_{13} are similar. To show monotonicity consider

$$\frac{R_{11}(U_{11}, U_{12}, U_{13})}{R_{12}(U_{11}, U_{12}, U_{13})} = \frac{e^{U_{11}}}{e^{U_{12}}}\left(\frac{1 + U_{11}^2 + U_{13}^2}{1 + U_{12}^2 + U_{13}^2}\right).$$

Again, since $f(U)$ is strictly increasing in U, $R_{11} > R_{12}$ (respectively, $R_{11} < R_{12}$) if and only if $U_{11} > U_{12}$ (respectively, $U_{11} < U_{12}$). The proof that monotonicity holds generally can be done in an analogous manner by considering R_{11}/R_{13} and R_{12}/R_{13}. Translation invariance is obviously not satisfied, and it is easily verified that symmetry does not generally hold. The novel feature of the example is that substitutability does not generally hold. For instance, $\partial R_{11}/\partial U_{13} = 2e^3/(3 + 2e + 3e^2)^2 > 0$ when $U_{11} = 1$, $U_{12} = 0$, and $U_{13} = -1$. ∎

2.7 QRE FOR CONTINUOUS GAMES

In many economic applications of game theory, action sets are continuous rather than finite. For instance, standard IO models of price or quantity competition assume that firms' prices (Bertrand) or quantities (Cournot) are nonnegative reals. Likewise, in the game-theoretic analysis of auctions, bids are typically assumed to be real valued. This section analyzes properties of logit QRE for continuous normal-form games $\Gamma = [N, \{A_i\}_{i=1}^n, \{u_i\}_{i=1}^n]$, where each action set A_i is a closed interval $A_i = [\underline{x}_i, \overline{x}_i]$. Existence of QRE (see proposition 2.10 below) generally requires that the action sets are bounded, that is, $\underline{x}_i$ and $\overline{x}_i$ are finite, since otherwise expected payoffs may diverge. But a bounded action set is not always necessary for existence and an example is given below where the action sets include all nonnegative real numbers, which is natural in some applications (e.g., auctions and pricing games).

In finite games, action labels are often abstract, for example, "tough" and "soft," and the game remains essentially the same if the order in which the actions are presented changes (it would simply correspond to permuting some rows/columns in the payoff matrix). In contrast, in continuous games, actions are naturally ordered. This has implications for the structural approach to QRE. While it is perfectly conceivable that "tough" and "soft" each get their own independent payoff shock, this is not plausible for a continuum of ordered actions.

For continuous games, one therefore follows the regular approach to QRE and assumes that a player's choice density, $f_i(x_i)$, is increasing in the expected payoff, $U_i(x_i)$, of action x_i. The equilibrium condition is that the distribution of choices that is determined by expected payoffs, for example, according to a logit choice rule, is the same as the distribution that is used to compute expected payoffs. Existence of QRE is easily established for the generalized Luce model in (2.18), now applied to continuous games. Define expected payoffs,

$$U_i(x_i) = \int_{\underline{x}_i}^{\overline{x}_i} u_i(x_i, x_{-i}) f_{-i}(x_{-i}) dx_{-i} \tag{2.19}$$

and choice densities,

$$f_i(x_i) = \frac{g(\lambda_i U_i(x_i))}{\int_{\underline{x}_i}^{\overline{x}_i} g(\lambda_i U_i(y)) dy}, \tag{2.20}$$

where $g(\cdot)$ is differentiable and strictly positive and increasing. For continuous games with the local payoff property, uniqueness of the symmetric logit equilibrium is implied by a variety of alternative monotonicity assumptions that arise in many applications.

Proposition 2.10 (Existence for continuous games)*: For all games in which players' expected payoffs are bounded and continuous in others' distribution functions, there exist differentiable equilibrium choice densities, $f_i(x_i)$, that solve (2.19) and (2.20).*

The proof is based on Schauder's fixed-point theorem for infinite-dimensional function spaces. For details see Anderson, Goeree, and Holt (2002).

While existence holds for general asymmetric games, the focus here will be on symmetric QRE for symmetric continuous games. In this case, one can drop the player-specific subscripts, and restrict attention to the logit QRE by assuming that the $g(\cdot)$ function in (2.20) is exponential. Differentiating (2.20) then yields a differential equation for the choice density,

$$f'(x) = \lambda f(x) U'(x). \tag{2.21}$$

This *logit differential equation* can be used to derive the choice density $f(x)$ keeping in mind that the expected payoff, $U(x)$, depends on $f(x)$ via (2.19). In general, the expected payoff $U(x)$ involves an integral over the entire range of choices and, hence, depends on the choice density at points different from x. But in many games of interest, the expected payoff $U(x)$ depends only on the density or distribution evaluated at x. For example, in a first-price auction, a prize V is awarded to the highest of N bidders, who must pay the winning bid b. In this case, the expected payoff of placing a bid b is given by $U(b) = (V - b)F(b)^{n-1}$. Since the logit differential equation (2.21) involves the marginal expected payoff, $U'(x)$, define a game having the *local payoff property* if the marginal expected payoff depends only on local variables, x, $f(x)$, $F(x)$, and possibly on an exogenous parameter, α (e.g., the value V in the first-price auction example). For games with the local payoff property one can thus write the marginal expected payoff as $U'(F, f, x, \alpha)$.

Proposition 2.11 (Uniqueness for symmetric continuous games)*: Any symmetric logit equilibrium for a game satisfying the local payoff property is unique if the expected marginal payoff, $U'(F, f, x, \alpha)$, is*

(i) strictly decreasing in x, or
(ii) strictly increasing in the common distribution function F, or

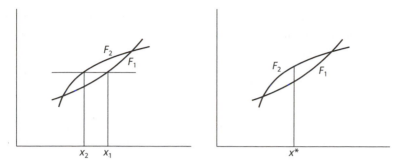

FIGURE 2.10. Horizontal lens proof (left panel) and vertical lens proof (right panel): configurations that yield a contradiction when $F_2 > F_1$.

(iii) independent of x and strictly decreasing in f, or
(iv) a polynomial expression in F, with no terms involving f or x.

Proof: For cases (i) and (ii), suppose, in contradiction, that there exist (at least) two symmetric logit equilibrium distributions, denoted by F_1 and F_2. Without loss of generality, assume $F_1(x)$ is lower on some interval, as shown in figure 2.10. Case (i) is based on a horizontal lens proof: any region of divergence between the distribution functions will have a maximum horizontal difference, as indicated by the horizontal line in the left panel of figure 2.10: $F_1(x_1) = F_2(x_2)$ with $x_1 > x_2$. The first-order and second-order conditions for the distance to be maximized at this height are that the slopes of the distribution functions be identical, that is, $f_1(x_1) = f_2(x_2)$, and that $f_1'(x_1) \geq f_2'(x_2)$. In case (i), $U'(F, f, x, \alpha)$ is decreasing in x, and since the values of the density and distribution functions are equal and $x_1 > x_2$, it follows that

$$U'(F_1(x_1), f_1(x_1), x_1, \alpha) < U'(F_2(x_2), f_2(x_2), x_2, \alpha),$$

but then the logit differential equation (2.21) implies that $f_1'(x_1) < f_2'(x_2)$, which yields the desired contradiction of the second-order condition for a maximum horizontal difference.

Case (ii) is proved with a vertical lens proof: if there are two symmetric distribution functions, then they must have a maximum vertical distance at x^* as shown in the right panel of figure 2.10. The first-order condition is that the slopes are equal, so the densities are the same at x^*. The second-order condition is that $f_2'(x^*) \leq f_1'(x^*)$. Under case (ii), $U'(F, f, x, \alpha)$ is increasing in F, so $U'(F_2, f_2, x^*, \alpha) > U'(F_1, f_1, x^*, \alpha)$, and it follows from (2.21) that $f_1'(x^*) < f_2'(x^*)$, which yields the desired contradiction. The proofs of cases (iii) and (iv) can be found in Anderson, Goeree, and Holt (2002). ∎

The logit differential equation (2.21) can be rewritten as the derivative of $\log(f_i(x))$ on the left-hand side and the derivative of $\lambda_i U_i(x)$ on the right-hand side, which can be integrated to yield

$$f(x) = \frac{e^{\lambda U(x)}}{\int_{\underline{x}}^{\overline{x}} e^{\lambda U(y)} dy}, \tag{2.22}$$

from which it is apparent that the logit equilibrium density is sensitive to all aspects of the expected payoff function, that is, choice propensities are affected by *magnitudes* of expected payoff differences, not just by the signs of the differences as in a Nash equilibrium. In particular, the logit predictions can differ sharply from Nash predictions when the costs of deviations from a Nash equilibrium are highly asymmetric, and when deviations in the less costly direction make further deviations in that direction even less risky, creating a kind of "snowball effect." These asymmetric payoff effects can be accentuated by shifts in parameters that do not alter the Nash predictions.

Since the logit equilibrium is a probability distribution, the comparative statics effects will be in terms of shifts in distribution functions. Our comparative statics effects pertain to shifts in the sense of first-degree stochastic dominance, that is, the distribution of decisions increases in this sense when the distribution function shifts down for all interior values of x. Assume that the expected payoff derivative, $U'(F, f, x, \alpha)$, is increasing in α, ceteris paribus. The next proposition shows that an increase in α raises the logit equilibrium distribution in the sense of first-degree stochastic dominance. Only monotonicity in α is required, since any parameter that decreases marginal profit can be rewritten so that marginal profit strictly increases in the new parameter. This comparative statics result will be used in the analysis of a variety of continuous games (e.g., the traveler's dilemma) in later chapters.

Proposition 2.12 (Comparative statics for symmetric continuous games): *Suppose the shift parameter, α, strictly raises (respectively, lowers) marginal expected payoffs, that is, $\partial U'(F, f, x, \alpha)/\partial \alpha > 0$ (respectively, < 0), for a symmetric game satisfying the local payoff property. Then an increase in α yields stochastically higher (respectively, lower) logit equilibrium decisions if $U'(F, f, x, \alpha)$ is*

(i) nonincreasing in x, or
(ii) nondecreasing in F.

Proof: The proofs are analogous to those of cases (i) and (ii) of proposition 2.11. Let $\alpha_2 > \alpha_1$ be the two parameter values and let the corresponding equilibrium distributions be represented by $F_2(x)$ and $F_1(x)$. The goal is to show that a higher parameter, α_2, generates higher choices in the sense of first-degree stochastic dominance, that is, $F_1(x) \geq F_2(x)$ for all x. Suppose instead that $F_1(x) < F_2(x)$ for some x as shown in the left panel of figure 2.10. For case (i), consider the line drawn at maximum horizontal difference. The associated choices, x_1 and x_2, yield the same values of their respective cumulative distributions: $F_1(x_1) = F_2(x_2)$. The first-order and second-order conditions for the horizontal distance to be maximized are $f_1(x_1) = f_2(x_2)$ and $f_1'(x_1) \geq f_2'(x_2)$. However, $U'(F, f, x, \alpha)$ is strictly increasing in α and decreasing in x so

$$U'(F_1(x_1), f_1(x_1), x_1, \alpha_1) < U'(F_2(x_2), f_2(x_2), x_2, \alpha_2),$$

but then the logit differential equation (2.21) implies $f_1'(x_1) < f_2'(x_2)$, a contradiction. Case (ii) follows similarly by considering the maximum vertical distance as in the right panel of figure 2.10. ∎

The above propositions are next illustrated with three simple examples of continuous games: a "continuation" of the asymmetric game of chicken, a continuous minimum-effort coordination game, and a Bertrand pricing game.

Example 2.8 (Example 2.3 continued: asymmetric game of chicken)*:* One natural class of continuous games arises when players in a finite normal-form game choose probabilities rather than actions. Consider, for instance, the asymmetric game of chicken in table 2.3 where Row picks a probability $p \in [0, 1]$ with which to play T. The chance of picking a particular p will be denoted $f_1(p)$, which depends on the expected payoff $U_1(p)$ via the logit choice rule. Similarly, Column chooses a probability $q \in [0, 1]$ with density $f_2(q)$. Row's expected payoff of choosing p is given by

$$U_1(p) = p \int_0^1 6(1-q)f_2(q)dq + (1-p) \int_0^1 (2-q)f_2(q)dq$$
$$= p(4 - 5\overline{q}) + (2 - \overline{q}),$$

where $\overline{q} = \int_0^1 q f_2(q)dq$. These expected payoffs map into choice probabilities using the familiar logit choice rule, that is, $f_1(p) = \exp(\lambda U_1(p))/\int_0^1 \exp(\lambda U_1(p'))dp'$, which yields

$$f_1(p) = \left(\frac{4 - 5\overline{q}}{e^{\lambda(4-5\overline{q})} - 1}\right) e^{\lambda p(4-5\overline{q})}.$$

Following the same steps for the Column player yields

$$f_2(q) = \left(\frac{12 - 13\overline{p}}{e^{\lambda(12-13\overline{p})} - 1} \right) e^{\lambda q(12 - 13\overline{p})}.$$

Notice that this model does not have the "local payoff" property discussed above; for example, Row's choice density evaluated at p depends on the entire $f_2(\cdot)$ function (and not just on functions evaluated at p). Nevertheless, the model can be easily solved by requiring that the expected values, $\overline{p}$ and $\overline{q}$, that determine the densities also result when the densities are used to calculate the expected values. This yields the fixed-point equations

$$\overline{p} = \frac{e^{\lambda(4-5\overline{q})}}{e^{\lambda(4-5\overline{q})} - 1} - \frac{1}{\lambda(4 - 5\overline{q})},$$

$$\overline{q} = \frac{e^{\lambda(12-13\overline{p})}}{e^{\lambda(12-13\overline{p})} - 1} - \frac{1}{\lambda(12 - 13\overline{p})}.$$

Define $G(x) = e^x/(e^x - 1) - 1/x$, which is a smoothly increasing distribution function on $\Re$, then the fixed-point equations can be written as

$$\overline{p} = G(\lambda(4 - 5\overline{q})), \qquad \overline{q} = G(\lambda(12 - 13\overline{p})),$$

resembling the logit equilibrium conditions when players choose actions T or S:

$$\sigma_{11} = F(\lambda(4 - 5\sigma_{21})), \qquad \sigma_{21} = F(\lambda(12 - 13\sigma_{11})),$$

where $F(x) = e^x/(e^x + 1)$ is the usual logistic distribution.

To summarize, a QRE model based on $F(\cdot)$ with players choosing probabilities results in the same *expected* choice probabilities for tough and soft as a QRE model based on $G(\cdot)$ with players choosing actions. The distribution function $G(\cdot)$ is more spread out than $F(\cdot)$, which is intuitive since noise is introduced for each of a continuum of possible choices. Qualitatively, however, the model where players choose probabilities yields very similar predictions to when they choose actions. In particular, for sufficiently high λ there are three logit QRE that limit to each of the Nash equilibria when $\lambda \to \infty$.

Example 2.9 (A continuous minimum-effort coordination game)*:*
Example 2.5 introduced a simple minimum-effort coordination game with two players and two possible effort levels, 1 and 2 (see table 2.5). This example considers a minimum-effort game with $n \geq 2$ players and a continuum of decisions. Each player $i = 1, \ldots, n$ selects a nonnegative effort,

$e_i \geq 0$, and payoffs are determined by the minimum effort minus the cost of player i's own effort:

$$u_i(e_1, \ldots, e_n) = \min(e_1, \ldots, e_n) - ce_i, \tag{2.23}$$

where the common cost parameter satisfies $\frac{1}{n} < c < 1$. This cost assumption ensures that any common effort is a Nash equilibrium, since a unilateral decrease will reduce the minimum by 1 but will result in cost savings of $c < 1$ only. Conversely, a unilateral effort increase above a common level will raise a player's cost but will not raise the minimum.[30] Since all of these equilibria are strict, standard refinements cannot be used to select one of them. More worrisome is the fact that the set of equilibria is not altered by noncritical changes in the cost of effort or the number of players, which contradicts the intuition that coordination on high-effort outcomes might be more difficult with more participants and a higher cost of effort.[31]

The QRE analysis of this game is based on the logit differential equation (2.21), so one needs a formula for the expected payoff derivative as a function of the distribution of the other players' effort choices. Let the distribution function for a player's effort decisions be $F(e)$, with density $f(e)$. One way to proceed would be to use this notation to obtain an expression for the expected value of the payoff function in (2.23), and then take the derivative with respect to a player's own effort, e_i. A quicker and more intuitive way to get the same result is to note that the cost of effort is linear in effort, so there will be a term in the expected payoff derivative that is the negative of the cost parameter. In addition, an increase in effort e_i will raise the minimum if all $n - 1$ other players' efforts are above e_i, which occurs with probability $(1 - F(e_i))^{n-1}$; so the derivative of the expected payoff is

$$U'(e) = (1 - F(e))^{n-1} - c.$$

This derivative is substituted into the formula for the logit differential equation (2.21) to obtain

$$f'(e) = \lambda f(e) \left((1 - F(e))^{n-1} - c \right). \tag{2.24}$$

---

[30] Note that this argument requires only that the common cost parameter is positive, i.e., $c > 0$. Attention is restricted to the $c > \frac{1}{n}$ case in order to ensure existence of QRE. Recall that when the action spaces are bounded, QRE exists generally. But when the possible effort choices are unbounded and the common cost parameter is low enough, $c < \frac{1}{n}$, existence of QRE fails because the logit QRE distribution puts all mass on the highest possible effort.

[31] Anderson, Goeree, and Holt (2001) showed that there is also a continuum of mixed-strategy Nash equilibria in which players randomize between two specific effort levels. These equilibria, however, have the perverse property that an increase in the cost of effort results in an *increase* in the probability of the high-effort decision.

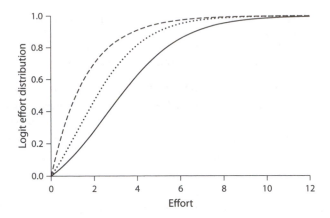

FIGURE 2.11. Effort distribution, $F(e)$, for the logit QRE model when $(n, c) = (2, 0.6)$ (solid line), $(n, c) = (2, 0.7)$ (dotted line), and $(n, c) = (100, 0.6)$ (dashed line).

Note that proposition 2.11(iv) implies that the solution will be unique. Moreover, proposition 2.12(i) implies that an increase in n or c will lower efforts in the logit QRE.

It is straightforward, albeit somewhat tedious, to verify that a "generalized logistic" distribution solves the logit differential equation

$$F(e) = 1 - \left(\frac{nc}{1 + (nc - 1) \exp(\lambda(n - 1)ce)} \right)^{1/(n-1)}.$$

Figure 2.11 shows the equilibrium distribution $F(e)$ for three sets of parameter values: the solid line corresponds to $(n, c) = (2, 0.6)$, the dashed line to $(n, c) = (100, 0.6)$, and the dotted line to $(n, c) = (2, 0.7)$. In all three cases, the precision parameter is $\lambda = 1$. Note that an increase in n or c produces lower effort choices in the sense of first-degree stochastic dominance, in accordance with proposition 2.12.

Two limit cases are of interest. First, as $nc \to 1$ the distribution limits to $F(e) \to 0$, which means that players coordinate on higher and higher efforts. This is the reason that, without an upper bound, logit QRE does not exist when $nc \le 1$. In contrast, when $nc > 1$ and $\lambda \to \infty$, the distribution limits to $F(e) \to 1$, which means that players coordinate on the lowest possible effort $e = 0$. In other words, the logit solution selects the "worst" equilibrium when $nc > 1$.

Example 2.10 (Bertrand pricing game)*:* The Bertrand model of price competition, one of the most important pillars of modern oligopoly theory, predicts cut-throat competition even with as few as two firms. When both firms have the same constant marginal costs, their incentives to capture more market sales will drive prices down to marginal costs, eliminating all profits. Uneasiness with this stark prediction stems from the absence of a cost of deviating from equilibrium. Why should firms exhibit rational equilibrium behavior if at the equilibrium they have no incentive to do so? In equilibrium, a firm's expected profit function is completely flat and any price (greater than or equal to the marginal cost) yields the same expected payoff. Moreover, if there is a slight chance that the rival will price above marginal cost, a firm is better off setting a higher (noncompetitive) price as well.

Consider a simple variant of the classic Bertrand duopoly game, where $n \geq 2$ competing firms simultaneously and independently choose prices for a homogeneous good produced at zero costs. Demand has a "box" structure: there is demand for one unit of the good for prices up to a reservation value of 1, and beyond that value demand falls to 0. Prices are therefore constrained to lie between 0 and 1. Under the homogeneity assumption, the lowest-price firm sells one unit of the good and the higher-price firms sell nothing. If m firms, with $1 < m \leq n$, tie for the lowest price then each of these sells a $\frac{1}{m}$ fraction of a unit. The Nash equilibrium prediction for this game is easy to derive: no strictly positive price, $p^* > 0$, can be an equilibrium price, since one firm could lower its price slightly below p^* and capture more market share at (almost) no cost. The unique Nash equilibrium is for each firm to charge a zero price, that is, $p^* = 0$.

Following Lopez-Acevedo (1997) and Baye and Morgan (2004), one can analyze the power Luce QRE for the Bertrand game. The power Luce QRE can be computed from a differential equation akin to (2.21), which determines logit QRE distributions. Recall that the power Luce formulation follows by applying the logit choice function to $\log(U_i)$ rather than U_i. This implies that U_i' in (2.21) is replaced by U_i'/U_i, assuming that payoffs are nonnegative, as is the case for the Bertrand model. The *Luce differential equation* is therefore

$$\frac{f_i'(p)}{f_i(p)} = \frac{\lambda_i U_i'(p)}{U_i(p)}. \tag{2.25}$$

If one considers symmetric power Luce QRE then the player-specific subscripts can be dropped and the expected payoff of charging a price p is simply

$$U(p) = p(1 - F(p))^{n-1},$$

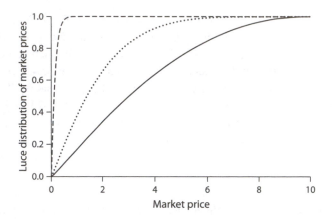

FIGURE 2.12. Market price distribution, $F_{\text{market}}(p)$, for the power Luce QRE model when $n = 2$ (solid line), $n = 4$ (dotted line), and $n = 10$ (dashed line).

where $F(p)$ is the common price distribution. Substituting into the Luce differential equation (2.25) yields

$$f' = \lambda f(p) \left(\frac{1}{p} - (n - 1) \frac{f(p)}{1 - F(p)} \right),$$

which, as in the case of the minimum-effort game, can be solved analytically:

$$F(p) = 1 - \left(1 - p^{\lambda+1} \right)^{1/(1-\lambda(n-1))}$$

is a solution for $p \in [0, 1]$ and $\lambda \in [0, \frac{1}{n-1})$ while $F(p) = 1$ is the solution for higher values of λ. In other words, for the Bertrand game, the power Luce QRE produces random behavior, $F(p) = p$, for $\lambda = 0$, as usual. But it converges to Nash equilibrium behavior, $F(p) = 1$, at a finite value $\lambda = \frac{1}{n-1}$, rather than only in the limit when $\lambda \to \infty$.

For the interesting range of λ values where behavior is neither fully random nor fully rational, the power Luce QRE exhibits intuitive comparative statics. It is instructive to consider the distribution of *market prices*, which follow from the individual price distributions in (2.10) by considering the minimum of n prices:

$$F_{\text{market}}(p) = 1 - \left(1 - p^{\lambda+1} \right)^{n/(1-\lambda(n-1))}.$$

Figure 2.12 shows the distribution of market prices when there are $n = 2$ (solid line), $n = 4$ (dotted line), and $n = 10$ (dashed line) firms. In contrast with the Nash equilibrium, power Luce QRE predicts that strictly positive prices occur in equilibrium and that prices fall, in the sense of first-degree stochastic dominance, with the number of firms.

3

Quantal Response Equilibrium in Extensive-Form Games

The previous chapter applied quantal response equilibrium (QRE) to normal-form games. Smoothed best responses were introduced to reflect that players are more likely to choose better strategies than worse strategies, but do not necessarily pick the best one with probability 1. The idea has its origins in statistical limited dependent variable models of *static* discrete choice (in economics and psychology) and stimulus/dosage response and bioassay (in biology and medical research).

There are also some applications of logit analysis to models of sequential choice, notably nested or conditional logit structures. For example, in Palfrey and Poole's (1987) analysis of voter turnout and strategic vote choice in US presidential elections, a voter is viewed as making a sequential choice. First, the voter decides whether to turn out or not, and if the decision is to turn out, then the voter decides which candidate to vote for.[1] This leads to a particularly simple nested formulation of logit choice where the first decision is made with some (payoff-related) noise, and then, if the decision is to vote, a second vote choice is made with some (payoff-related) noise. One can also extend models of discrete consumer choice in this same way. For example, for a decision about buying an automobile, the decision maker might first decide whether to buy a car or a truck and then decide which model to purchase.

This chapter extends this logic of sequential choice to extensive-form games. The formulation of the model is necessarily more complicated because timing and information now play a direct role in the decision maker's choice. This can have interesting and unanticipated consequences. For example, in the simple voting

---

[1] To some this may seem like the reverse timing of most voters' reasoning, but like in any sequential choice problem, the optimal decision is reached by first looking at the end of the game tree and solving backward. In this application, this means the turn-out decision is made with the voter correctly anticipating which candidate he would vote for if he votes at all.

problem, it may make a difference whether the voter first decides who to vote for or first decides whether to vote at all. And it could be even more complicated if the voter first had to register to vote, and had to accumulate information about the election and candidates.

Such subtle differences in modeling the game normally have no effect on standard equilibrium analysis, due to the *principle of strategic invariance*: extensive-form games with the same strategic form produce the same behavior. This principle lies at the very foundation of virtually all applications of game theory. It has been argued vigorously by some (e.g., Kohlberg and Mertens, 1986, p. 1011) that "the reduced normal form captures all the relevant information for decision purposes," leading to the traditional view that any good theoretical equilibrium concept for noncooperative games must satisfy invariance. However, it is known from experimental data (and common sense) that this invariance principle does not hold universally.

A simple example of invariance failure arises in the sequential version of the battle-of-the-sexes game. Two players can each choose either A or B. If they both choose A, then player 1 receives a large prize and player 2 a small prize. If they both choose B, player 1 receives the small prize and player 2 receives the large prize. If they choose differently, neither receives a prize. There are three Nash equilibria, two pure equilibria where both players choose the same, and one mixed equilibrium. In the one-shot simultaneous play version of the game (in the laboratory), one typically sees choice frequencies that correspond to a logit version of the mixed-strategy Nash equilibrium. However, suppose player 1 chooses first, and then player 2 chooses, but player 2 is not shown player 1's choice until after both have made a choice. In this variation of the game, the choice frequencies are mostly (A, A), reflecting intuition. Both versions of the game are strategically equivalent, but clearly choices depend on the actual extensive-form representation of the game, perhaps due to focal-point effects.

As seen in the previous chapter, an interesting property of QRE is that systematic deviations from Nash equilibrium are often predicted even without introducing systematic features to the error structure (i.e., with independently and identically distributed payoff disturbances). The systematic effects of the statistical disturbances are (indirect) equilibrium phenomena that arise because of the strategic responses of players to the noisy environment, not because of some kind of bias or game misspecification.

Perhaps an even more important property of QRE, which is explored in this chapter, is that strategically equivalent games may have different equilibria. In our statistical theory of decision making, the compelling arguments for invariance fall apart. In fact, it is easy to construct examples where QRE predicts systematic differences in play depending on which "equivalent" version of a game is played.

Moreover, the intuition behind these differences is very sensible, and supported by experimental data. Thus, QRE provides a completely general framework for studying violations of strategic equivalence. And while it cannot account for all known violations, it does explain some.

While the breakdown of strategic equivalence may be viewed as a "bug" by traditional game theorists,[2] QRE's sensitivity to classically inessential details of the extensive form can be viewed as a "feature." Understanding the breakdown of invariance is important for understanding the behavioral foundations of strategic decision making and the understanding of dynamic individual choice. From a cognitive standpoint, choosing from all available strategies at the beginning of a game (a dynamic choice problem) is much different than thinking about the problem only as a collection or sequence of smaller choices, where little or no thought is given to "off-the-path" decision nodes.

Besides discriminating between different versions of a game that have equivalent strategic-form representations, QRE also makes different predictions depending upon whether the game is played in "agent" normal form, the more traditional normal form, or the reduced normal form of the same game. The intuition for why QRE predicts "representation dependence" is that the error structure introduces private information in particular ways that depend on the exact details of how the game is actually played. Because of the specific statistical predictions that are implied by the dependence, many aspects of QRE can be examined directly by applying standard maximum-likelihood techniques to data generated in controlled laboratory experiments. For computational reasons, this book focuses most of its attention on a version of QRE that is similar in spirit to the agent model of how an extensive-form game is played, and so it is called the *agent quantal response equilibrium* (AQRE).

The logit AQRE model implies a unique selection from the set of Nash equilibria. This selection is defined by the connected component of the logit AQRE correspondence, as with the logit solution characterized in chapter 2. This selection thus generates a unique prediction of a strictly positive probability distribution over play paths for every value of the logit response parameter, for almost every finite extensive-form game. As demonstrated in chapter 2, this selection may be inconsistent with trembling-hand perfection as applied to the reduced normal form of the game, as well as all other refinements.

In signaling games, there is a well-known multiple equilibrium problem to which a great deal of effort has been directed by modern game theorists,

---

[2] Kohlberg and Mertens (1986), for instance, defend their Nash refinement criterion, "stability," partly on the grounds that it satisfies strategic invariance.

generating a vast literature on the general problem of deductively based refinements. Several experiments have been conducted to test whether the refinements developed to distinguish these equilibria are useful in helping to predict which equilibria (if any) are more likely to be played. This is a natural setting in which to apply AQRE, since this equilibrium concept does not have the problem of trying to define behavior off the equilibrium path. This equilibrium concept also makes multiple equilibrium predictions, but the logit AQRE has a natural refinement that generically selects a unique equilibrium in a way that is much different from traditional (deductive) refinement arguments. In addition to predicting the patterns of play "on the Nash equilibrium path," this equilibrium concept also predicts patterns of play "off the Nash equilibrium path." Some of the anomalies (vis-à-vis more standard theories) uncovered in these experiments have to do with behavior off the equilibrium path, and logit AQRE can account for these.

In centipede game experiments (McKelvey and Palfrey, 1992) frequent violations of Nash predicted play are observed. This, and the fact that observed outcomes are substantial Pareto improvements over Nash play, can be rationalized by a model in which some of the players have altruistic preferences. Using arguments based on reputation building (Kreps et al., 1982) it can be shown that frequent violations of Nash behavior (similar to what is observed) can be accounted for even if altruistic players are "rare." While this explanation of the data does account pretty well for some of the salient features of that particular data set, it is clearly ad hoc. The explanation involves the invention, or assumption, of a "deviant type" who systematically violates Nash behavior in exactly the direction observed in the data. A preferable explanation would be able to account for this data without resorting to such "ad-hoc-ery." The AQRE provides a framework for doing exactly that, and has the desirable feature of being applicable to arbitrary games without necessitating the invention of systematically deviant types that are tailored to the specific peculiarities of the game.

Accordingly, the centipede data is reexamined using the one-parameter logit AQRE model to show that it accounts for the same qualitative violations of Nash behavior as the five-parameter altruism model in McKelvey and Palfrey (1992).[3] The logit AQRE model also outperforms competing models when applied to data from much different centipede-type experiments where all outcomes are Pareto optimal (Fey, McKelvey, and Palfrey, 1996).

---

[3] Zauner (1999) reexamined the centipede data using a similar model, but with different distributional assumptions about the error structure.

3.1 REGULAR QRE FOR EXTENSIVE-FORM GAMES

It is instructive to begin by describing four possible ways to define QRE in extensive-form games, depending on how the games are represented. Importantly, *all four* alternatives yield outcome-equivalent results for standard Nash equilibrium, but with QRE there are subtle differences. As in the definition of QRE for normal-form games, this is explained using the reduced-form approach.

1. *Normal-form representation.* In this case, strategies are defined as complete plans of action. In the reduced-form approach, quantal response functions are defined the same as described in chapter 2, that is, as functions mapping i's expected payoffs of strategies into a probability distribution over those same strategies. So a quantal response equilibrium of the game's normal form will be a profile of mixtures over complete plans of action, one for each player.

2. *Reduced normal-form representation.* In this case, the normal form of the game is first reduced by eliminating equivalent strategies (plans of action) for all players. Two strategies for player i are equivalent if and only if the expected payoffs for all players are the same, for all profiles of strategies for players other than i. That is, all duplicate strategies are condensed into a single "representative" strategy. A quantal response equilibrium of the game's reduced normal form is a profile of mixtures over a maximal set of nonequivalent plans of action, one for each player.

3. *Agent normal-form representation.* In this case, additional (fictitious) players, called *agents*, are added to the game. At each of player i's information sets, replace player i with a new "player-agent" who has the same payoffs as i in every terminal node. If H_i denotes the collection of all of player i's information sets, then what was originally an n- player game is now a game with $\sum_{i=1}^{n} |H_i|$ player-agents. Each player-agent then has a quantal response function that maps expected payoffs of the actions available at a (single) information set into a probability distribution over those actions. The expected payoffs are computed ex ante, given the mixed strategies of all the other player-agents in the game. A quantal response equilibrium of the game's agent normal form is a profile of mixtures over complete plans of action at every information set, one for each player-agent.

4. *Agent extensive-form representation (behavioral strategies with player-agents).* This is the representation used for the applications in this book, and it is also the approach proposed in McKelvey and Palfrey (1998). Behavioral strategies are mappings that assign to each information

set a mixture over the actions available to the player at that information set. Each player then has a quantal response function that maps expected payoffs of the actions available at each of his information sets into a probability distribution over those available actions. There is a similarity to the agent normal-form representation, but the payoffs are computed differently in the behavioral strategy approach, conditional on arriving at a particular information set, and based on continuation values, rather than ex ante. That is, expected payoffs for player i at an information set $h \in H_i$ are computed at an interim stage, considering the mixed strategies only at information sets that follow h. This approach also implies a stochastic version of sequential rationality, as proved later in this chapter.

For all four of these approaches, regular QRE is defined as a fixed point of quantal response functions that satisfy interiority, continuity, monotonicity, and responsiveness. The differences between the game representations are illustrated with a simple example of the "centipede game" (McKelvey and Palfrey, 1992, 1998; Zauner, 1999).

Example 3.1 (Centipede game)*:* Rosenthal (1981) originally studied a class of two-person games which have come to be known as "centipede games" (Binmore, 1987; Aumann, 1995). The version considered here can be described as follows. There are two piles of money, a larger one containing \$4 and a smaller one containing \$1. Player 1 moves first and can either pass (P) or terminate (T) the game. If he terminates the game, he receives the larger pile and his opponent receives the smaller one. If he passes, both piles of money double in size to \$8 and \$2. Then it's the second player's move to either pass or terminate and receive the larger pile. If she passes, both piles of money double in size again, to \$16 and \$4. Then it is player 1's move again to either pass or terminate and receive the larger pile. If he passes, both piles of money double in size to \$32 and \$8. Then it is player 2's move again to either pass or terminate and receive the larger pile. If she passes, both piles double in size one last time to \$64 and \$16, she receives the smaller pile, the original first mover receives the larger pile, and the game is over. The game tree is given in figure 3.1. Player 1 has two information sets, labeled 1a and 1b. Player 2 also has two information sets, labeled 2a and 2b.

1. *Normal-form representation.* In this simple four-move centipede game, each player has two information sets and two available moves at each information set. The strategy set for both players is $S = \{TT, TP, PT, PP\}$, where TT is the strategy where a player terminates at both information sets, and so forth. The payoffs of the game

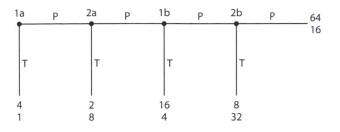

FIGURE 3.1. Game tree for a four-move centipede game.

TABLE 3.1. Normal form for the four-move centipede game in figure 3.1.

	TT	TP	PT	PP
TT	4, 1	4, 1	4, 1	4, 1
TP	4, 1	4, 1	4, 1	4, 1
PT	2, 8	2, 8	16, 4	16, 4
PP	2, 8	2, 8	8, 32	64, 16

are represented in matrix form in table 3.1. The first thing to notice is that TT and TP are equivalent strategies. Looking slightly ahead, these strategies will be merged into a single strategy, T, in the reduced normal form. But in this representation the strategies are retained, and referred to as "clones" of each other.[4]

The approach of chapter 2 can be applied to the normal-form game in table 3.1 to obtain logit QRE strategies $\{\sigma_i^*(TT), \sigma_i^*(TP), \sigma_i^*(PT), \sigma_i^*(PP)\}$ for $i = 1, 2$ as functions of λ. It is straightforward to convert these strategies into *conditional take probabilities,* $\{q_{1a}, q_{2a}, q_{1b}, q_{2b}\}$, which have a more intuitive interpretation. The conditional take probability at an information set is the probability the game is terminated by the player moving at that information set, given that no earlier player terminated the game: $q_{ia} = \sigma_i^*(TT) + \sigma_i^*(TP)$ and $q_{ib} = \sigma_i^*(PT)/(\sigma_i^*(PT) + \sigma_i^*(PP))$ for $i = 1, 2$. These logit QRE conditional take probabilities are shown in the upper-left panel of figure 3.2.

Note that for the normal-form representation, logit QRE does not limit to the subgame-perfect outcome: while player 1 terminates the game at the first opportunity (information set 1a) both players pass with probability $\frac{1}{7}$ at later nodes. This explains why the conditional take probability q_{1a} limits to 1 while q_{1b}, q_{2a} and q_{2b} limit to $\frac{6}{7}$. In terms of normal-form game strategies, player 1's strategy $\sigma_1^*(TT) + \sigma_1^*(TP) = 1$ makes player 2

---

[4] The term "cloned strategies" is used in Camerer, Ho, and Chong (2004).

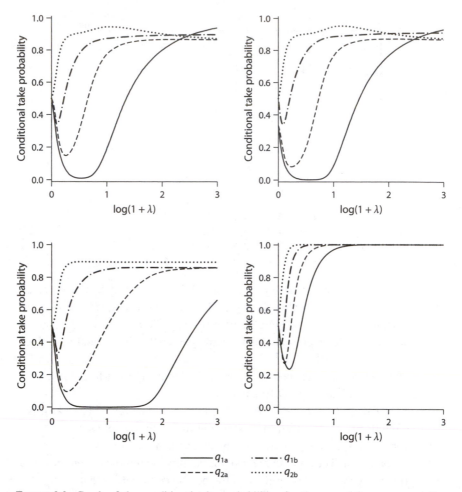

FIGURE 3.2. Graph of the conditional take probabilities for the normal-form representation (upper-left panel), reduced normal-form representation (upper-right panel), agent normal-form representation (lower-left panel), and agent extensive-form (behavioral strategy) representation (lower-right panel) of the four-move centipede game in figure 3.1.

indifferent. Likewise, player 2's strategies $\sigma_2^*(TT) + \sigma_2^*(TP) = \frac{6}{7}$, $\sigma_2^*(PT) = \frac{6}{49}$, and $\sigma_2^*(PP) = \frac{1}{49}$ make player 1 indifferent. In other words, logit QRE selects a non-subgame-perfect mixed-strategy Nash equilibrium.

2. *Reduced normal-form representation.* In the reduced normal form of the game, cloned strategies are collapsed into a single strategy. Therefore, the reduced normal form is a 3×3 game with payoffs shown in table 3.2.

TABLE 3.2. Reduced normal form of the four-move centipede game.

	T	PT	PP
T	4, 1	4, 1	4, 1
PT	2, 8	16, 4	16, 4
PP	2, 8	8, 32	64, 16

TABLE 3.3. Agent normal form of four-move centipede game.

		T_{2a} T_{2b}	T_{2a} P_{2b}	P_{2a} T_{2b}	P_{2a} P_{2b}
T_{1a}	T_{1b}	4, 1	4, 1	4, 1	4, 1
T_{1a}	P_{1b}	4, 1	4, 1	4, 1	4, 1
P_{1a}	T_{1b}	2, 8	2, 8	16, 4	16, 4
P_{1a}	P_{1b}	2, 8	2, 8	8, 32	64, 16

Now the logit QRE is computed as the solution to a smaller system of equations. The implied conditional take probabilities, graphed in the upper-right panel of figure 3.2, are different from those in the normal-form game shown in the upper-left panel, although they exhibit the same qualitative pattern. For instance, when $\lambda = 0$ and behavior is completely random, $\sigma_i^*(T) = \frac{1}{3}$ while previously $\sigma_i^*(TT) + \sigma_i^*(TP) = \frac{1}{4} + \frac{1}{4} = \frac{1}{2}$. However, the differences disappear for higher values of λ and when $\lambda \to \infty$, logit QRE for the reduced normal-form representation selects the same mixed-strategy Nash equilibrium as under the (nonreduced) normal-form representation.

3. *Agent normal-form representation.* The agent normal form is a four-player game since there are four information sets, and the four agents are labeled 1a, 1b, 2a, and 2b, following the labels in figure 3.1. Each player-agent has two strategies: pass and terminate. The game is represented by the four matrices in table 3.3. Agent 2b chooses a column in the matrices (same column for all four matrices) while agent 1b chooses a row in the matrices (same row for all matrices). Agent 2a chooses either the left or right pair of matrices and agent 1a chooses either the upper or lower pair of matrices. For ease of notation, the payoffs are written as pairs, with the first payoff for agents 1a and 1b and the second payoff for agents 2a and 2b.

The logit equilibrium of the $2 \times 2 \times 2 \times 2$ game specifies the termination probabilities for the agents, which coincide with the "conditional" take probabilities. That is, in the agent normal-form representation of the game, $q_{ia} = \sigma_{ia}^*(T)$ and $q_{ib} = \sigma_{ib}^*(T)$ for $i = 1, 2$. These conditional take probabilities are shown in the lower-left panel of figure 3.2. While convergence is slower under this representation of the game, the conditional take probabilities share the same qualitative features as in the previous two representations. In particular, the same mixed-strategy Nash equilibrium is selected when $\lambda \to \infty$.

4. *Behavioral strategy representation.* As for the *agent* normal-form representation, the analysis is carried out for each information set, but the analysis exploits the timing of the extensive form, rather than the strategic (normal) form. Thus, the logit equilibrium of the game again specifies the termination probability for each agent using expected payoffs that are given by continuation payoffs rather than ex ante expected payoffs. In other words, the logit QRE for the behavioral strategy representation is similar to that of the agent normal-form representation, but with one important difference. In the agent normal-form representation, the difference in expected utility between terminate and pass for a player at an information set is *deflated* by a factor equal to the probability of reaching that information set. This deflation factor will be higher for later information sets, because it is the product of all the pass probabilities in earlier information sets. Because the payoffs in the behavioral strategy approach are the actual (undeflated) continuation values, this will imply a different QRE solution than the agent normal form.

The conditional take probabilities (equivalently the behavioral strategies) of the logit AQREs are shown in the lower-right panel of figure 3.2. Note that under this representation, convergence is fast and the logit AQRE limits to the subgame-perfect equilibrium where agents choose to terminate the game at every decision node. Because of its intuitive interpretation and superior limit properties, the agent extensive-form (behavioral strategy) representation is used for the formal definition of structural AQRE.

3.2 STRUCTURAL AQRE FOR EXTENSIVE-FORM GAMES

3.2.1 Notation

Let I be a finite set of $n + 1$ *players*, one of whom (player "0") is designated *chance*, let X be a finite set of *outcomes*, and let A be a finite set of *actions*. Let $\boldsymbol{\Gamma} = (\Gamma, Q)$ be a *topological tree* with Γ the set of *nodes*, and $Q \subseteq \Gamma \times \Gamma$ a binary relation on Γ representing *branches*, and let Q^* be the transitive closure of Q. If vQv' then v *immediately follows* v' (or v' *immediately precedes* v). If vQ^*v' then v *follows* v' (or v' *precedes* v). Write $Q(v)$ and $Q^*(v)$ for the set of immediate followers and the set of followers of $v \in \Gamma$, respectively. Q is an asymmetric, acyclic binary relation on Γ such that every node $v \in \Gamma$ follows at most one node, and such that there is exactly one node, $v^* \in \Gamma$, which follows no node. The element v^* is called the *root node*. A node $v \in \Gamma$ is *terminal* if $Q(v) = \emptyset$, otherwise it is *nonterminal*. Let Γ^t and Γ^0 represent the set of terminal and nonterminal nodes of Γ, respectively.[5]

Define an extensive-form game $\mathcal{G}(N, X, A, \Gamma)$ by defining an *outcome function* $\psi : \Gamma^t \to X$, a *partition function* $P : \Gamma^0 \to I \times \mathcal{Z}$, an *index function* $\phi : \Gamma \to A$ which is one-to-one on $Q(v)$ for each $v \in \Gamma^0$, and which satisfies $\phi(Q(v)) = \phi(Q(v'))$ whenever $P[v] = P[v']$, and a *probability function* $\Lambda : \Gamma^0 \to [0, 1]$ which is a probability function on $Q(v)$ for each $v \in \Gamma^0$ with $P_1[v] = 0$.

For any $v \in \Gamma^0$, define $A(v) = \phi(Q(v))$ to be the set of *actions* available at v. For each $v \in \Gamma^0$ define $h(v) = \{v' \in \Gamma^0 : P[v] = P[v']\}$ to be the *information set* containing v. Let $H = \{h : h = h(v) \text{ for some } v \in \Gamma^0\}$ be the set of all information sets. If $h = h(v)$ write $P[h] = P[v]$ and $A(h) = A(v)$. Write $H_i = \{h \in H : P_1[h] = i\}$ as the set of information sets for player i. If $P[v] = (i, j)$, write $h_i^j = h(v)$. An action $a \in A(h)$ is said to *precede* a node $v \in \Gamma$ if there is an $x \in h$ and $y \in Q(x)$ with $v \in Q^*(y)$ and $\phi(y) = a$.

An extensive-form game $\mathcal{G}$ is said to have *perfect recall* if for every player $i \in I$, all information sets $h, h' \in H_i$, every action $a \in A(h)$, and all nodes $x, y \in h'$, then a precedes x if and only if a precedes y. In a game of perfect recall, a player remembers what he previously knew and what he previously did. Assume that $\mathcal{G}$ has perfect recall.

A *behavioral strategy* for player $i \in I$ is a function $\sigma_i : H_i \to \mathcal{M}(A)$ satisfying $\sigma_i(h) \in \mathcal{M}(A(h))$ for all $h \in H_i$. Here $\mathcal{M}(C)$ indicates the set of all probability measures over the set C. Again, for $h \in H_i$ use the shorthand $\sigma_{ih} = \sigma_i(h)$ and $\sigma_{ij} = \sigma(h_i^j)$. For each $a \in A(h)$, $\sigma_{iha} = \sigma_{ih}(a)$ denotes the probability of action a. Let Σ_i denote the set of all behavioral strategies for player i, and $\Sigma = \prod_{i \in N} \Sigma_i$ be the set of behavioral profiles. If $v \in h \in H_0$ is a chance node, and $v' \in Q(v)$, then define σ by $\sigma_h(v') = \Lambda_v(v')$. Define $\Sigma_{-i} = \prod_{j \neq i} \Sigma_j$, and write elements of Σ in the form $\sigma = (\sigma_i, \sigma_{-i})$ to focus on a particular player $i \in I$.

Let Σ° denote the interior of Σ, that is, totally mixed behavioral strategies. Each behavioral strategy profile $\sigma \in \Sigma^\circ$ determines a strictly positive *realization probability* $\rho(v|\sigma)$ for each node $v \in \Gamma$, as follows. If $v = v^*$ is the root node, then define $\rho(v|\sigma) = 1$. Otherwise, find v' with $v \in Q(v')$. Since $v' \in \Gamma^0$, $v' \in h \in H_i$

---

[5] This section includes material from McKelvey and Palfrey (1998).

for some $i \in I$, define $\rho(v|\sigma) = \rho(v')\sigma_h(v')$. For any $h \in H$, define $\rho(h|\sigma) = \sum_{v \in H} \rho(v|\sigma)$. Also, for any $h \in H$, define $Q^*(h) = \cup \{Q^*(v) : v \in h\}$. Define a *conditional realization probability*, $\rho(v|h, \sigma)$ on $Q^*(h)$, by $\rho(v|h, \sigma)\rho(h|\sigma) = \rho(v|\sigma)$. Note that for any $\sigma \in B^\circ$, $\rho(v|\sigma)$ defines a strictly positive probability measure over the terminal nodes. Similarly, $\rho(v|h, \sigma)$ defines a probability measure over $\Gamma^t \cap Q^*(h)$.

For any $h \in H$, $i \in N$, and $\sigma \in \Sigma^\circ$, define the conditional expected payoff function $U_i : \Sigma^\circ \to \Re$ by

$$U_i(\sigma|h) = \sum_{v \in \Gamma^t \cap Q^*(h)} \rho(v|h, \sigma) u_i(v).$$

For any interior behavioral strategy $\sigma \in \Sigma^\circ$, any i, and any information set $h_i^k \in H_i$, denote by $U_{ijk}(\sigma)$ the conditional expected payoff to i of playing action $a_j \in A(h_i^k)$ at h_i^k with probability 1, and following σ_i in all subsequent information sets.

3.2.2 Agent Quantal Response Equilibrium

Structural AQRE is modeled in a way that is compatible with the behavioral strategy representation. Each agent of each player has an additive payoff disturbance that is added to the continuation payoff for each possible action at that agent's information set. It is called "agent" QRE because each player-agent acts independently, as in the agent normal form, without knowledge of any other agents' payoff disturbances.

For each i, h_i^k, $a_j \in A(h_i^k)$, let ε_{ijk} be a random variable, which represents i's *payoff disturbance* from choosing a_j at information set h_i^k. For any $\sigma \in \Sigma^\circ$, let

$$\widehat{U}_{ijk} = U_{ijk}(\sigma) + \varepsilon_{ijk},$$

where ε is private information to the players. So player i observes ε_{ijk} but the other players (or the econometrician) do not. Let ε_{ik} be the vector of i's errors at h_i^k, let ε_i be the entire error vector across i's information sets, and let $\varepsilon = (\varepsilon_1, \ldots, \varepsilon_n)$ be the vector of errors for all players. Assume ε is an absolutely continuous random vector (with respect to Lebesgue measure), distributed according to a joint distribution with full-support density function $f(\varepsilon)$. Also assume that the ε_{ik} are statistically independent and that $E(\varepsilon_{ijk})$ exists for all $a_j \in A(h_i^k)$. Any probability density function, f, satisfying the above assumptions is called *admissible*.

At each information set, h_i^k, player i selects an action a_j that maximizes $\widehat{U}_{ijk}$. For any action $a_j \in A(h_i^k)$ and any vector of expected payoffs $U_i = (U_i, \ldots, U_{J^i})$, define

$$D_{ijk}(U_i) = \{\varepsilon : U_{ijk} + \varepsilon_{ijk} \geq U_{ij'k} + \varepsilon_{ij'k} \quad \forall a_{j'} \in A(h_i^k)\}$$

and

$$R_{ijk}(U_i) = \int_{D_{ijk}(U_i)} f(\epsilon) d\epsilon.$$

Definition 3.1: *For any extensive-form game $\mathcal{G}$ and error structure $f(\varepsilon)$, a behavioral strategy $\sigma^* \in \Sigma$ is an **agent quantal response equilibrium** (**AQRE**) if it is a fixed point in Σ of $R \circ U$. It is a behavioral strategy profile $\sigma^* \in \Sigma^\circ$ such that for all $i \in I$, h_i^k, $a_j \in A(h_i^k)$, we have $\sigma_{ijk}^* = R_{ijk}(U_i(\sigma^*))$.*

That is, σ^* is a quantal response equilibrium if, when each player i is choosing their best response at every information set h_i^k, taking σ_{-ij}^* and ε_{ijk} as given, this generates a behavioral strategy that is exactly the same as σ^*. To emphasize a simple point, this structural approach does not imply that players are boundedly rational in any sense of the term. By "choosing their best response," what is meant is that player i chooses $a_j \in A(h_i^k)$ if and only if $\varepsilon \in D_{ijk}(U_i)$, that is, if and only if private payoff disturbances are such that $\widehat{U}_{ijk} \geq \widehat{U}_{ij'k}$ for all $a_{j'} \in A(h_i^k)$. But to an outsider who only observes U_i this might appear to be suboptimal choice behavior by a boundedly rational player.

This definition assumes the "agent model" of play in an extensive-form game, where different information sets of a player are played out by different agents, all of whom share the same payoff function. In the agent model, each agent i_k simply chooses the maximum of $\widehat{U}_{ijk}$ at information set h_i^k and acts independently of the other agents of the same player.[6] For this reason the equilibrium defined above is called the *agent quantal response equilibrium* (AQRE). If the disturbances are i.i.d., then choice behavior in an AQRE will mimic the "reduced-form" version introduced in chapter 2, regular QRE, where players are "soft optimizers" relative to U_i and choose suboptimal choices (relative to U_i) in a payoff-responsive way. This gives rise to the agent form variation on regular quantal response equilibrium, where $R(U)$ follows a smooth function such as logit. Applications to extensive-form games typically follow the *regular AQRE* approach.

---

[6] This is not to be confused with the agent *normal-form* QRE discussed in the previous section, where payoffs are computed ex ante rather than by continuation values.

Proposition 3.1 (Existence of AQRE): *For any admissible f, an AQRE exists.*

Proof: Define $\phi = R \circ U : \Sigma \mapsto \Sigma$. Note that ϕ is continuous. Extend ϕ to have domain $\overline{\Sigma}$ by

$$\phi_{ij}(b) = \mathrm{co}\left\{ \lim_{t \to \infty} \phi_{ij}(\sigma_t) : \{\sigma_t\} \subseteq \Sigma^\circ \text{ and } \lim_{t \to \infty} \sigma_t = \sigma \right\}.$$

Then $\overline{\Sigma}$ is compact, convex, and $\phi : \overline{\Sigma} \rightrightarrows \overline{\Sigma}$ is convex valued and upper hemicontinuous. By Kakutani's fixed-point theorem, ϕ has a fixed point, $\sigma^* \in \overline{\Sigma}$. It is easy to show there can be no fixed point on the boundary. Hence, there is a $\sigma^* \in \Sigma^\circ$ with $\phi(\sigma^*) = \sigma^*$. ∎

The above proposition guarantees existence for AQRE, and the full support assumption guarantees that the equilibrium behavioral strategies place positive probability on every available action in every information set of every player. The above model can be generalized by dropping the independence assumption, or by specifying an observation function $\mathcal{O}(h_i^k)$, where for each i, h_i^k, $\mathcal{O}(h_i^k)$ specifies a signal that i receives about ε at information set h_i^k, rather than just assuming i observes only U_{ik} at h_i^k. While these changes would require some additional notation, the basic ideas and properties of the quantal response model would not change. Some models of $\mathcal{O}(h_i^k)$ other than the one focused on here might be interesting to explore.

3.3 LOGIT AQRE

The logit response function at information set h_i^k is given by

$$\sigma_{ijk} = \frac{e^{\lambda U_{ijk}(\sigma)}}{\displaystyle\sum_{a_{j'} \in A(h_i^k)} e^{\lambda U_{ij'k}(\sigma)}}, \tag{3.1}$$

where the equilibrium (expected) continuation values, U, are as defined earlier in the chapter. Given $\lambda \in [0, \infty)$, a *logit AQRE* is any solution to this set of equations (one equation for each action in each information set of each agent).

Define the *logit AQRE correspondence* $\sigma^* : \mathcal{R}_+ \rightrightarrows \Sigma^\circ$ as the mapping from $\lambda \in [0, \infty)$ to the set of logit AQRE behavioral strategies when the precision equals λ:

$$\sigma^*(\lambda) = \{\sigma \in \Sigma^\circ : \text{equation (3.1) is satisfied for all } i, j, k\}.$$

The logit AQRE correspondence has several useful properties. Proposition 3.1 has already established that $\sigma^*(\lambda) \neq \emptyset$ for all $\lambda \geq 0$, that is, at least one logit

AQRE exists. A second property is that limit points of $\sigma^*(\lambda)$ as $\lambda \to \infty$ are behavioral strategy Nash equilibria of the game. Both of these properties follow from the same arguments as in chapter 2. A third property is that limit points of $\sigma^*(\lambda)$ are not only Nash equilibrium behavioral strategy profiles, but are sequential equilibrium strategy profiles:

> ***Proposition 3.2*** (Limit points of logit AQREs are sequential equilibria): *For every finite extensive-form game, every limit point of a sequence of logit AQREs when $\lambda \to \infty$ corresponds to a sequential equilibrium of the game.*

Proof: Let σ^*_∞ be a limit point of $\sigma^*(\lambda)$. Then there exist consistent beliefs μ^*_∞ (i.e., assignments of a probability distribution over the nodes at each information set that satisfy the Kreps–Wilson (1982) consistency condition) such that, under those beliefs, σ^*_∞ specifies optimal behavior for every continuation game. That is, σ^*_∞ is sequentially rational given μ^*_∞ and μ^*_∞ is consistent with σ^*_∞. This is proved below.

Take any sequence $\{\lambda_t\}_{t=1}^\infty$ and $\{\sigma^*_t\}_{t=1}^\infty$ such that $\lim_{t \to \infty} \lambda_t = \infty$ and $\lim_{t \to \infty} \sigma^*_t = \sigma^*$ and σ^*_t is a logit equilibrium for λ_t for each t. First, note that for all t, for all i, k, and for all information sets h_i^k, we have $\sigma^*_{ijkt} > 0$ for all $a_j \in A(h_i^k)$. Consequently every node is reached with positive probability. Therefore, by Bayes' rule, for each t, σ^*_t uniquely defines a set of beliefs, μ^*_t, over the nodes of every information set. Moreover, since μ_t varies continuously with σ_t, there is a unique limit μ^*_∞. Therefore μ^*_∞ are consistent beliefs with respect to σ^*_∞. What remains to be shown is that $\sigma^*_\infty(h)$ is optimal given $\mu^*_\infty(h)$ for all h. (If so, then the assessment, $(\sigma^*_\infty, \mu^*_\infty)$, is a sequential equilibrium, so that σ^*_∞ is a sequential equilibrium strategy.) Suppose not, then there exists $\varepsilon > 0$ such that for some h_i^k there is some pair of actions $\{a_j, a_{j'}\} \in A(h_i^k)$ such that $\sigma^*_{ijk\infty} > 0$ but $U_{ij'k}(\sigma^*_\infty) - U_{ijk}(\sigma^*_\infty) > \epsilon$. But then there must exist $T > 0$ such that

$$U_{ij'k}(\sigma^*_t) - U_{ijk}(\sigma^*_t) \geq \frac{\epsilon}{2} \quad \forall t \geq T.$$

Therefore,

$$\frac{\sigma^*_{ijkt}}{\sigma^*_{ij'kt}} \leq e^{-\lambda_t(\epsilon/2)} \quad \forall t \geq T.$$

So,

$$\sigma^*_{ijk\infty} \equiv \lim_{t \to \infty} \sigma^*_{ijkt} \leq \lim_{t \to \infty} \sigma^*_{ij'kt} e^{-\lambda_t(\epsilon/2)} = 0,$$

which contradicts $\sigma^*_{ijk\infty} > 0$. ∎

The next result establishes uniqueness of the AQRE for games of perfect information.

Proposition 3.3 (Uniqueness of logit AQRE in games of perfect information)*: For every finite extensive-form game of perfect information and any $\lambda \in [0, \infty)$, $\sigma^*(\lambda)$ is unique.*

Proof: This is proved by construction of $\sigma^*(\lambda)$. Since the game has perfect information, all information sets are singletons, and the predecessors of each terminal node all lie in different (singleton) information sets. Consider one such (singleton) information set, h_i^k. For each $a_j \in A(h_i^k)$, there exists a unique outcome $\nu_{a_{ijk}} \in \Gamma^t$ such that $\rho(\nu_{a_{ijk}} | h_i^k, \sigma_j) = 1$, where σ_j is the behavioral strategy profile σ, with σ_{ik} replaced by the pure action a_j, so $U_{ijk}(\sigma_j) = u_i(\nu_{a_{ijk}})$. This *uniquely* determines i's action probabilities of logit AQRE at h_i^k by

$$\sigma_{ijk}^* = \frac{\exp(\lambda u_j(\nu_{a_{ijk}}))}{\sum\limits_{a_{j'} \in A(h_i^k)} \exp(\lambda u_i(\nu_{a_{ij'k}}))}.$$

Then the expected utility to an agent who has reached h_i^k but has not chosen an action at h_i^k is $U_{ik}(\sigma) = \sum_{a_j \in A(h_i^k)} \sigma_{ijk}^* u_i(\nu_{a_{ijk}})$. Notice that $U_{ik}(\sigma)$ is therefore uniquely determined, for all i and for all immediate predecessors of Γ^t. Next consider the immediate predecessors of the immediate predecessors of Γ^t. Repeating the above argument yields a unique $\sigma^*(h')$ for all such h' such that $Q|(Q(h')) \subseteq \Gamma^t$. Since the game is finite, one can iterate this argument until σ^* is defined (uniquely) at each (singleton) information set. ∎

A fifth property of the logit AQRE correspondence is that for generic games there exists a unique connected path in the graph of logit AQRE correspondence which includes σ_0^* (i.e., the equilibrium when $\lambda = 0$). This unique connected path defines a solution $\overline{\sigma}_\lambda^*$ for all values of $\lambda \geq 0$, and furthermore there is a unique limit point $\overline{\sigma}^* = \lim_{\lambda \to \infty} \overline{\sigma}_\lambda^*$. Call $\overline{\sigma}^*$ the *logit AQRE solution* of the game.

Proposition 3.4 (Properties of the logit AQRE correspondence)*: For almost all finite extensive-form games,*

1. *the logit AQRE correspondence, $\sigma^*(\lambda)$, is a one-dimensional manifold;*
2. *for almost all λ, $\sigma^*(\lambda)$ consists of a finite and odd number of logit AQRE;*
3. *there is a unique branch (the* principal branch*) of the graph of $\sigma^*(\lambda)$ connected to the centroid of the game at $\lambda = 0$; $\sigma^*(\lambda)$ consists of a unique element for sufficiently small positive values of λ;*

4. *the principal branch of the logit AQRE correspondence selects a unique*
   component from the set of sequential equilibria of the game, in the sense
   that if $\{(\lambda_t, \sigma_t^*)\}_t$ and $\{(\lambda_t', \sigma_t'^*)\}_t$ are sequences along the principal branch
   with $\lim_{t\to\infty} \lambda_t = \lim_{t\to\infty} \lambda_t' = \infty$, and if σ^* is a limit point of $\{\sigma_t^*\}_{t=1}^\infty$
   and σ'^* is a limit point of $\{\sigma_t'^*\}_{t=1}^\infty$, then σ^* and σ'^* are in the same
   component of the set of sequential equilibria of the game.

Proof: The proof is slightly more involved than that for the normal-form logit
QRE correspondence, but the basic logic is similar. A detailed proof is
provided in McKelvey and Palfrey (1998, Appendix).   ∎

Note that the example in table 2.2 of the previous chapter demonstrates that
the converse of proposition 3.4 is not true. That is, there are sequential equilibria
that are not approachable by a sequence of logit AQREs. Thus, the set of limit
points of logit AQRE provides a refinement of sequential equilibrium. Whenever
there is a unique connected path selection of the logit AQRE correspondence, this
gives a refinement of the AQRE-approachable sequential equilibria which is an
even stronger refinement (in fact, an essentially unique selection) of sequential
equilibrium. As illustrated later in the book, it offers a quite different refinement
from traditional belief-based refinements such as the intuitive criterion,[7] and
provides a better predictor of real behavior than the intuitive refinement in some
signaling games studied in laboratory experiments.

Another feature of logit AQRE is that it makes predictions about the relative
likelihood of *all* different play paths. In contrast, other refinements tend to predict
that only a subset of the play paths can occur, while the others are supposed to
occur with probability 0. Because each play path arises with a (unique) positive
probability in every logit AQRE, player "beliefs off the equilibrium path" are
uniquely pinned down.

Next, consider the application of logit AQRE to the one-shot chain-store-
paradox game. This shows how again the invariance principle is violated. The way
the game is actually played, that is, as a game where players choose strategies
simultaneously versus a two-stage game in which the incumbent first observes
whether an entrant has entered before choosing his strategy, will systematically
affect the predicted pattern of play using the logit AQRE. Furthermore, some of
the violations implied theoretically by logit AQRE mirror findings reported in the
experimental literature (e.g., Schotter, Weigelt, and Wilson, 1994).

---

[7] That is, there exist games in which there is a unique intuitive equilibrium which is different from
the unique logit AQRE selection.

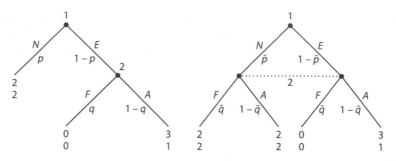

FIGURE 3.3. Chain-store-paradox game: player 1, the entrant, chooses to enter (E) or not (N) and player 2, the incumbent, chooses to fight (F) or acquiesce (A).

TABLE 3.4. Normal form for the chain-store-paradox game.

	F	A
N	2, 2	2, 2
E	0, 0	3, 1

Example 3.2 (Chain-store-paradox game)*:* This second example of strategic invariance violations (besides the centipede game example) is included for three reasons. First, it is related to the well-known ultimatum bargaining game; in fact, it has been dubbed the "mini-ultimatum" game (Gale, Binmore, and Samuelson, 1995).[8] Second, there are some experimental data from this and related games that clearly shows empirical violations of invariance that are consistent with QRE. Third, the game was originally introduced to illustrate subgame perfection and to demonstrate how some of the implications of perfect rationality and the refinement of subgame perfection may be suspect. In the normal-form version of this game, the *imperfect* equilibrium is *not* generally refined away by QRE-based refinements. Moreover, for some variations of the game, it is the unique logit solution of the game. The extensive form is illustrated in the left panel of figure 3.3. The normal-form representation of this game is given in table 3.4.

The story is that there is a potential entrant (player 1) into a market that is currently monopolized (by player 2). If player 1 chooses not to enter (N), both players continue to earn profits from their current activities, which are given by 2 and 2 respectively. If player 1 chooses to enter (E), then player 2

---

[8] This game can also be interpreted as a discrete (binary) version of an ultimatum game in which the offerer may offer either a 50/50 split or a 75/25 split of the pie. The unfair split (75/25) may be either accepted or rejected by the other player. The fair split is automatically accepted. Gale, Binmore, and Samuelson (1995) call it the "ultimatum mini-game," and study the stability of its Nash equilibria under the replicator dynamic.

can either fight (F) which results in a costly (to both sides) price war with profits reduced to 0 each. Or, player 2 can acquiesce (A), in which case player 1 earns some additional profits (3 instead of 2) and player 2, now having to share the market, earns lower, but still positive, profits (1 instead of 2). The logit QRE correspondence for the normal form of table 3.4 is equivalent to the logit AQRE correspondence of the extensive-form game in the right panel of figure 3.3, which is a strategically invariant transformation of the game in the left panel.[9]

One can compare the logit AQRE for the game in the left panel of figure 3.3 to that of the right panel, referred to as the simultaneous play version of the game.[10] The logit AQRE conditions for the game in the left panel are

$$p^* = \frac{1}{1 + e^{\lambda(1-3q^*)}}, \qquad q^* = \frac{1}{1 + e^{\lambda}}, \tag{3.2}$$

where p^* is the equilibrium probability of not entering and q^* is the equilibrium probability of fighting. Likewise, the equilibrium conditions for the extensive-form game in the right panel are

$$\tilde{p}^* = \frac{1}{1 + e^{\lambda(1-3\tilde{q}^*)}}, \qquad \tilde{q}^* = \frac{1}{1 + e^{\lambda(1-\tilde{p}^*)}}. \tag{3.3}$$

The solutions for (p^*, q^*) and $(\tilde{p}^*, \tilde{q}^*)$ as functions of λ are shown in the top and bottom panels respectively of figure 3.4. While the two solution sets are qualitatively similar, there are several points that can be illustrated with this example. First, in the sequentially played game there is a unique AQRE for every value of λ, since q^* does not depend on p^*. This is generally the case in games of perfect information, as was established in proposition 3.3. However, this is *not* true in the game where player 2 chooses a strategy without having observed player 1's move. In this case there is an additional QRE component for large values of λ, which converges to the *imperfect* Nash equilibrium $\tilde{p}^* = 1$, $\tilde{q}^* = \frac{1}{2}$; see the bottom panels of figure 3.4. Thus, the imperfect entry deterrence equilibrium component is more plausible in the simultaneous play version of the chain-store game since it is approachable by a sequence of AQREs.

---

[9] The right panel of figure 3.3 can be interpreted as a version of the game where player 2 has to decide to fight or acquiesce before knowing whether player 1 has decided to enter. For example, it might be that some preparations for a possible price war need to be made in advance.

[10] This is equivalent to comparing logit AQRE as a function of λ in the first case to the (normal-form) logit equilibrium of the game in table 3.4.

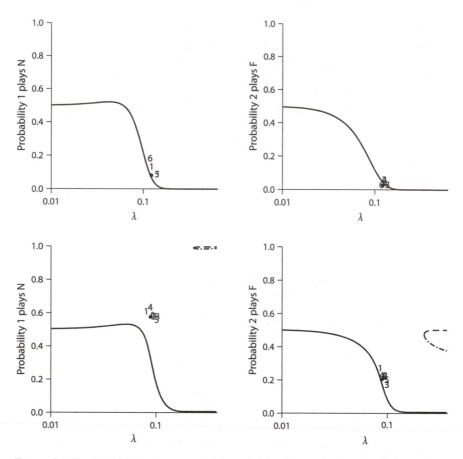

FIGURE 3.4. Logit AQRE for the sequential (top panels) and strategic (bottom panels) versions of the chain-store-paradox stage game studied by Schotter, Weigelt, and Wilson (1994).

There is another reason why the subgame-imperfect outcome (no entry) is more likely to be observed in the second version of the game. In the logit AQRE that converges to the perfect equilibrium, $\tilde{p}^*$ and $\tilde{q}^*$ are both higher than p^* and q^*, respectively, for *all* positive values of λ. To see this, note from (3.2) and (3.3) that for any value of $\tilde{p}^* \in (0, 1)$ and for any $\lambda > 0$, we have $q^* < \tilde{q}^*$, as $1 - \tilde{p}^* < 1$. Since $q^* < \tilde{q}^*$, it follows that $p^* < \tilde{p}^*$. The intuition is simple. In the simultaneous version of the game, the probability that F is suboptimal is less than 1, since it is only suboptimal if player 1 chooses E. This reduces the difference in expected payoffs between F and A (relative to the sequential version), leading to a higher probability that player 2 will choose F. This in turn leads to a higher probability player 1 will choose N. This gives the following proposition.

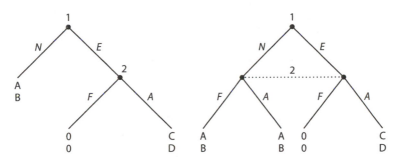

FIGURE 3.5. A 2 × 2 game of perfect information and its strategic version.

Proposition 3.5: *Entry occurs less frequently in the strategic version of the chain-store-paradox game and the incumbent more frequently fights, for all* $\lambda > 0$, $p^* < \tilde{p}^*$, *and* $q^* < \tilde{q}^*$.

This result can be generalized to the class of games illustrated in figure 3.5. Suppose that $D > 0$. In the strategy version of the game, player 2's (dominated) choice of F is less costly than in the sequential version of the game. Therefore $q^* < \tilde{q}^* < \frac{1}{2}$ when $D > 0$. Similarly, $q^* > \tilde{q}^* > \frac{1}{2}$ if $D < 0$. In general, $|q^* - \frac{1}{2}| > |\tilde{q}^* - \frac{1}{2}|$, which implies that deviations from the subgame-perfect Nash equilibrium prediction should always be *greater* when the game is played in the strategy version. How this affects the moves by player 1 is more subtle, and depends on whether C and D have the same or different signs. E is a relatively more attractive option in the strategy version if and only if $C > 0$ and $\tilde{q}^* > q^*$ or $C < 0$ and $\tilde{q}^* < q^*$. This requires $DC < 0$. Similarly E is a relatively less attractive option if and only if $DC > 0$. Therefore $p^* > \tilde{p}^*$ if and only if $CD < 0$.

Schotter, Weigelt, and Wilson (1994) conducted an experiment using a setup with $A = 4$, $B = 3$, $C = 6$, and $D = 2$. Some subjects played the game in the left panel of figure 3.5 while others played the game on the right and yet other subjects played the simultaneous-move game with the matrix form representation. Schotter, Weigelt, and Wilson (1994) found that entry was deterred more often with the two-strategy versions of the game than with the sequential version of the game. Furthermore, the second mover chose to fight more often in the strategy version of the game.

It is a straightforward exercise to fit the logit AQRE to the data from the sequential-move experiment and the normal-form QRE to the data from the simultaneous-move experiment.[11] In the sequential-move game, for any logit

---

[11] Details of the estimation procedures are explained in chapter 6.

TABLE 3.5. Schotter, Weigelt, and Wilson (1994) experimental results. The observed choice frequencies in the data are denoted by $\bar{p}, \bar{q}$ and the estimated equilibrium choice frequencies are denoted by $\hat{p}, \hat{q}$.

		$\bar{p}, \bar{q}$	$\hat{p}, \hat{q}$
	N	0.08	0.07
Sequential	E	0.92	0.93
version	F	0.03	0.05
	A	0.97	0.95
	N	0.57	0.47
Strategic	E	0.43	0.53
version	F	0.20	0.31
	A	0.80	0.69

AQRE parameter, λ, one solves for the equilibrium frequencies of moves at each information set. The estimated value of $\hat{\lambda}_{\text{seq}}$ is then the value of λ such that equilibrium frequencies, $(\hat{p}_{\text{seq}}, \hat{q}_{\text{seq}})$, at that parameter value best fit the actual choice frequencies in the sequential-move data, using maximum likelihood as the measure of fit. In the simultaneous-move game the estimated value of $\hat{\lambda}_{\text{sim}}$, and the fitted equilibrium choice probabilities, $(\hat{p}_{\text{sim}}, \hat{q}_{\text{sim}})$, are obtained similarly, using the observed choice frequencies in the simultaneous-move data.

Table 3.5 compares the estimated equilibrium frequencies and actual data frequencies from the Schotter, Weigelt, and Wilson (1994) experiment. Figure 3.4 superimposes the observed choice frequencies on a graph of the equilibrium correspondences of the two games. The upper panels show the AQRE choice frequencies in the sequential-move game as a function of λ, with the left panel showing the probability the entrant chooses N and the right panel the probability the incumbent chooses F, conditional on entry. The small numbers in each panel show the observed choice frequency graphed against the estimated value of λ for different experience levels. The bottom two panels are similar, except they are for the simultaneous-move game. Note that another logit equilibrium converging to the imperfect Nash equilibrium is picked up at very high values of λ, but the logit solution of the game converges to the perfect equilibrium (E, A). The data are consistent with the comparative static prediction of proposition 3.5. The observed entry rates and fight rates are both significantly higher in the simultaneous-move version of the game.

Several versions of these games were replicated in a later experiment by Coughlan, McKelvey, and Palfrey (1999). That paper investigated the invariance question, and also explored QRE comparative static predictions about the effect of changing the relative payoffs to entry and no entry, and the opportunity for potential entrants to establish group reputations.

That experiment confirmed invariance effects, as well as some other QRE comparative static predictions.

3.4 AQRE ANALYSIS OF THE CENTIPEDE GAME

A laboratory experiment of the four-move centipede game shown in figure 3.1 was reported in McKelvey and Palfrey (1992). It was run as a sequential game, and they developed a model of play based on two-sided incomplete information, altruism, learning, and heterogeneous prior beliefs, where a player's altruism parameter was private information. Players could be one of two privately known types. *Altruists* were assumed to maximize the expected sum of payoff for the two players and *nonaltruists* were assumed to maximize their own expected payoff. Errors were also introduced, and the ϵ-perturbed (sequential) Nash equilibrium for the resulting game was solved.[12] One of the key predictions of that model is that the equilibrium probability with which a player chooses T after k passes is higher than the probability the player chooses T after $k - 1$ passes, for $k = 1, 2, 3$, a pattern found in the data (see the first column of table 3.6). The resulting five-parameter model explained the essential qualitative features of the data and fit the move frequencies observed in the data fairly well, including time trends.

As an alternative to the five-parameter model, one can instead analyze and fit to the data a more parsimonious model: the one-parameter logit AQRE model.[13] Referring to figure 3.1, for $i = 1, 2$, denote player i's behavioral strategies by (p_{ia}, p_{ib}), where

$$p_{1a} = \text{Prob}[1a \text{ chooses } T],$$

$$p_{1b} = \text{Prob}[1b \text{ chooses } T \text{ (if reached)}],$$

$$p_{2a} = \text{Prob}[2a \text{ chooses } T \text{ (if reached)}],$$

$$p_{2b} = \text{Prob}[2b \text{ chooses } T \text{ (if reached)}].$$

At the first move, player 1a evaluates the expected payoffs of T and P as follows:

$$U_{1a}(T) = 4,$$

$$U_{1a}(P) = 2p_{2a} + (1 - p_{2a})(16p_{1b} + (1 - p_{1b})(8p_{2b} + 64(1 - p_{2b}))).$$

---

[12] Kreps (1990) independently analyzed a one-sided incomplete information model of a centipede game, with homogeneous prior beliefs and no errors to illustrate reputation building in the sequential equilibrium.

[13] A similar analysis is reported in Fey, McKelvey, and Palfrey (1996) using an experiment that employs a constant-sum version of the centipede game.

TABLE 3.6. Estimated take probabilities in the four-move centipede game.
(Source: McKelvey and Palfrey, 1998.)

Take probabilities	Data	Logit AQRE	Zauner model
T	0.071	0.246	0.214
T/P	0.383	0.315	0.301
T/PP	0.644	0.683	0.677
T/PPP	0.754	0.938	0.939

The logit formula then implies the following take probability at the first node:

$$p_{1a} = \frac{e^{4\lambda}}{e^{4\lambda} + e^{2p_{2a}+(1-p_{2a})(16p_{1b}+(1-p_{1b})(8p_{2b}+64(1-p_{2b})))}}.$$

The expected payoffs of T and P at subsequent moves can be evaluated similarly. For example, at the final move, player 2b has payoffs $U_{2b}(T) = 32$ and $U_{2b}(P) = 16$. Applying the logit rule to the expected payoffs yields the following expressions for the remaining *conditional* take probabilities:

$$p_{2a} = \frac{e^{8\lambda}}{e^{8\lambda} + e^{\lambda(4p_{1b}+(1-p_{1b})(32p_{2b}+16(1-p_{2b})))}},$$

$$p_{1b} = \frac{e^{16\lambda}}{e^{16\lambda} + e^{\lambda(8p_{2b}+64(1-p_{2b}))}},$$

$$p_{2b} = \frac{e^{32\lambda}}{e^{32\lambda} + e^{16\lambda}}.$$

This system of four equations and four unknowns is solved recursively (p_{2b}, then p_{1b}, then p_{2a}, then p_{1a}) and has a unique solution. There are several interesting features of the equilibrium correspondence, which is graphed in the lower-right panel of figure 3.2. First, for most values of λ, the equilibrium take probabilities increase as the game progresses. That is, $p_1 < q_1 < p_2 < q_2$, which was a key qualitative feature in the data. Second, for *very* low values of λ, p_{1a}, p_{2a}, and p_{1b} are all *decreasing* in λ. That is, the equilibrium does not converge monotonically toward the Nash equilibrium as a function of λ.

The best-fitting value of λ was estimated using standard maximum-likelihood techniques similar to the Schotter, Weigelt, and Wilson (1994) data for the chain-store stage game in the previous section. The results are reported in table 3.6. These AQRE estimates mirror similar results reported in Zauner (1999), using a probit specification for the AQRE model. The estimated take probabilities are given in the last column of table 3.6. The estimates are very similar, as both models share two basic properties: they capture the feature in the data that take probabilities increase as the end of the game is approached, and they overestimate the take probabilities on the first and last moves. But as pointed out in Zauner (1999), neither model includes a heterogeneity parameter, which produced a significant improvement in fit in the models estimated by McKelvey and Palfrey (1992). The inclusion of a heterogeneity parameter could significantly improve the fits of these two models. The next chapter addresses the general question of how to incorporate heterogeneity into quantal response equilibrium models.

4

Heterogeneity

Players differ in skill, which has implications for the degree to which they make errors. Low-skill hitters in baseball often swing at bad pitches, beginning skiers frequently fall for no apparent reason, and children often lose at tic-tac-toe. At the other extreme, there are brilliant chess players, bargainers, and litigators who seem to know exactly what move to make or offer to decline. From a QRE perspective, these skill levels can be modeled in terms of variation in error rates—in the structural QRE approach—or in responsiveness of quantal response functions—in the regular QRE approach. High-skill players do not make errors very often and are much more likely to make optimal decisions than suboptimal ones; the opposite being true for low-skill players. By the same token, error rates can also vary with the difficulty of the task. Beginning skiers fall less frequently on bunny slopes than on intermediate slopes. And it is easier to master tic-tac-toe than Go or chess.

The original formulation of structural QRE proposed by McKelvey and Palfrey (1995) allows for heterogeneity in the additive random-utility error terms for different players. For each player, i, errors are distributed according to a well-behaved distribution function F_i, but there is no requirement that F_i be the same for all i. For example, each player might have i.i.d., extreme value disturbances, resulting in logit choice quantal response functions, but the precision parameter, λ_i, could be different for each i, with higher λ_i corresponding to higher skill. Similarly, the formulation of regular QRE in Goeree, Holt, and Palfrey (2005) allows quantal response functions to differ across players, provided they are "regular." Heterogeneity in other dimensions, for example, propensities or aversions, could be (and has been) incorporated in an analogous manner, which enhances the usefulness of this approach. The heterogeneous quantal response equilibrium (HQRE) that will be introduced in the first part of the chapter preserves the equilibrium property that choice distributions match the belief distributions over others' actions, even if there is extra variation in choice distributions due to individual differences, for example, in skill levels. HQRE is especially useful in situations for which it is desirable to model or estimate persistent behavioral differences between people in a parsimonious manner.

This equilibrium belief property can be relaxed by letting the perceived distribution of others' skill levels be biased, for example, if people with higher skills expect others to also have high skills. In other words, a model could specify subjective beliefs over others' precisions in a manner that will lead to a subjective quantal response equilibrium (SQRE).

Alternatively, subjective beliefs could be *downward looking* if players cannot imagine that others are more skilled than they are, so people with low precisions expect others to have even lower precisions. One such model involves a truncation, for example if a person with precision λ_i considers the distribution of precisions to be truncated at that level, so that the density of precisions would be scaled up by dividing by $F(\lambda_i)$ so that it integrates to 1 on the interval $[0, \lambda_i]$, which leads to a consideration of truncated quantal response equilibrium (TQRE).

The most widely used downward-looking models of nonequilibrium beliefs involve introducing heterogeneity in terms of "levels" of strategic sophistication. In particular, increasing degrees of rationality correspond to a higher number of levels of iterated thinking: A "level-0" person does not think but makes decisions that are randomly determined in a manner that assigns equal probabilities to all options. A "level-1" person then makes a best response to level-0 randomness, a "level-2" makes a best response to the behavior of a level-1 player, etc. This "level-k" model has the property that each person thinks that the other players are exactly one level below them. A more general approach, known as the cognitive hierarchy (CH) model specifies a probability distribution over the possible levels that are below the player's own level, which would correspond to level-k if all mass were on the next lower level. These models have parameters that determine the proportions of players of each type, parameters which can be (and have been) estimated in the process of analyzing data from laboratory experiments. In particular, a consideration of nonequilibrium beliefs is particularly appealing for games that are played only once, for which it is not possible for players to refine their beliefs about others' decisions.

The final model to be considered is one of noisy introspection (NI), in which players who cannot learn by observing others' decisions must rely on learning by introspection, about what the other players might do, what the other players think other players might do, etc. It is easy to see that this process of iterated introspection would not go very far before it is blurred by noise. The noisy introspection model typically specifies lower levels of precision for higher-order beliefs (e.g., about what "he thinks I think"). This increasing noise for more layers of iterated thinking ensures that belief distributions will exhibit more noise than choice distributions, so noisy introspection is also a model of nonequilibrium beliefs, although it does not involve a downward-looking perspective. The NI model can be parameterized and estimated in empirical applications (e.g., data

TABLE 4.1. A game where Row has a strictly dominant strategy and Column does not.

	L	R
U	5, 4	4, 7
M	4, 6	4, 15
D	6, 4	5, 3

from games played once) for which observed data patterns are systematically biased away from Nash (and even QRE) predictions.

4.1 SKILL AND ROLE HETEROGENEITY

4.1.1 Systematic Heterogeneity Depending on Player Roles

It is conceptually useful to distinguish between two different kinds of skill heterogeneity. The first kind of heterogeneity, which we call *systematic*, is across player *roles*, and applies to asymmetric games. For example, the decision problem in one of the player roles may be less challenging (i.e., require less skill) than for the other player roles. In the game shown in table 4.1, the Row player has a strictly dominant strategy to play D, but the Column player's optimal strategy depends on Row's strategy.

If the variance of the error term (or equivalently the responsiveness of the quantal response function) is driven by the complexity of the decision problem, then it is reasonable to expect the Row player to have a relatively low variance error term (highly responsive quantal response function) and the Column player to have a higher variance error term (less responsive quantal response function).

4.1.2 Idiosyncratic Heterogeneity

The second kind of heterogeneity, which we call *idiosyncratic*, occurs at an individual level, irrespective of the player role. For example, one can think of the variance of the error term, or the responsiveness of the quantal response function as an idiosyncratic skill parameter. Some individuals may be concentrating harder, or have more experience with a game, or have greater cognitive or strategic acuity than others. The effect of (and the modeling of) idiosyncratic heterogeneity applies also in symmetric games, where fundamental heterogeneity due to role asymmetries is absent.

These ideas are formalized by Rogers, Palfrey, and Camerer (2009) as *heterogeneous quantal response equilibrium* (HQRE). The key assumption is that different individuals playing a game, possibly in equivalent roles, have different logit quantal response functions. The next section develops the HQRE formulation more generally, with players' quantal response functions being drawn from a commonly known distribution of regular quantal response functions. In the tradition of Bayesian game theory, a player's idiosyncratic "skill" type corresponds to a quantal response function (and possibly a role). Players all know the distribution from which the other players' idiosyncratic skill types are drawn, but do not know the exact skill type of each player in the game. In principle, the HQRE model with idiosyncratic heterogeneity can also allow the distribution of skill types to depend on player roles in a systematic way as described in the previous section.

There are two additional possible sources of heterogeneity that the QRE framework is equipped to deal with. The first, and perhaps most obvious, is unobserved heterogeneity of parameters that govern the underlying preferences of the players, such as risk aversion, altruism, inequality aversion, impatience, and so forth. This can be particularly important when applying the QRE framework to the structural estimation of alternative models using data from laboratory game experiments. Accordingly, we will postpone a detailed discussion of this until chapter 6, but note here that incorporation of this category of heterogeneity is relatively straightforward.

The third source of heterogeneity that has received considerable attention by theorists and experimentalists is heterogeneity of beliefs about the quantal response functions of other players. The equilibrium approach is to impose rational expectations on all the players, but allow for either commonly known differences in the quantal response functions for the various players, or to allow for incomplete information about the quantal response functions in a Harsanyi common prior framework. The HQRE model presented in the next section is formulated in a general way to allow for this kind of belief heterogeneity. A richer source of belief heterogeneity is to allow players to have incorrect beliefs about the quantal response functions of the other players. Rogers, Palfrey, and Camerer (2009) formulate a general approach to modeling this within the QRE framework, called *subjective quantal response equilibrium* (SQRE). The SQRE formulation has its roots in earlier models of QRE with incorrect beliefs that were used to analyze behavior in specific laboratory experiments.

4.2 HETEROGENEOUS QUANTAL RESPONSE EQUILIBRIUM

This section develops the more general version of skill heterogeneity where players have incomplete information about the quantal response functions of the other. Let $\Gamma = [N, \{A_i\}_{i=1}^n, \{u_i\}_{i=1}^n]$ denote a normal-form game and $R = (R_1, \ldots, R_n)$ be a collection of regular quantal response functions, one for each player. Let $\lambda_i \in [0, \infty)$ denote player i's *type*, so that a player type simply corresponds to the precision of the quantal response function governing i's stochastic choices as a function of expected payoffs, $U_i = (U_{i1}, \ldots, U_{iJ_i})$. We assume player i with type λ_i chooses action $a_{ij} \in A_i$ with probability $p_{ij}(\lambda_i) = R_{ij}(\lambda_i U_i)$.

Player types are drawn independently from possibly different, commonly known type distributions over $[0, \infty)$, and we denote player i's type distribution by F_i.[1] From an outsider's perspective, the probability that player i chooses $a_{ij} \in A_i$ is then equal to

$$\sigma_{ij} = \int_0^\infty p_{ij}(\lambda_i) dF_i(\lambda_i) = \int_0^\infty R_{ij}(\lambda_i U_i) dF_i(\lambda_i).$$

The type-dependent strategies, $p_{ij}(\lambda_i)$, are assumed to be common knowledge in HQRE, as are the type distributions, $F_i(\lambda_i)$. Hence, the induced mixed strategies, σ_{ij}, are common knowledge. Let $p = \{p_{ij}(\cdot)\}_{i \in N, j \in A_i}$ be the collection of commonly known type-dependent strategies. Then player i's expected utility from choosing strategy a_{ij} can be written as

$$U_{ij}(p) = \sum_{a_{-i} \in A_{-i}} \left(\prod_{k \neq i} \sigma_k(a_k) \right) u_i(a_{ij}, a_{-i}).$$

Definition 4.1: *The strategy profile p^* is a **heterogeneous quantal response equilibrium** of the normal-form game Γ if $p^*_{ij}(\lambda) = R_{ij}(\lambda U_i(p^*))$ for all $i \in I$, $j \in A_i$, and $\lambda \geq 0$.*

4.3 SKILL HETEROGENEITY WITHOUT RATIONAL EXPECTATIONS

4.3.1 Egoistic Beliefs about Opponents

The QRE framework was first extended to allow for λ-heterogeneity by McKelvey, Palfrey, and Weber (2000) in an experimental study of asymmetric matching pennies games. The baseline game A in their study is a minor variation of the asymmetric matching pennies game introduced in chapter 2 (see table 2.1) with $X = 9$. The behavior in this game was compared with that in two variations: game B is the same as game A, except the column payoffs are quadrupled, and game C is the same as game A, except both players' monetary payoffs are quadrupled.[2]

---

[1] This formulation allows heterogeneity across player roles, as well as idiosyncratic heterogeneity.
[2] The design also included a fourth game, which was not based on game A.

These payoff magnitude manipulations do not change the Nash equilibrium of the game, but do change the logit equilibrium correspondences. The quantal response equilibrium correspondence for game C is a rescaled version of the quantal response equilibrium of game A, because quadrupling the payoffs is the same as quadrupling the λ of both players, as in the HQRE model.[3] Game B's QRE correspondence is the same as the QRE correspondence of game A, except with the λ parameter of the Column player being four times that of the Row player, which would be a form of structural heterogeneity as described earlier in this chapter.

The experiment was designed so that QRE predicted specific payoff magnitude effects across these three games, assuming a homogeneous model of λ. Specifically, λ is assumed constant across player roles (Row versus Column), individual subjects, and across games.[4] None of the expected payoff magnitude effects were significant; in fact, there was relatively little variation in aggregate behavior across the games.

Examination of the individual level data indicated significant amounts of heterogeneity in individual choice behavior. This observation led to the development and estimation of a model of skill heterogeneity. In this model, each individual is assumed to have a λ_i and believes that all other subjects have the same $\lambda_j = \lambda_i$. Because each player assumes the other player shares the same skill level, beliefs are called *egoistic*. The mean and variance of the distribution of λ_i was estimated and this model was found to fit the data significantly better than the homogeneous QRE model. However, that model of heterogeneity is not the same as HQRE, because each player assumes that all players are homogeneous ($\lambda_j = \lambda_i$), when in fact different players have different skill types.

4.3.2 Unconstrained Point Beliefs about Opponents' Skill

Weizsäcker (2003) takes this approach an important step further by allowing players to have incorrect (degenerate) beliefs about the skill types of the other player, and, as in McKelvey, Palfrey, and Weber, (2000), the analysis and theoretical development is limited to two-player games. Specifically, each player has a skill type, λ_i, and a belief about the skill type of the other player, $\widehat{\lambda}_j$. The predicted behavior of player i is given by $p_i^*(\lambda_i, \widehat{\lambda}_j)$. Weizsäcker (2003) estimates a distribution of λ_i and $\widehat{\lambda}_j$ in the population, using data from earlier

---

[3] This equivalence between increasing payoffs by a scale factor and increasing λ by a scale factor requires an assumption that utility is linear in monetary payoffs. Goeree, Holt, and Palfrey (2003) reexamine these data as well as additional data from asymmetric matching pennies games, allowing for concave utility (risk aversion).

[4] The analysis also assumed the logit form of stochastic choice, as well as the rational expectations equilibrium assumption that players acted as if λ were common knowledge.

papers by Stahl and Wilson (1994, 1995), and the estimation supports the notion that individuals believe they are more skilled than the other players in the game.[5]

This kind of directional heterogeneity with arbitrary beliefs about the other players' quantal response function includes some interesting special cases that correspond to different levels of strategic sophistication, as defined in Stahl and Wilson's (1994, 1995) model:

1. If $\lambda_i = 0$, then i is a *level*-0 player (completely random), and $\widehat{\lambda}_j$ does not affect i's choice probabilities.
2. If $\lambda_i = \infty$ and $\widehat{\lambda}_j = 0$, then i is a *level*-1 player, who best responds to a uniform mixed strategy by j.
3. If $\lambda_i \in (0, \infty)$ and $\widehat{\lambda}_j = 0$, then i is a *quantal responding level*-1 player, who quantal responds to a uniform mixed strategy by j, according to the precision parameter λ_i.
4. If $\lambda_i = \infty$ and $\widehat{\lambda}_j = \infty$, then i is a *Nash* player, who plays a Nash equilibrium strategy.
5. If $\lambda_i = \widehat{\lambda}_j = \lambda \in (0, \infty)$, then i is a *QRE* player, who plays a QRE strategy corresponding to the precision parameter λ.

4.4 QRE WITHOUT RATIONAL EXPECTATIONS

4.4.1 Subjective Heterogeneous QRE

For general normal-form games, the subjective heterogeneous quantal response equilibrium (SQRE) allows expectations about choice probabilities to be inconsistent with the actual choice frequencies of the other players.[6] Indeed, allowing for incorrect beliefs seems to better explain the data from the simple games studied in McKelvey, Palfrey, and Weber (2000) and Weizsäcker (2003). Models with this property might be particularly useful in explaining behavior in one-shot games, or complex games in which learning or other forces have not enabled beliefs to fully equilibrate to actual choices. However, the particular form of inconsistencies allowed in SQRE still permits it to be thought of as an equilibrium model: choice probabilities conditional on type are common knowledge; it is only the perceived distribution of types that varies across players.

More specifically, let $\lambda_i \in [0, \infty)$ denote player i's type and replace the rational expectations assumption by an assumption of *subjective expectations*.

---

[5] This type of egocentric overconfidence is a property of most models based on levels of strategic sophistication.

[6] Subjective AQRE can be defined analogously for games in extensive form.

According to this model, the equilibrium strategies, which map types into choice probabilities of all players, are common knowledge in equilibrium, but players may have different beliefs about the type distributions. Denote the conditional subjective beliefs of player i about player k's type by the distribution $F_i^k(\lambda_k|\lambda_i)$. Typically, one assumes that each F_i^k has support contained in $[0, \infty)$, a smooth density function, f_i^k, and finite moments, but this is not necessary for most purposes. For example, the models of McKelvey, Palfrey, and Weber (2000) and Weizsäcker (2003) assume that individuals have point beliefs about the other players. In the McKelvey, Palfrey, and Weber model, i's point belief about player k is assumed to be perfectly correlated with i's type, and in Weizsäcker's model there is no such restriction except that $F_i^k(\lambda_k|\lambda_i)$ is degenerate. For the general case studied here, the density f_i^k could be an exponential or log-normal distribution and could in principle be independent of λ_i, although beliefs generally may depend on a player's own type.[7] As shown below, this framework provides a general way to link heterogeneous QRE with alternative approaches that model belief heterogeneity by assuming different levels of strategic sophistication (e.g., Stahl and Wilson, 1995; Camerer, Ho, and Chong, 2004).

Formally, let $\Gamma = \left[N, \{A_i\}_{i=1}^n, \{u_i\}_{i=1}^n\right]$ denote a normal-form game and let ΔA_i denote the set of probability distributions over A_i and let $\Delta A = \Delta A_1 \times \cdots \times \Delta A_n$ denote the product set. For any subjective belief about the mixed-strategy profile, $\widehat{\sigma} \in \Delta A$, the (subjective) expected payoff to player i from using action $a_{ij} \in A_i$ is

$$U_{ij}(\widehat{\sigma}) = \sum_{a_{-i} \in A_{-i}} \left(\prod_{k \neq i} \widehat{\sigma}_k(a_k)\right) u_i(a_{ij}, a_{-i}).$$

The probability that player i with type λ_i chooses action a_{ij} given i's beliefs about the mixed-strategy profile, $\widehat{\sigma}$, equals

$$p_{ij}(\lambda_i) = R_{ij}(\lambda_i U_i(\widehat{\sigma})). \tag{4.1}$$

Call any measurable function $p_i : [0, \infty) \to \Delta A_i$ a *strategy* for player i.

Next, consider the beliefs that another player, $k \neq i$, has about i's choice probabilities given player k's subjective belief $F_i^k(\lambda_i|\lambda_k)$ about player i's type λ_i. Because the subjective beliefs about player i's type may differ across players, so may their beliefs about player i's choice probabilities. Assume that any differences in their beliefs about i's mixed strategy stem from differences in beliefs about the distribution of λ_i. That is, the strategy profile, $p = (p_1, \ldots, p_n)$, is assumed to be common knowledge, although the structure of beliefs about the

---

[7] A special case is when $F_i^k(\lambda_k|\lambda_i) = F_k(\lambda_k)$ for all $i, k, \lambda_i, \lambda_k$. In this case subjectivity is absent, so HQRE is a special case of SQRE.

distribution of types departs from the conventional common-prior assumption. Given i's strategy, $p_i(\cdot)$, the prior belief of player k with type λ_k that player i chooses action a_{ij} (i.e., before λ_i is drawn) is therefore

$$\sigma_{ij}^k(p_i|\lambda_k) = \int_0^\infty p_{ij}(\lambda_i) f_i^k(\lambda_i|\lambda_k) d\lambda_i. \tag{4.2}$$

Given $\sigma_{-i}^i(p_{-i}|\lambda_i)$, the expected payoff for player i with type λ_i of choosing a_{ij} is given by

$$U_{ij}(\sigma_{-i}^i(p_{-i}|\lambda_i)) = \sum_{a_{-i} \in A_{-i}} \left(\prod_{k \neq i}^n \sigma_k^i(a_k|\lambda_i) \right) u_i(a_{ij}, a_{-i}). \tag{4.3}$$

In an SQRE with logit response functions, equations (4.1), (4.2), and (4.3) must all be satisfied simultaneously.

> **Definition 4.2:** *The strategy profile p^* is a **subjective quantal response equilibrium** of the normal-form game Γ if $p_{ij}^*(\lambda) = R_{ij}(\lambda U_i(\sigma_{-i}^i(p_{-i}^*|\lambda)))$ for all $i \in I$, $j \in A_i$, and $\lambda \geq 0$.*

This definition reflects that in an SQRE players have rational expectations about strategies (i.e., a player's behavior conditional on the player's type λ), but may have different beliefs about the distribution of mixed strategies induced by different beliefs about the distribution of λ types. That is, the beliefs in SQRE are *subjective* and do not necessarily come from a common prior. We refer to this as an *equilibrium* in exactly the same sense as a Bayesian equilibrium. In standard definitions of Bayesian equilibrium (see, for example, Geanakoplos, 1994, p. 1461), beliefs are type contingent and there is no requirement for a common prior model of beliefs.[8]

4.4.2 Truncated QRE and Cognitive Hierarchy

SQRE, as defined above, is a very general framework in that there is little that constrains the extent of heterogeneity and subjective beliefs. This section explores several plausible ways to impose further constraints on SQRE, which are relevant from both an empirical and a purely theoretical standpoint.[9] As noted in the last section, one special case of SQRE is HQRE, where rational expectations

---

[8] In fact, the entire theoretical analysis of this section could be done using Bayesian equilibrium, because QRE is itself a particular form of Bayesian equilibrium with i.i.d. payoff disturbances that follow an extreme value distribution. Heterogeneity in λ simply means that the distribution of payoff disturbances, and in particular, the variances of those disturbances, can vary across players.

[9] This section contains material from Rogers, Palfrey, and Camerer (2009).

are imposed, and a limiting case of HQRE is QRE, where players have rational expectations and there is no heterogeneity. More surprising is the fact that the cognitive hierarchy model of Camerer, Ho, and Chong (2004) is a limiting case of a class of SQRE models where beliefs are "downward looking."

This is done by introducing *truncated rational expectations*: players act as if they are not aware of the existence of types who are more rational than some maximum upper bound, and this upper bound may depend on their own type. Given their truncated beliefs, they form expectations by integrating over their perceived type distribution.

Denote the upper bound on player i's imagined types by $\theta_i(\lambda_i)$, where $\theta_i(\cdot)$ is commonly known. Assume that $\theta_i(\lambda_i)$ is strictly increasing and uniformly continuous in λ_i and for each i there exists $\bar{\theta}_i$ such that $\theta_i(\lambda_i) \leq \bar{\theta}_i \lambda_i$ for all λ_i. The beliefs of player k with type λ_k about λ_{-k} are rooted in the true distribution, F_{-k}, but normalized to reflect the missing density. That is, for $\lambda_k > 0$, the subjective beliefs of player k about player i's type is given by $F_i^k(\lambda_i|\lambda_k) = F_i(\lambda_i)/F_i(\theta_k(\lambda_k))$ for $\lambda_i \in [0, \theta_k(\lambda_k)]$ and $F_i^k(\lambda_i|\lambda_k) = 1$ for $\lambda_i > \theta_k(\lambda_k)$. In other words, the density, $f_i(\lambda_i)$, is scaled up by a factor $1/F_i(\theta_k(\lambda_k))$ on the relevant range so that it integrates to 1 on that range. This is truncated HQRE, or *TQRE*. Note that as $\theta_i(\lambda_i) \to \infty$ for all i, the upper bound on λ is lifted and the model converges to the standard HQRE model.

The truncation, $\theta_i(\lambda_i)$, can be interpreted as type λ_i of player i's *imagination*. Since $\theta_i(\lambda_i)$ is finite, this is a model of bounded imagination, in the sense that for any type λ_i of player i, all types λ_{-i} that exceed the threshold $\theta_i(\lambda_i)$ are unimaginable (i.e., i assigns zero probability to all such higher types). Notice that since $\theta_i(\lambda_i)$ is increasing, then players who are more skillful in the sense of payoff responsiveness (i.e., higher λ_i) necessarily also have more accurate expectations, in the sense that their beliefs are closer to the true distribution $F_i(\lambda_i)$. Types for which $\theta_i(\lambda_i) \approx 0$ are almost completely unimaginative in the sense that they believe all other players are nearly random. Hence these very low types will act approximately as if they are applying the principle of insufficient reason to form expectations about the other players' strategy choices (as do level-1 types in the cognitive hierarchy model), and then quantal respond to these beliefs. If $\theta_i(\lambda_i) \leq \lambda_i$, then we say that players are *self-limited*, because they cannot imagine types with higher λ than their own. A proof of existence of TQRE appears in Rogers, Palfrey, and Camerer (2009).

There are a number of reasons why truncated beliefs might be a plausible way to constrain belief heterogeneity. One rationale is that players with a low value of λ who can imagine players with higher λ, and compute what those other players will do, might want to switch to the higher-type behavior. There is also a significant body of evidence from the psychology literature indicating that people are often overconfident about their relative skill.[10] A third rationale is computational complexity: if there are cognitive costs to computing expected payoffs, those costs increase as players have more other types to consider. The benefits from more imagination—the expected payoff differential from imagining what a wider range of types will do— are likely to fall as λ rises, so the truncated expectations assumption can be seen as a reduced-form model of cost-benefit calculations which lead players to ignore information that is both hard to process and not too costly to ignore.

TRUNCATION AND HETEROGENEITY IN CH

Truncation of beliefs in a similar way to TQRE (albeit relative to beliefs about the distribution of a much different parameter) is the central feature of the cognitive hierarchy (CH) model of Camerer, Ho, and Chong (2004). CH introduces heterogeneity of player types of a much different kind than HQRE. In CH there is a discrete distribution $f(k)$ of players who do k steps of thinking, so k indexes *strategic sophistication*. The choice probabilities for a k-step player i choosing strategy j are $p_{ij}(k)$. A 0-step player randomizes uniformly over a (finite) number of strategies J_i, so $p_{ij}(0) = 1/J_i$ for all j. Note that these players do not form beliefs or even attend to their payoffs; their presence is just assumed to get a hierarchical process started in a simple way. Also note that these players are equivalent to $\lambda = 0$ players in the QRE approach.

Players who do $k \geq 1$ steps of thinking form truncated beliefs about the fraction of h-step types according to $g_k(h) = f(h)/\sum_{n=0}^{k-1} f(n)$ for all $h < k$ and $g_k(h) = 0$ for all $h \geq k$. In this specification, players do not imagine that any others are at their level (or higher), so, in the notation of the TQRE, they effectively have $\theta(\lambda) < \lambda$. All positive-step thinkers best respond given their beliefs, so in a two-player game,[11] $p_{ij}(k) = 1$ if and only if[12] $a_{ij} = \text{argmax}_a \sum_{h=0}^{k-1} g_k(h) \sum_{m=1}^{J_{-i}} p_{im}(h) u_i(a, a_{-im})$. The expected choice probabilities for player i implied by the CH model are given by $p_{ij} = \sum_{k=0}^{\infty} p_{ij}(k) f(k)$.

Camerer, Ho, and Chong (2004) assume $f(k)$ is Poisson and estimate the mean of the distribution using data from more than 100 normal-form games. Other types of hierarchical models have been explored as well. Nagel (1995)

---

[10] Kahneman and Tversky (1973) first studied overconfidence, and much work has followed, e.g., Camerer and Lovallo (1999) and Santos-Pinto and Sobel (2005).

[11] The expressions are more cumbersome to write out with n-player games because the probabilities of other players' types have a multinomial distribution with many terms. Roughly speaking, CH models become hard to compute as the number of players increases, while QRE models, which require finding a fixed point, become more difficult to compute as the number of strategies increases.

[12] If more than one action is a best response they are assumed to randomize equally across all best responses.

and Stahl and Wilson (1995) were the first to use strategic hierarchies to study dominance-solvable "beauty contest" games and matrix games, respectively. In Nagel's approach, k-step players think all others do $k - 1$ steps of reasoning (i.e., $g_k(h) = 1$ if $h = k - 1$ and 0 otherwise). Stahl and Wilson's limited-step types have the same one-step-below beliefs as in Nagel, but they also permit equilibrium types and "worldly" types who maximize against the empirical distribution of play. When applied to data, these models typically further assume that players use quantal responses rather than best responses.[13]

DIFFERENCES AND SIMILARITIES BETWEEN CH AND TQRE

The general form of TQRE is different from CH in three distinct ways. First, the maximum "imagined" type of other players could be equal to, greater than, or less than a player's own type, and this could be a second source of heterogeneity, whereas in CH and related approaches the imagination parameter for all players is strictly less than 1.[14] Second, levels of rationality are indexed by λ in TQRE, rather than k, so that types correspond to increasing payoff responsiveness rather than strategic sophistication. Third, in TQRE, all types exhibit some degree of randomness in response, reflecting the stochastic choice modeling. In CH all players with $k \geq 1$ best respond, so the only source of stochastic choice behavior is buried in the 0-level types.

In spite of these major differences between the two models, there are a number of important similarities between the TQRE and CH approaches. First, a central feature of both models is heterogeneity in types. Second, both models incorporate stochastic behavior. Third, they share an important type in common: the bottom of the food chain, $k = 0$ or $\lambda = 0$, and these lowest types are in the support of the beliefs of *all* (other) types. Fourth, both models assume there is a limit to the rationality of the other players, and this limit is monotonically increasing in type. Fifth, in both approaches, there is heterogeneity of beliefs as well as heterogeneity of types, and these are correlated: higher types have more accurate beliefs. These beliefs move in the direction of rational expectations about $f(\lambda)$ (or $f(k)$) as λ (or k) increases.[15] Finally, *all* players are overconfident in the sense that

---

[13] Recent applications of this approach include Costa-Gomes and Crawford (2006) and Crawford and Iriberri (2007a,b). The approach is also used to analyze Swedish lottery and experimental data by Östling et al. (2011) and box office reviews of unreviewed movies by Brown, Camerer, and Lovallo (2007).

[14] The Stahl and Wilson (1995) and Costa-Gomes and Crawford (2006) specifications include other types that do not correspond to levels in the thinking hierarchy. If the maximum imagined type is always less than one's own type, then the model's solution can be computed recursively, as in CH. However, if $\theta = 1$, so that players are aware that others share their level of thinking, the model must be solved using fixed-point methods.

[15] Full convergence (in λ) to rational expectations would require $\lim_{\lambda \to \infty} \theta_i(\lambda) = \infty$.

they underestimate the gamesmanship (be it sophistication or responsiveness) of other players.

THE FORMAL CONNECTION BETWEEN TQRE AND CH

By placing two restrictions on TQRE, for any CH model there exist distributions of types in TQRE that lead to behavioral predictions that are essentially equivalent to CH. By *essentially* equivalent, we mean (i) that the equivalence is in terms of approximations that can be made arbitrarily close and (ii) that the approximating equilibria in TQREs are unique.

To make this approximation, first consider distributions such that the set of λ values is discrete, $L_\gamma = \{0, \gamma, 2\gamma, \ldots, k\gamma, \ldots\}$, with grid size γ. A player of type k is called a *level-k* player, and has response parameter $\lambda = k\gamma$. Fix the distribution over k, so that the probabilities of types are $f = \{f(0), f(1), \ldots, f(k), \ldots\}$. This is simply an HQRE specification with discrete λ types.

The first restriction is $\theta_i(\lambda) = \frac{k}{k+1}\lambda$ for all i, $\lambda \in L_\gamma$. That is, players recognize only (and all) lower types, but otherwise have correct beliefs about the distribution. Level-0 players randomize uniformly, for any value of γ. Level-1 players quantal respond using $\lambda = \gamma$, assuming all other players are type 0. Level-2 players quantal respond using $\lambda = 2\gamma$, assuming all other players are type 0 or type 1 , with perceived probabilities $f(0)/(f(0) + f(1))$ and $f(1)/(f(0) + f(1))$, respectively. Higher-level types are defined analogously.

Let $\gamma \to \infty$, so that *all* $k > 0$ types have unboundedly large values of λ, and choice behavior approaches best response. By effectively removing the stochastic choice component (except for the random $k = 0$ types), this converges to a generalized version of CH in which the type probabilities have the probability distribution $\{f(0), f(1), \ldots, f(k), \ldots\}$. The second parametric assumption is that $f(k)$ follows a Poisson distribution, although this can be easily generalized, and less restrictive models have been used in other papers.

The formal connection between TQRE and CH is asymptotic in γ.[16] In particular, for almost all games and almost all values of τ, the Poisson distribution parameter, the aggregate choice probabilities implied by the γ-TQRE model converge to the aggregate choice probabilities of CH. This equivalence is stated more formally, as follows.

Fix τ. Denote the CH choice probability that level k of player i chooses action j by p_{ijk}^τ, and denote the γ-TQRE choice probability (and f distributed Poisson with parameter τ) that type $\lambda = \gamma k$ of player i chooses action j by p_{ijk}^γ. Denote the expected CH choice probability of player i choosing action j by

---

[16] For finite values of γ the model can be viewed alternatively as either a downward-looking version of SQRE or a quantal response version of CH.

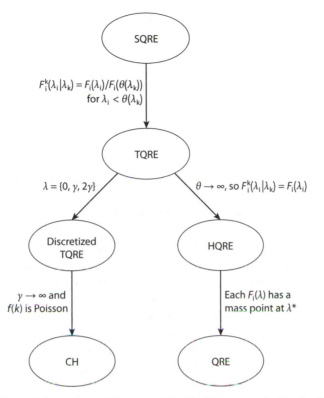

FIGURE 4.1. Several alternative models as special or limiting cases of subjective HQRE. The relationships among the models are depicted in a "family tree."

$\overline{p}_{ij}^{\tau} = \sum_{k=0}^{\infty} p_{ijk}^{\tau} f(k)$ and the expected γ-TQRE choice probability of player i choosing action j by $\overline{p}_{ij}^{\gamma} = \sum_{k=0}^{\infty} p_{ijk}^{\gamma} f(k)$. Define $\Delta^{\tau,\gamma} = \sum_{i=1}^{n} \sum_{j=1}^{J_i} (\overline{p}_{ij}^{\tau} - \overline{p}_{ij}^{\gamma})^2$.

Proposition 4.1 (CH models as a limiting case of TQRE)*: Fix τ. For almost all finite games Γ and for any $\varepsilon > 0$, there exists $\overline{\gamma}$ such that $\Delta^{\tau,\gamma} < \varepsilon$ for all $\gamma > \overline{\gamma}$.*

The proof can be found in Rogers, Palfrey, and Camerer (2009). As illustrated in figure 4.1, one can identify a "family tree" generated from SQRE by imposing additional restrictions. When all subjectivity takes the form of truncation at a player's $\theta_i(\lambda_i)$, we have TQRE. From TQRE there are two branches to follow. If $\theta \to \infty$, then subjectivity vanishes, producing the rational expectations version, HQRE. From there, a limiting distribution that places all mass at one value of λ corresponds to standard QRE. Following the other branch from TQRE corresponds to discretizing TQRE so that λ takes on a countable set of values, $L_\gamma = \{0, \gamma, 2\gamma, \ldots, k\gamma, \ldots\}$. Sending $\gamma \to \infty$, and assuming a Poisson distribution on $F(k)$ then yields the standard CH model. As will be discussed later in the book, this particular TQRE model can be a useful two-parameter specification for estimation purposes, where the two parameters are γ and τ (the Poisson parameter).

THE FORMAL CONNECTION BETWEEN SQRE AND LEVEL-K MODELS

Level-k models differ from CH by assuming that a level-k player thinks all other players are level-$(k - 1)$ players. Level-0 players can be modeled in various ways, but we adopt here the neutral assumption that level-0 players are completely random. Just as there is a formal connection between TQRE and CH, there is a similar formal connection between SQRE and level-k models. The main difference is that in level-k models, players do not have truncated beliefs based on some "true" distribution of types. Hence in games where the true distribution of types is concentrated on the lowest few levels (say, $k < 5$), level-k models have the built-in feature that more sophisticated types have relatively less accurate beliefs about the true distribution of types than less sophisticated types.

As before, first consider distributions such that the set of λ types is discrete, $L_\gamma = \{0, \gamma, 2\gamma, \ldots, k\gamma, \ldots\}$, with grid size γ, and fix the distribution over k, so that the probabilities of types are $f = \{f(0), f(1), \ldots, f(k), \ldots\}$. Consider the SQRE model where the belief of type k is concentrated on type $(k - 1)$ rather than the CH assumption that k's beliefs are distributed according to the conditional distribution f_k over types $j = 0, 1, \ldots, k - 1$, derived from f. For very large values of γ, each k type is approximately *best responding* to beliefs that all other players are type $k - 1$.

For finite values of γ this provides a quantal response version that combines level-k heterogeneity with skill heterogeneity. Specifically, higher level-k-type players will also be higher skill (λ) types. A number of studies have estimated quantal response versions of level-k models with considerable success fitting data (see footnote 13). In fact, that is the approach taken in the original analysis of Stahl and Wilson (1995), although they allow for some additional type categories. However, those models generally assume only level-k heterogeneity, estimating a distribution over k assuming a λ parameter that is common to all levels. It would be interesting to compare whether the more parsimonious two parameter (γ, λ) model described here would fit the data as well.

It is important to remember that level-k and other models with nonrational expectations are probably best suited for situations like one-shot games, in which players are not able learn much about the distributions of others' decisions. There may be some games for which these models do provide good predictions even if players interact repeatedly via random matchings, but care should be used in these cases. For example, consider a Bertrand duopoly game with zero costs in which 1 unit is sold to the firm with the lowest price on a range of feasible prices between 0 and 100 (step function demand). A level-0 player would choose randomly in this range, for an average of 50. A level-1 player (who expects the other person to randomize) would realize that a price P in this range will result in a sale with probability $(100 - P)/100$. This results in an expected profit function that is a quadratic in price, which is maximized at 50. A level-2 person, who best responds, would choose a price just below 50, for example, at 49. Similarly, higher-level players would also be bunched together in this middle price range. In contrast, observed prices in experiments with repetition and rematching tend to be relatively low in the range, closer to marginal cost, but prices do depend on the number of competitors (Dufwenberg and Gneezy, 2000). The point is that, with experience, players are able to learn what prices to expect, and hence, expectations are not as biased as would be predicted by level-k models in this game.

4.5 NOISY INTROSPECTION

The cognitive hierarchy and level-k models provide one approach to introducing imperfect strategic thinking into game theory. These are downward-looking models of strategic sophistication. A different kind of strategic sophistication involves the degree to which players can navigate the steps of thinking associated with common knowledge. In deciding one's strategy in game theory, it is not just a matter of blindly forming beliefs about how one's opponents are likely to play, because to do so in a sophisticated way also requires developing the intricate strategic reasoning behind how they will play. To make a reasoned prediction about whether my (rational) opponent will choose U or D, in principle I need to introspect as to what my opponent might think would be the better strategy. But that depends on the beliefs my opponent holds about the strategies I might play. There are potentially infinite steps of strategic thinking along the lines of "I think that my opponent will choose X," "I think that my opponent thinks that I will choose Y," "I think that my opponent thinks that I think my opponent thinks that I think. . . ."[17]

---

[17] This introspective approach based on higher-order beliefs is a more general notion of strategic sophistication than the level-k approach, where the behaviors of strategic types are generated by iterating best responses to the next lower level, starting from some well-specified "naive" model of behavior (level 0). There, a player's behavior corresponds to a single level. In the introspective

The *noisy introspection* (NI) model developed by Goeree and Holt (2001, 2004) takes a quantal response approach and further assumes that choice behavior associated with these higher orders of beliefs will be increasingly random with higher orders of the "I think that you think" reasoning. Intuitively it should be more difficult to predict behavior several steps ahead in this chain of logic than just the first step of such reasoning.

Consider first the special case of the NI model for symmetric two-player games with two possible actions, a_1 and a_2. In this case a player's choice can be simply represented by a single probability, for example, the probability of choosing a_1, and player-specific subscripts can be suppressed. A player's expected payoffs, $U_1(q)$ and $U_2(q)$, depend on the player's belief about the other's play, that is, the probability, q, with which the other player chooses a_1. In turn, the expected payoffs determine the player's choice probability via a logit choice rule[18]

$$\sigma^{\lambda_0}(q) = \frac{e^{\lambda_0 U_1(q)}}{e^{\lambda_0 U_1(q)} + e^{\lambda_0 U_2(q)}}, \tag{4.4}$$

where λ_0 is the precision associated with a player's decision. This rule determines a player's choice probability in terms of the player's "first-order" belief, q, about the other's decision. In a QRE, the requirement is that this first-order belief matches the other's choice probability, that is, $\sigma^{\lambda_0}(q) = q$, which turns (4.4) into a fixed-point equation. Equilibrium consistency is *not* imposed in the NI model, but instead the other's decision will depend on the other's belief about the player's own decision, as discussed next.

A player's first-order belief B^1 is modeled as the other's logit quantal response, $\sigma^{\lambda_1}(B^2)$, given the player's second-order belief, B^2. In other words, the thought process that produces a player's first-order belief about what the other will do is modeled as the other's logit quantal response given what the player thinks the other thinks that the player will do. This second level of introspection is likely to be somewhat imprecise, so we assume that the precision λ_1 associated with this step is less than the precision parameter λ_0 that determines a player's choice probabilities.

This naturally leads us to consider higher-order beliefs, denoted $B^0, B^1, B^2, \ldots$, where

- B^0 represents a player's choice probability as a function of B^1 beliefs: $B^0 = \sigma^{\lambda_0}(B^1)$;

---

approach, each player may consider many levels, or depths, of reasoning about own and others' beliefs about choice behavior. Below we show that level-k is a specific case of the NI model.

[18] The NI model is extended below to general quantal response functions and arbitrary normal-form games.

- B^1 represents a player's (first-order) belief about the other's choice based on the other's beliefs: $B^1 = \sigma^{\lambda_1}(B^2)$;
- B^2 represents a player's (second-order) belief about the other's belief about the player's choice: $B^2 = \sigma^{\lambda_2}(B^3)$, where B^3 and higher-order beliefs are by iterative substitutions; so
- B^k represents a player's kth-order beliefs, defined as the limit

$$B^k = \lim_{n \to \infty} \sigma^{\lambda_k} \circ \sigma^{\lambda_{k+1}} \circ \cdots \circ \sigma^{\lambda_n}(q). \tag{4.5}$$

Successive iterations become increasingly harder so the associated precision parameters fall, that is, $\lambda_0 \geq \lambda_1 \geq \cdots \geq \lambda_\infty = 0$. The zero limit ensures that the kth-order beliefs defined above are independent of the starting point q, which can thus be chosen arbitrarily. In other words, the assumption that the sequence of precision parameters converges to 0 implies that higher-order beliefs become more and more diffuse, that is, players effectively start out reasoning from a uniform prior. Since the NI model allows for an arbitrary nonincreasing sequence of precision parameters, it is quite general and includes many other models of boundedly rational behavior in (one-shot) games as special cases. One major distinction between these models is whether they assume homogeneous behavior or allow for individual heterogeneity.

An example that allows for individual heterogeneity is the "NI-k model" proposed by Goeree, Louis, and Zhang (2015) where players differ in the number of noisy introspective steps they make. In particular, an NI-k player is characterized by the sequence of precision parameters that takes the form $\lambda_0 = \cdots = \lambda_{k-1} = \lambda$ and $\lambda_k = \cdots = \lambda_\infty = 0$. Note that the level-$k$ model is a limit case of the NI-k model when better responses (with finite λ) are replaced by best responses (with $\lambda = \infty$). For instance, if $\lambda_0 = \infty$ and $\lambda_k = 0$ for $k \geq 1$, then B^1 is completely random, and the model predicts that players will best respond to random behavior by the opponent, that is, players are level-1 types. Higher-level types are constructed similarly; for example, for $k > 1$ a level-k type corresponds to the sequence $\lambda_0 = \cdots = \lambda_{k-1} = \infty$ and $\lambda_k = \cdots = \lambda_\infty = 0$. To summarize, the level-k model is obtained by considering (degenerate) non-increasing sequences of the precision parameters that involve only ∞ (best response) and 0 (uniform behavior).

Goeree, Louis, and Zhang (2015) apply the NI-k model to data from one-shot play in variants of the "11–20" money request game (Arad and Rubinstein, 2012). The 11–20 game consists of a nonstrategic part—players are paid the amount they request, which has to be an integer between 11 and 20—and a strategic part—players receive a bonus of 20 if the amount they request is 1 less than what their opponent requests. Arad and Rubinstein (2012) propose the game as a "litmus test" for level-k thinking, since a naive level-0 player will optimize only the nonstrategic part of the game and choose 20, level-1 will best respond to 20 and choose 19, level-2 will choose 18, etc.

Goeree, Louis, and Zhang (2015) consider variants of the basic game by arranging the numbers 11–20 on a line, not necessarily in increasing order but always with 20 in the rightmost position, and players receive the bonus when they pick the number directly to the left of their opponent's choice. If the numbers are increasing from left to right then the game is identical to the baseline game. However, consider an "extreme" variant with numbers arranged in a declining pattern: 19, 18, . . . , 11, ending with 20. In terms of positions, the level-k predictions for this extreme variation are identical to those in the baseline game since neither the obvious choice for the naive player nor the best-response structure have changed: level-0 will choose the rightmost number (20), level-1's best response is one to the left (11), level-2 chooses one to the left of that (12), and so forth. The data, however, are dramatically different: the modal choice in the extreme variation is 19, which would correspond to a level-9 player under the continued assumption that the level-k model applies, which is unrealistic.

Goeree, Louis, and Zhang (2015) show that under the NI-k model such behavior is realistic for $k = 1$ and higher. The intuition is that when others' behavior is noisy, the cost of picking 19 (instead of 20) is only 1 while it yields a strictly positive chance of the bonus (a choice of 20 never yields the bonus). To summarize, because behavior is noisy and because subjects "rationally" expect behavior to be noisy, the NI-k model dramatically outperforms the level-k model, which predicts no better than Nash in variations of the 11–20 game. The improvement is not merely in terms of "fit," but rather in terms of reproducing main qualitative features of the data (e.g., modal decisions and histograms of observed actions).

A parsimonious approach to modeling homogeneous players is to assume a geometrically declining sequence for all players, that is, $\lambda_k = \lambda t^k$, where $t \in [0, 1)$ measures how fast precision falls with successive iterations. This model reproduces logit QRE in the limit when $t \to 1$, assuming (4.5) converges in this limit.[19] To see this, note that if the process in (4.5) converges with a constant sequence of precision parameters, then applying σ^λ one more time does not change the outcome, that is, $\sigma^\lambda(B^k) = B^k$, which means the B^k probabilities constitute a logit QRE. Goeree and Holt (2004) apply this homogeneous model to 37 matrix games studied by Guyer and Rapoport (1972) and show that it fits

---

[19] Without the assumption that $\lambda_\infty = 0$, the NI process in (4.5) may not converge. For instance, in a matching pennies game, a sequence of (nearly) best responses will "cycle" forever. For λ small enough, however, the process (4.5) would converge also for a matching pennies game.

the one-shot data better than logit QRE, and the improvement is particularly noticeable for the asymmetric coordination and chicken games considered.

So far the NI model has been defined only for symmetric games. We next generalize the model to arbitrary finite games, drawing a parallel with the notion of rationalizability.

4.5.1 Noisy Rationalizability

Consider a normal-form game $\Gamma = [N, \{A_i\}_{i=1}^n, \{u_i\}_{i=1}^n]$, as defined in section 2.1. Each player $i = 1, \ldots, n$ chooses an action from the set A_i with J_i elements denoted by a_{ik}. Let $A = \prod_i A_i$ be the Cartesian product of strategy spaces and, for each i, let P_{-i} denote the projection from A to $A_{-i} = \prod_{k \neq i} A_k$, that is, the set of actions of player i's rivals.

Rationalizability is based on the idea of iteratively eliminating those strategies that are never best responses for any (consistent) set of beliefs (Bernheim, 1984; Pearce, 1984). The set of rationalizable strategies can be constructed by defining $\Lambda : A \to A$ as $\Lambda(A) = \prod_i f_i(P_{-i}(A))$, where $f_i(a_{-i})$ is player i's best response to her opponents' strategy a_{-i}. Bernheim shows that the set of rationalizable strategies can be obtained by recursively applying the Λ-mapping. In other words, the rationalizable strategies, RS, are given by the limit set

$$\text{RS} = \lim_{n \to \infty} \Lambda^n(A). \tag{4.6}$$

Noisy rationalizable outcomes can be defined analogously by replacing players' best-response functions, $f_i(\cdot)$, by better-response functions, $\sigma_i^\lambda(\cdot)$. Since smoothed better responses are interior, they are defined over the set of mixed strategies, Σ. Suppose player i's first-order belief about rivals' play is $B_i^1 \in \Sigma_{-i}$. Then the expected payoff of choosing action a_{ik} is given by $U_{ik}(B_i^1)$ and the probability of selecting a_{ik} equals $\sigma_{ik}^\lambda(B_i^1) = R(\lambda U_{ik}(B_i^1))$, where $R(\cdot)$ is some regular quantal response function defined in section 2.1. The J_i-dimensional vector, σ_i^λ, with elements σ_{ik}^λ for $k = 1, \ldots, J_i$, defines a mapping from Σ_{-i} to Σ_i. Define $\sigma^\lambda(\Sigma)$ to be the Cartesian product $\prod_i \sigma_i^\lambda(P_{-i}(\Sigma))$. It is this mapping that replaces the Λ-mapping used by Bernheim (1984) in the construction of rationalizable strategies.

Recall that the notion of rationalizability assumes perfectly rational decision making at any level of introspection, that is, irrespective of the number of iterations. For this reason, the rationalizable strategies follow from applying the same Λ-mapping recursively in (4.6). In contrast, we assume that higher levels of introspection become increasingly more noisy. The set of noisy rationalizable strategies, NRS, is obtained by recursively applying the σ^λ-mapping to Σ, using a lower precision at every step:

$$\text{NRS} = \lim_{n \to \infty} \sigma^{\lambda_0} \circ \sigma^{\lambda_1} \circ \cdots \circ \sigma^{\lambda_n}(\Sigma), \tag{4.7}$$

where the $\{\lambda_n\}_{n=0}^{\infty}$ form a nonincreasing sequence with $\lambda_\infty = 0$. The latter assumption implies that the set of noisy rationalizable outcomes consists of a single element of Σ, since σ^0 maps Σ into a single point, corresponding to uniform belief probabilities for all players. To summarize, while the set of rationalizable strategies generally consists of more than one point, the noisy rationalizable (mixed) strategy is always unique, even in games with multiple Nash equilibria.

So far we have implicitly assumed that the iterative process in (4.7) converges. The next proposition, which summarizes the main result of this section, establishes that this is indeed the case (for a proof, see Goeree and Holt, 2004).

Proposition 4.2 (Convergence of introspective beliefs)*: If $\lambda_0, \lambda_1, \lambda_2, \ldots$ is a sequence of nonincreasing precision parameters that converges to 0, then the sequence $\sigma^{\lambda_0}(p^0)$, $\sigma^{\lambda_0}(\sigma^{\lambda_1}(p^0))$, $\sigma^{\lambda_0}(\sigma^{\lambda_1}(\sigma^{\lambda_2}(p^0)))$, $\ldots$ converges to a unique point, the noisy introspection strategy, independent of the starting point, $p^0 \in \Sigma$.*

Example 4.1 (Example 2.3 continued: asymmetric game of chicken)*:* As an application, consider again the example of the asymmetric game of chicken shown in table 2.3. Figure 4.2 shows the set of actions (B^0), first-order beliefs (B^1), and second-order beliefs (B^2) when we use geometrically declining precision parameters, that is, $\lambda_k = \lambda t^k$, and vary $\lambda \in [0, \infty)$ and $t \in [0, 1)$. The entire set that combines the light, medium, and dark gray areas shows the possible actions; the medium and dark area combined show the first-order beliefs; and the dark area shows the second-order beliefs.

Note that the lower boundaries of the set of actions, first-order beliefs, and second-order beliefs are identical and correspond to the logit QREs that arise in the limit when $t \to 1$. The intuition is given above: if the noisy introspection process converges (as it does for the asymmetric game of chicken) for a constant sequence, $\lambda_k = \lambda$ for $k \geq 0$, then the B^k must be a fixed point of the σ^{λ}-mapping, that is, they correspond to logit QRE. Also note that the set of higher kth-order beliefs gets smaller with k, which is intuitive since beliefs get more dispersed, that is, closer to the $(0.5, 0.5)$ outcome. Indeed, for large k, the set of kth-order beliefs essentially consists of a single point, $(0.5, 0.5)$, for all $t \in [0, 1)$ and $\lambda \in [0, \infty)$, plus the curve of logit QRE for $t = 1$ and $\lambda \in [0, \infty)$.

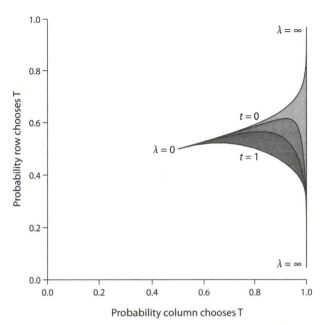

FIGURE 4.2. The set of noisy rationalizable outcomes for the asymmetric game of chicken in table 2.3 for $t \in [0, 1)$ and $\lambda \in [0, \infty)$. The dark gray set shows the set of second-order beliefs, B^2, the medium and dark gray sets combined show first-order beliefs, B^1, and the entire set (light, medium, and dark gray) shows actions, B^0. The upper boundary of this entire set corresponds to quantal responses to uniform beliefs ($t = 0$) while the lower boundary coincides with the set of logit QRE, which arise in the limit $t \to 1$.

Importantly, note that the set of all noisy introspection outcomes is *not* the full unit square, that is, it is not true that the two-parameter (λ, t) model can simply explain anything. Interestingly, the noisy introspection model predicts only the outcomes in which the Column player, who has the larger advantage in an asymmetric outcome, is more likely to play tough than soft. Depending on the degree of sophistication of Row as measured by t, Row may be more likely to choose tough as well (e.g., when t is close to 0) or Row may be more likely to succumb and play soft (i.e., when t is close to 1). These predictions are in line with experimental results but contrast with those of logit QRE, which allows for both asymmetric outcomes where either Row or Column plays tough and the other player accommodates, as well as a mixed outcome; see figure 2.4.

4.5.2 Noisy Introspection in Extensive-Form Games

Noisy introspection is defined above only for games in strategic form, but can be extended to finite games in extensive form analogously to how QRE is extended to AQRE for extensive-form games. This approach has been applied by Kübler and Weizsäcker (2004) to analyze data from information cascade games. An important simplifying feature of those games is that each player makes only a single decision, in sequence, and observes the choices of all previous players, and payoffs depend only on one's own action and an underlying state of the world. In games with this structure, it is only necessary to consider chains of higher-order beliefs of length less than or equal to the number of players minus one. This observation is illustrated in the example that follows.

Consider a three-player two-state, two-action, two-signal information cascade game, where the prior on state 1 is $\frac{1}{2}$ and signal informativeness is $q > \frac{1}{2}$, as analyzed in detail in chapter 7.[20] Assume the payoff for a correct action is 1 and the payoff for an incorrect action is 0. The (logit) noisy introspection model in the three-stage example works in the following way. Denote by π_i the posterior belief of player i that the state is A, conditional on the actions of previous players and i's own signal. Then i chooses action A with probability

$$\sigma_A^i(\pi_i) = \frac{e^{\lambda_0 \pi_i}}{e^{\lambda_0 \pi_i} + e^{\lambda_0(1-\pi_i)}}.$$

The problem, of course, is determining how i's posterior, π_i, is derived. In the standard logit equilibrium model of the cascade game, it is assumed player i assumes that all previous players' choice probabilities were determined by the logit formula, using the *same* λ_0. The noisy introspection model is more complicated, and specifies a (possibly) different value of λ_k for the kth-order beliefs.[21] Thus, for the second mover, π_2 is derived from Bayes' rule, conditioning on player 2's signal and player 1's action, assuming that player 1's probability of choosing action A was given by

$$\sigma_A^1(\pi_1) = \frac{e^{\lambda_1 \pi_1}}{e^{\lambda_1 \pi_1} + e^{\lambda_1(1-\pi_1)}}. \tag{4.8}$$

Things get more complicated for constructing the belief of the third mover, player 3. Player 3 also assumes the choice probabilities for both players 1 and 2 are given by the logit formula with precision parameter λ_1. However, player 3 believes that player 2's beliefs about player 1's choice probabilities are given by applying λ_2 (second-order belief precision parameter) rather than λ_1. That is,

[20] Kübler and Weizsäcker (2004) also allow for endogenous information acquisition, in the sense that each player has the option of buying a signal before making a choice. The methodology of applying NI is basically the same.

[21] A natural assumption, consistent with the normal-form model of noisy introspection, would be that $\lambda_0 \geq \lambda_1 \geq \lambda_2 \geq \cdots$, but such an assumption is not formally required. If $\lambda_0 = \lambda_1 = \lambda_2 = \cdots = \lambda$, then the model reduces to the logit AQRE with precision parameter λ.

player 2 thinks that π_2 is derived conditioning on player 2's signal and assuming player 1's probability was the one given in (4.8), *except for substituting λ_2 for λ_1 in that formula*. Obviously, the specification of higher-order beliefs beyond the second order is not necessary. For longer games with more players, this logic is extended in the natural way, so that one must specify up to $n - 1$ higher-order beliefs if there are n players, and a precision parameter corresponding to each. Notice that this approach to modeling quantal response behavior in cascade games sneaks in some heterogeneity, because player k draws inferences from the history of play in a fundamentally different way than does player $k' \neq k$.

In principle, there is no reason the sequences of precision parameters need to be the same for all players. Just as players could have different "skill" levels in the sense of higher or lower payoff responsiveness (corresponding to λ_0 in the noisy introspection model), they could form expectations about higher-order beliefs differently as well. That is, the precision parameters could vary across individuals, so the sequences could be indexed by i as well as by the order of beliefs, as $\lambda_0^i = \{\lambda_0^i, \lambda_1^i, \lambda_2^i, \ldots\}$. Even more complex kinds of heterogeneity are possible as well, because player i's kth-order beliefs about player j's beliefs about player j' could be constructed based on precision parameter $\lambda_k^{ijj'}$, while player i's kth-order beliefs about player j's beliefs about player j'' could be constructed based on precision parameter $\lambda_k^{ijj'} \neq \lambda_k^{ijj''}$. However, it is hard to imagine developing a general and parsimonious way to represent all these different kinds of heterogeneity with respect to the construction of higher-order beliefs.

5
Dynamics and Learning

The first part of this chapter explores *dynamic QRE* models where strategies are conditioned on observed histories of past decisions and outcomes of stage games. A key aspect of managing the complexity of infinitely repeated stage games is to consider strategies that respond to particular aspects of prior experience, for example, whether a prior "defect" decision has been observed in a prisoner's dilemma. For this reason, histories are defined in a manner that partitions data into decision-relevant equivalence classes. Players' quantal responses are functions of expected payoffs in the current stage plus discounted continuation values determined by history-contingent beliefs about future outcomes. The noise inherent in QRE models ensures that players have well-defined beliefs associated with histories that are not on the Nash equilibrium path. These beliefs are "rational" in the sense that the history-contingent distributions of action profiles are consistent with the distributions determined by equilibrium interactions over time.

Section 5.1 defines *recursive quantal response equilibrium* for infinitely repeated games, which are characterized by quantal subgame perfection (the continuation strategies specify a quantal response equilibrium in every proper subgame). Section 5.2 develops the QRE generalization of the notion of a Markov perfect equilibrium for dynamic and stochastic games, by requiring strategies to be measurable with respect to payoff-relevant states. The concept of a *Markov quantal response equilibrium* is defined and illustrated in a simple stochastic game with two payoff-relevant states. In one state the players play a prisoner's dilemma; in the other, a matching pennies game.

Off-path beliefs are almost surely influenced by a player's prior experience during a process of adjustment to equilibrium. The second part of this chapter considers models in which players are *learning* about others' behavior via a process in which they may update and respond to current beliefs in a noisy (quantal) manner. Some of these models necessarily involve less forward-looking expectational rationality than is implicit in a dynamic QRE formulation.

The most extreme simplification involves no forward- or backward-looking behavior at all, that is, an evolutionary process with noisy adjustments at each point of time that enhance the probabilities associated with actions that are more profitable at that point in time, based on current distributions of actions. In section 5.3, we specify a simple evolutionary process in which decisions are adjusted locally in the direction of increasing payoffs, subject to a white noise (Wiener) process, and show that any steady state will be a logit equilibrium.

The final section explores learning models that involve quantal responses to beliefs formed by processing information from finite (but possibly long) histories of prior or observed action profiles. The formulation permits consideration of a wide variety of exogenous or even *endogenous* (e.g., least squares) learning rules. For example, a learning rule could weight each prior observation equally (e.g., fictitious play), or it could exhibit recency effects whereby beliefs are more sensitive to more recent observations. These and other learning models have been used in laboratory experiments to explain time patterns and directions of convergence for out-of-equilibrium data in laboratory experiments.

Recency effects in forecasting behavior can introduce additional noise into the data, since with random matching and stochastic choices, different players will see different finite histories, and hence, will have different beliefs and quantal response choice probabilities. Note that histories determine beliefs via the learning rule being used, and the beliefs that are generated from this learning determine the choice probabilities for each history, which in turn, determine how that particular history gets updated. A steady-state equilibrium can be defined in terms of a stationary distribution over (truncated) histories, which will be called a *stochastic learning equilibrium*. The extra noise produced by differences in individual histories and recency effects can cause the stochastic learning equilibrium choice distributions to be "flatter" than would be the case with QRE for a stationary game. With very long memories and fictitious play (no recency), players would be able to learn the nature of any stationary state distribution of decisions, and the equilibrium condition in this limiting case is analogous to a quantal response equilibrium. This general approach can be used to study interesting dynamic phenomena such as price cycles or coordinated alternating decisions in a game with a pair of asymmetric Nash equilibria.

5.1 QRE IN INFINITELY REPEATED GAMES

Repeated games are applied widely in the social sciences, often as a workhorse model of the possibility of enduring relationships to support cooperative behavior that would not be possible in a one-shot game. Finitely repeated games are a special case of extensive-form games for which the AQRE model was fully developed earlier in this book. Infinitely repeated games are another story, and

TABLE 5.1. A prisoner's dilemma game.

	D	C
D	4, 4	10, 2
C	2, 10	8, 8

the purpose of this section is to show how QRE can be extended to analyze such "supergames."

Before proceeding with the formal development of QRE for supergames, we offer the following rationale for why QRE may offer new insights into the structure of equilibria, and the stability of equilibria for these games.

Example 5.1 (Repeated prisoner's dilemma game)*:* Consider the prisoner's dilemma stage game in table 5.1. In this game, the unique Nash equilibrium is (D, D). Also, it is well known that in any finitely repeated game with T repetitions, playing (D, D) in every history is the only subgame-perfect equilibrium of the game. However, if the game is infinitely repeated and if players are sufficiently patient, then there exist many subgame-perfect Nash equilibria whose equilibrium paths involve players choosing (C, C) in every period. Such equilibria are supported by strategies that ensure lower long-run payoffs off the equilibrium path.

   There are at least two conceptual problems with this. First, if there is any noise in the play (as there would be in an experiment or in most other applications), then play will certainly fail to be on the equilibrium path at some point. One then must face the question about whether the specified strategies off the path are plausible. There are many reasons why they might not be. In particular, there is an extensive literature on renegotiation-proofness requirements for off-path strategies.[1] Ex post, players may not be willing to implement draconian punishment strategies, such as the "grim trigger" strategy of playing (D, D) forever in response to a single deviation. Moreover, if play is noisy, then the off-path punishment strategies themselves may have random components. Second, even in a world with no error, there is an epistemological question about how players could possibly ever come to know what will be played off the equilibrium path if (in equilibrium) off-path histories are never observed.

QRE addresses both issues. First, all histories are reached with positive probability, and behavior in the continuation game from every history takes into account that every future history is also reachable. Strategies do not gratuitously

---

[1] See, e.g., Bernheim and Ray (1989) and Farrell and Maskin (1989).

encode behavior for information sets that are impossible to reach, although in principle one can imagine that players who start the game with incorrect or incomplete beliefs about behavior following some histories might eventually be able to learn.[2] Questions related to learning are addressed in the second part of the chapter.

5.1.1 Notation

Formally, let $I = \{1, \ldots, n\}$ be the (finite) set of players, $A = A_1 \times \cdots \times A_n$ be the (finite) set of stage-game action profiles, where $A_i = \{a_{i1}, \ldots, a_{iJ_i}\}$ is player i's set of feasible stage-game actions, and $u = \{u_i\}_{i=1}^n$ is the collection of stage-game payoff functions, where $u_i : A \to \Re$. A *history* of the game at period t is any sequence of action profiles, $h^t = (a^0, \ldots, a^{t-1})$. Histories are defined recursively. If the history at time t is $h^t = (a^0, \ldots, a^{t-1})$ and action profile a is implemented in period t, then we define $h^{t+1} = (a^0, \ldots, a^{t-1}, a) \equiv (h^t, a)$. Player i's (behavioral) strategy, σ_i, assigns to each history a probability distribution over A_i. Each player evaluates payoffs as the discounted sum of stage-game payoffs, so, for any infinite sequence of actions $a^\infty = (a^0, \ldots, a^t, \ldots)$, player i's long-run payoff is the discounted sum of stage-game payoffs, which we write as $u_i(a^\infty) = \sum_{t=1}^\infty \delta^t u_i(a^t)$.[3]

Given any strategy profile σ, and any history h^t, this implies a unique sequence of probability distributions over future action profiles. For each $\tau > t$ we denote by $\rho(a^\tau | h^t, \sigma)$ the probability of action profile a in period $\tau > t$. Thus, define for any history h^t the continuation value of the game under strategy profile σ as

$$Z_i^\sigma(h^t) = \sum_{a \in A} \left\{ \prod_{j=1}^n \sigma_j(a_j; h^t) \right\} u_i(a) + \sum_{\tau=1}^\infty \delta^\tau \left\{ \sum_{a^{t+\tau} \in A} \rho(a^{t+\tau} | h^t, \sigma) u_i(a^{t+\tau}) \right\}.$$

(5.1)

The expected utility to player i from choosing action a in history $h^t \in H_k$, given the strategy profile σ and the continuation values defined by (5.1), is given by

$$U_i^\sigma(a, h^t) = \sum_{a_{-i} \in A} \left\{ \prod_{j \neq i}^n \sigma_j(a_j, h^t) \right\} \left(u_i(a) + \delta Z_i^\sigma(h^t, a) \right).$$

A (logit) QRE in this repeated game is then defined for any given precision parameter λ as follows.[4] A strategy profile σ^* is a logit QRE of the repeated

---

[2] This latter point has been explored; see, e.g., Fudenberg and Kreps (1993).

[3] The assumption that all players share the same discount factor is easily generalized.

[4] QRE for repeated games can be defined similarly for any regular quantal response function.

game if and only if, for all $i \in I$, for all $a_i \in A_i$, for all t, and for all t-histories h^t,

$$\sigma_i^*(a_i, h^t) = \frac{e^{\lambda U_i^{\sigma^*}(a_i, h^t)}}{\sum_{a_i' \in A_i} e^{\lambda U_i^{\sigma^*}(a_i', h^t)}}. \tag{5.2}$$

One feature that is built into the definition of QRE for repeated games is (quantal) subgame perfection. That is, the quantal response functions implicit in the equilibrium system of equations in (5.2) are defined conditional on histories, taking into account the continuation strategies of all players. Thus, in every proper subgame, the continuation strategies of the players are required to form a quantal response equilibrium of that subgame. Also recall that, just as in AQRE, all subgames are reached with positive probability so there is no issue about off-path behavior. An existence theorem is trivial for these games, and follows from the existence of QRE in the stage game. Suppose that σ^* is a logit QRE of the (finite) stage game. Then it is also an equilibrium where, at every history, players use that one-shot QRE strategy.

5.1.2 A Finite Recursive Form for Repeated Games

In a repeated game, the set of histories is enormous. For practical reasons, recursive QRE is defined by first specifying a finite partition of the set of histories, and then considering only strategies that are measurable with respect to that partition.[5] Denote such a partition of histories by $\mathbf{H} = \{H_1, \ldots, H_K\}$. For any history, h^t, denote by $\eta(h^t)$ the element of the partition to which h^t belongs. That is, $\eta(h^t) = H_t$ if and only if $h^t \in H_t$.[6] Given a partition, $\mathbf{H}$, σ_i is required to be measurable with respect to $\mathbf{H}$ for all i. Formally, we define

$$\sigma_i \text{ is } \mathbf{H}\text{-measurable} \Leftrightarrow \eta(h^t) = \eta(h^{t'}) \Rightarrow \sigma_i(h^t) = \sigma_i(h^{t'}) \forall i, h^t, h^{t'}.$$

That is, this measurability restriction on strategies requires that each player use the same probability distribution over an action set for any two histories that are in the same equivalence class.

As a very simple example, one could partition histories so that two histories are in the same equivalence class if and only if the last action profile of that history is the same. That is, for two histories, h^t and $g^{t'}$, $\eta(h^t) = \eta(g^{t'})$ if and only if $a^{t-1} = a^{t'-1}$. For example, in the infinitely repeated prisoner's dilemma game, this would divide the histories into five equivalence classes: the null history,

---

[5] The equivalence classes defined by the partition could be called "states," but this could create confusion with the definition of states in Markov perfect equilibrium defined in the next section, where states are directly tied to the game's payoff function (see the next section).

[6] This approach is related to the finite automata approach to studying complexity and bounded rationality in repeated games.

h^0, and four history classes, depending on whether the last action profile was $(D, D), (D, C), (C, D)$, or (C, C). The measurability restriction then implies that a player must have the same probability of choosing D in period 5 after the history (CC, CC, DD, CD) as the probability he chooses D in period 3 after (DD, CD).[7]

In keeping with the *recursive* nature of repeated games, it only makes sense to consider partitions of histories such that the continuation game depends on history in a way that is consistent with the partition. For example, if two histories are in the same partition, then the new histories that result from appending the action profile observed in the current stage should also be in the same partition. Formally, for a partition **H** to be *admissible*, a partition must be measurable with respect to actions, for the same reason we require strategies to be **H**-measurable. This admissibility condition on **H** is formally stated as follows:

$$\mathbf{H} \text{ is an } admissible \ partition \Leftrightarrow \eta(h^t) = \eta(h^{t'})$$

$$\Rightarrow \eta(h^t, a) = \eta(h^{t'}, a) \quad \forall a, h^t, h^{t'}.$$

If a partition were inadmissible it would mean that there is relevant information about the continuation game that is not encoded in the partition. Consider the following example of an inadmissible partition. Let two histories, h^t, $h^{t'}$ be in the same equivalence class if and only if $a^{t-2} = a^{t'-2}$. Then, in the infinitely repeated prisoner's dilemma game, as in the earlier example, this would divide the histories into six equivalence classes: the null history, h^0, the *set* of one period histories, $h^1 = \{(C, C), (D, D), (D, C), (C, D)\}$, and four history classes defined on those histories, h^t, such that $t > 1$, depending on whether the $t - 2$ action profile two periods ago was $(D, D), (D, C), (C, D)$, or (C, C). Now, let $t = 2$ and consider the two histories $h^2 = (DD, CD)$ and $\widehat{h}^2 = (DD, CC)$. Then, $\eta(h^2) \neq \eta(\widehat{h}^2)$ but for any action profile in period 3, a, we will have $\eta(h^2, a) = \eta(\widehat{h}^2, a)$. This means that the value of the continuation game at h^2 will in general be different from the value of the continuation game at $\widehat{h}^2$ so that some payoff-relevant information is not incorporated in the partition of histories. In a real sense, such partitions are not internally consistent.

---

[7] Many subgame-perfect equilibria of the repeated prisoner's dilemma game have this structure, such as grim trigger ("cooperate in the first period and then play grim trigger thereafter"). The grim trigger strategy is measurable with respect to the simpler partition of histories into two equivalence classes. One equivalence class is the set of all histories where the three action profiles $(D, D), (D, C)$, (C, D) have never been played, and the other equivalence class is the set of all histories where one of the three action profiles $(D, D), (D, C), (C, D)$ has been played at least once. Essentially all strategies that have been identified specifically in the repeated game constructions are based on finite partitions of the set of histories.

Given any admissible partition, $\mathbf{H}$, and any $\mathbf{H}$-measurable strategy profile of the players, $\sigma^{\mathbf{H}}$, this implicitly defines a collection of *value functions*, one for each player.[8] Denote these by $V^{\sigma^{\mathbf{H}}} = \{V_i^{\sigma^{\mathbf{H}}}\}_{i=1}^n$, where $V_i^{\sigma^{\mathbf{H}}} : \mathbf{H} \to \Re$ specifies, for each equivalence class defined by the partition of the set of histories, the discounted payoff of the continuation game from any history in that equivalence class. Thus, for each i, $V_i^{\sigma^{\mathbf{H}}}(H_k)$ is the value of the continuation game starting from any history $h^t \in H_k$, given $\sigma^{\mathbf{H}}$. The value function associated to $\sigma^{\mathbf{H}}$ is defined recursively, for any $H_k \in \mathbf{H}$, as

$$V_i^{\sigma^{\mathbf{H}}}(H_k) = \sum_{a \in A} \left\{ \prod_{j=1}^n \sigma_j^{\mathbf{H}}(a_j, H_k) \right\} u_i(a) + \delta \sum_{l=1}^K \rho(H_l; \sigma^{\mathbf{H}}) V_i^{\sigma^{\mathbf{H}}}(H_l), \qquad (5.3)$$

where ρ is the probability distribution over $\mathbf{H}$ implied by $\sigma^{\mathbf{H}}$, given the current history lies in H_k, and the players are playing the mixed action profile $\sigma^{\mathbf{H}}(H_k)$. Applying the one-shot-deviation principle in this recursive structure, an $\mathbf{H}$-equilibrium is then defined as any strategy profile, σ^*, such that for all i, for all $H_k \in \mathbf{H}$, and for all mixtures over i's stage-game actions, α_i,

$$V_i^{\sigma^*}(H_k) \geq \sum_{a \in A} \left\{ \alpha_i(a_i) \prod_{j \neq i} \sigma_j^*(a_j, H_k) \right\} u_i(a) + \delta \sum_{l=1}^K \rho(H_l; \alpha_i, \sigma_{-i}^*) V_i^{\sigma^*}(H_l).$$

RECURSIVE QRE FOR REPEATED GAMES

Fix λ, and fix some repeated game in recursive form defined by $\langle I, A, U, \delta, \mathbf{H} \rangle$. Given any mixed $\mathbf{H}$-measurable strategy profile, σ, we define $V_i^\sigma(\cdot)$ for each i, recursively, exactly as in equation (5.3). Define the expected utility to player i from choosing strategy a in history $h^t \in H_k$, given the strategy profile σ and its implied value function $V_i^\sigma(\cdot)$, by

$$U_i^\sigma(a, H_k; V_i^\sigma, \sigma_{-i}) = \sum_{a_{-i} \in A} \left\{ \prod_{j \neq i} \sigma_j(a_j, H_k) \right\} u_i(a) + \delta \sum_{l=1}^K \rho(H_l; a_i, \sigma_{-i}) V_i^\sigma(H_l).$$

A logit quantal response equilibrium is a profile of Markov strategies, σ^*, such that, for all $i \in I$, for all $a_i \in A_i$, and for all $s \in \mathbf{S}$,

$$\sigma_i^*(a, H_k) = \frac{e^{\lambda U_i^{\sigma^*}(a, H_k; V_i^{\sigma^*}, \sigma_{-i}^*)}}{\sum_{a' \in A_i} e^{\lambda U_i^{\sigma^*}(a', H_k; V_i^{\sigma^*}, \sigma_{-i}^*)}}, \qquad (5.4)$$

where $V_i^{\sigma^*}$ is given by the unique solution of (5.3).[9]

---

[8] Without the admissibility and measurability restrictions, these value functions would not be well defined.

[9] In this simple recursive, finite, bounded framework with discounting, existence of a value function associated with a Markov strategy σ^* is guaranteed from basic principles of dynamic programming.

Example 5.2 (Example 5.1 continued: repeated prisoner's dilemma game):
Going back to the prisoner's dilemma example, one can map the recursive
logit QRE correspondence for different values of δ and for different
specifications of admissible partitions. The simplest partition is, of course, the
null partition, where **H** consists of a single element. That is, strategies are
constant across histories. As noted earlier, in any repeated game, infinite
repetition of a stage-game QRE is a QRE of the repeated game. Because of
the dominance structure of the repeated game, there is a unique stage-game
logit QRE for any value of λ.

The simplest nontrivial partitions divide the set of all histories into just two
equivalence classes of subhistories. Let H_1 be the set of all histories in which
there has been at least one prior instance of a strategy other than CC, and let
H_2 be the set of all other possible histories.[10] Call H_1 and H_2 the two *phases*
of the game, referred to as the noncooperative and cooperative phases. First,
observe that this partition is admissible. Characterizing a *symmetric* logit
QRE of the game is straightforward, as it requires solving for only two
mixing probabilities: q_1, the probability of choosing C in the noncooperative
phase, and q_2, the probability of choosing C in the cooperative phase.
Formally, as in QRE for normal-form games, the logit equilibrium (q_1^*, q_2^*) is
a fixed point. One can think of computing the fixed point in the following
way. Start with an initial value of (q_1^0, q_2^0). This initial strategy implies value
functions, V_1 and V_2, for the two phases, as defined in (5.3). Using these value
functions, one obtains a new strategy profile, (q_1^1, q_2^1) from equation (5.4),
which implies new value functions, V_1 and V_2, etc.

For this prisoner's dilemma example, the equations for V_1 and V_2 and the
equations for (q_1, q_2) are

$$V_1 = 8(q_1)^2 + 12q_1(1 - q_1) + 4(1 - q_1)^2 + \delta V_1$$

or

$$V_1 = \frac{4 - 6q_1}{1 - \delta},$$

and

$$V_2 = 8(q_2)^2 + 12q_2(1 - q_2) + 4(1 - q_2)^2 + \delta(q_2)^2 V_2 + \delta(1 - (q_2)^2)V_1,$$

and the logit equations mapping (V_1, V_2) into choice probabilities (q_1, q_2) are
obtained from (5.4).

---

[10] This example was explored in Nayyar (2009), which also established a folk theorem for QRE
when the discount factor is close to 1 and λ approaches ∞.

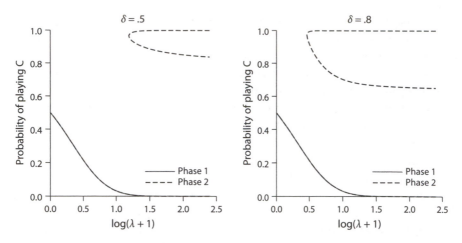

FIGURE 5.1. Recursive logit QRE for the repeated prisoner's dilemma game in table 5.1.

This mapping from (q_1^0, q_2^0) to (q_1^1, q_2^1) is single valued and continuous. One can then use computational methods to find the set of fixed points of this mapping. In principle, there could be many recursive logit QRE, depending on the value of λ. There always exists at least one recursive QRE, which is the history-independent QRE that is a noisy version of always defect, and the strategy is the same in H_1 and H_2. This is isomorphic to the unique QRE for the null partition.[11]

For low enough values of λ there will be a unique fixed point. For high enough values of λ, however, there exist recursive QREs with the property that $q_1^* < q_2^*$, and $V(H_1) < V(H_2)$, provided the players are patient enough.[12] These recursive QREs converge to the cooperative "grim trigger" equilibrium as λ tends to infinity. Because the logit equilibrium correspondence is smooth, new equilibria will arise in pairs as λ is varied, so in addition to this cooperative equilibrium, there will also be a logit equilibrium that converges to a mixed-strategy equilibrium of the repeated game. Figure 5.1 shows the (symmetric) recursive logit QRE correspondence values of $q_1^*(\lambda)$ associated with this partition for this prisoner's dilemma game for two different values of δ. The logit solution of the game is the noncooperative branch of the correspondence that converges to (D, D).

---

[11] More generally, infinite repetition of a stage-game QRE is always a recursive QRE for any admissible partition.

[12] If the players are so impatient that grim trigger is not supported as a strict subgame-perfect Nash equilibrium, then it cannot be supported as a recursive logit equilibrium in the limit. In this game it requires $\delta > \frac{1}{3}$.

5.2 QRE IN DYNAMIC AND STOCHASTIC GAMES

In dynamic and stochastic games, many of the equilibrium constructions that can be found in the repeated game "folk theorem" literature can be adapted to support an indeterminate number of equilibria. These constructions are by their very nature quite complicated, both analytically, and in terms of more formal mathematical complexity. Markov perfect equilibrium (MPE; see Maskin and Tirole, 1989) is a widely accepted and commonly applied refinement of subgame-perfect equilibrium for this class of games. For most applications, MPE drastically reduces the size of the strategy space that needs to be considered in the equilibrium analysis. With subgame-perfect equilibrium, strategies can depend in arbitrary ways on the history of play, which opens up the possibility for strategies to encode a system of punishments and rewards that can support many sequences of action profiles in the game. MPE restricts attention to those subgame-perfect equilibria in which each player's strategy depends only on a well-defined set of *payoff-relevant states*. Thus, continuation strategies can differ across two different histories only if the two histories lead to two different payoff-relevant states, in the sense that the continuation games defined by the two histories are identical in terms of the mapping from future actions to payoffs.

To illustrate the powerful reduction in the strategy space that obtains compared to subgame-perfect equilibrium, consider an infinitely repeated game with perfectly observable actions, so that all histories are public and common knowledge. In that case, there is only a single payoff-relevant state (past outcomes have no effect on payoffs in the current stage game), which means that the strategies composing a Markov perfect equilibrium cannot depend on history in any way. For example, in the prisoners' dilemma, there is only one unique MPE: that is, always defect. More generally, the only MPE in such games involves the infinite repetition of one-shot Nash equilibrium strategies.

To define a Markov quantal response equilibrium (MQRE), we first define MPE for a simple framework and consider only infinite horizon dynamic games with a simple recursive structure. Assume that there is a finite number of states that are deterministic functions of histories, action sets are finite and history independent, actions are perfectly observed, there is a finite number of players, and payoff functions are given by the discounted sum of payoffs in each period of the game. Also assume a logit response function.[13]

---

[13] This relatively simple framework is easily generalized to arbitrary response functions, alternative specifications of the long-run payoff function, action sets that could be history dependent, and a stochastic mapping from histories to states.

5.2.1 Markov Perfect Equilibrium

Formally, let $I = \{1, \ldots, n\}$ be the set of players, $A = A_1 \times \cdots \times A_n$ be the set of stage-game action profiles, where $A_i = \{a_{i1}, \ldots, a_{iK_i}\}$ is player i's set of feasible actions in any period, and $S = \{s_1, \ldots, s_S\}$ is the set of payoff-relevant states. Denote by $T : S \times A \to S$ the state transition function that maps the current state and current stage action profile into a new state, and $U = \{u_i\}_{i=1}^{n}$ is the set of stage-game payoff functions, where $u_i : A \times S \to \Re$. In this setup, without loss of generality, for the purposes of MPE or MQRE, restrict strategies to those that depend only on the current state. That is, for each player i consider only subgame-perfect equilibrium behavior strategies (possibly mixed) that are measurable in the payoff-relevant state, that is, they take the form $\sigma_i : S \to A_i$. Given any infinite sequence of actions and states, $a^\infty = (a^0, \ldots, a^t, \ldots)$, $s^\infty = (s^0, \ldots, s^t, \ldots)$, player i's long-run payoff is the discounted sum of stage-game payoffs, denoted by $u_i(a^\infty, s^\infty) = \sum_{t=1}^{\infty} \delta^t u_i(a^t, s^t)$.[14]

Given the recursive structure, any Markov strategy profile of the players, σ, defines a collection of *value functions*, one for each player. We denote these by $V^\sigma = \{V_i^\sigma\}_{i=1}^n$, where $V_i^\sigma : S \to \Re$. Thus, for each i, $V_i^\sigma(s)$ is the value of the continuation game starting from state s, if all players are adopting state-contingent strategies as given by σ. The value function is defined recursively, with some abuse of notation, as

$$V_i^\sigma(s) = \sum_{a \in A} \left\{ \prod_{j=1}^{n} \sigma_j(a_j, s) \right\} u_i(a, s) + \delta \sum_{l=1}^{S} \rho(s_l; \sigma, s) V_i^\sigma(s_l), \qquad (5.5)$$

where $\rho(s_j; \sigma, s)$ is the probability distribution of states under the transition function T, given the current state is s and the players are mixing according to $\sigma(s)$. Applying the one-shot-deviation principle in this recursive structure, an MPE is then defined as any strategy profile, σ^*, such that for all i, for all s, and for all mixtures over player i's stage-game actions, α_i,

$$V_i^{\sigma^*}(s) \geq \sum_{a \in A} \left\{ \alpha_i(a_i) \prod_{j \neq i} \sigma_j^*(a_j, s) \right\} u_i(a, s) + \delta \sum_{l=1}^{S} \rho(s_l; \alpha_i, \sigma_{-i}^*, s) V_i^{\sigma^*}(s_l).$$

Example 5.3 (Two-state example): As an illustration, consider the following simple two-state, two-player, two-action game. Each player has two stage-game actions: C or D in state s and H or T in state s'. The payoff functions in the two states, s and s', are given by the two payoff matrices

---

[14] For ease of notation, we assume here that all players share the same discount factor, but this is easily generalized.

TABLE 5.2. Payoff functions for a two-state example.

$u(s)$	D	C		$u(s')$	H	T
D	$(4, 4)$	$(10, 2)$		H	$(2, 0)$	$(0, 2)$
C	$(2, 10)$	$(8, 8)$		T	$(0, 2)$	$(2, 0)$

shown in table 5.2, and players have the same discount factor, δ. The transition function is given by

$$T(s', a) = s \quad \forall a \in A,$$
$$T(s, (C, C)) = s,$$
$$T(s, a) = s' \quad \forall a \neq (C, C).$$

In this game, players are either playing a prisoners' dilemma (state s) or they are playing a matching pennies game (state s'). The transition function says that if you and your partner cooperate when you are in the prisoner's dilemma, (C, C), then you continue to play a prisoner's dilemma game tomorrow, but switch to the matching pennies game tomorrow if either of you defects. If you are playing the matching pennies game today, then you will be playing a prisoner's dilemma tomorrow. First, observe that there is a unique one-shot Nash equilibrium in state s', and the transition function at s' is independent of a. Thus both players play their unique stage-game strategy in s', which is to mix uniformly, that is, each player plays $(\frac{1}{2}, \frac{1}{2})$.

Now consider the two symmetric pure stage-game strategies the players could be adopting in s, either (C, C) or (D, D). It is easy to see that the Markov strategy $\sigma^* = [(D, D), (\frac{1}{2}, \frac{1}{2})]$ is a Markov equilibrium for all values of $\delta \in [0, 1)$. In state s, if your opponent is always defecting, then no matter what you do, you will always play a matching pennies game tomorrow, so the best you can do is defect in response. If you are in the matching pennies game, then no matter what your opponent does, you will be stuck in the prisoner's dilemma game tomorrow, so you play the one-shot Nash equilibrium. Formally, one can solve for the value functions, writing V_s^{DD} and $V_{s'}^{DD}$:

$$V_s^{DD} = 4 + \delta V_{s'}^{DD}, \qquad V_{s'}^{DD} = 1 + \delta V_s^{DD},$$

which implies

$$V_s^{DD} = \frac{4 + \delta}{1 - \delta^2}, \qquad V_{s'}^{DD} = \frac{1 + 4\delta}{1 - \delta^2}.$$

If one player were to change and cooperate in the prisoner's dilemma in some period, then the payoff to that deviator would be $2 + \delta V_{s'}^{DD} < 4 + \delta V_{s'}^{DD}$ for all $\delta \in [0, 1)$. As noted above, since both players are mixing $(\frac{1}{2}, \frac{1}{2})$ in state s', any deviation in state s' yields no change in payoffs.

First consider the possibility that both players cooperate in the prisoner's dilemma state, that is, they play the strategy profile (C, C). Can this behavior also be supported as part of an MPE (with the other part necessarily involving uniform mixing in state s')? The answer is yes, for sufficiently high δ. As before, we can compute the value functions corresponding to this joint strategy profile, which we denote by V_s^{CC} and $V_{s'}^{CC}$:[15]

$$V_s^{CC} = 8 + \delta V_s^{CC}, \qquad V_{s'}^{CC} = 1 + \delta V_s^{CC},$$

which implies

$$V_s^{CC} = \frac{8}{1 - \delta}, \qquad V_{s'}^{CC} = 1 + \frac{8\delta}{1 - \delta}.$$

If one player defects in the prisoner's dilemma game in some period, then the payoff to that defector would be $10 + \delta V_{s'}^{CC} = 10 + \delta[1 + \frac{8\delta}{1-\delta}] > \frac{8}{1-\delta}$ if and only if $\delta < \frac{2}{7}$. Thus, cooperation in the prisoner's dilemma state can be supported as an MPE outcome in this game provided $\delta \in [\frac{2}{7}, 1)$.

There is a third possibility for a symmetric MPE in this game, where both players mix between cooperation and defection in the prisoner's dilemma state. Let q be the probability that a player cooperates in state s.[16] A mixed equilibrium is characterized by three equations: one equation defining V_s^{qq} as a function of q; one equation defining $V_{s'}^{qq}$ as a function of q; and a third equation, specifying that a player is indifferent between cooperating and defecting in state s. These three equations are

$$V_s^{qq} = q^2(8 + \delta V_s^{qq}) + (1 - q)^2(4 + \delta V_{s'}^{qq}) + 2q(1 - q)(6 + \delta V_{s'}^{qq}),$$
$$V_{s'}^{qq} = 1 + \delta V_s^{qq},$$

and

$$q(8 + \delta V_s^{qq}) + (1 - q)(2 + \delta V_{s'}^{qq}) = q(10 + \delta V_{s'}^{qq}) + (1 - q)(4 + \delta V_{s'}^{qq}).$$

This system of equations has a solution in the $(0, 1]$ interval if and only if $\delta > \frac{2}{7}$, and the mixing probability will be decreasing in δ. The equilibrium

---

[15] Of course, if the initial state, s^0, happens to be s', then the equilibrium path will have the matching pennies game in the first period, but all subsequent periods will be cooperation in the prisoner's dilemma.

[16] A natural conjecture is that the MPEs are generically odd in this class of games with finite states and actions. However, we are not aware of a general result that identifies sufficient conditions for this property in stochastic games.

solution, for values of $\delta > \frac{2}{7}$ is

$$q^*(\delta) = -\frac{1}{4} + \sqrt{\frac{1}{16} + \frac{1+\delta}{3\delta}}.$$

5.2.2 Logit Markov QRE

Defining quantal response equilibrium for dynamic games with a recursive structure was developed independently by Breitmoser, Tan, and Zizzo (2010) and Battaglini and Palfrey (2012) as a framework for analyzing experimental data for such dynamic games. Breitmoser, Tan, and Zizzo (2010) consider an experiment based on Hörner's (2004) theoretical model of perpetual R&D races while Battaglini and Palfrey (2012) study dynamic voting games. The latter application is covered in chapter 8 as an illustration of MQRE.

In an MPE, one can think of the equilibrium as a coupling of the equilibrium strategy σ^* and the equilibrium value functions V^{σ^*}, because the value function, V^{σ^*}, is much like an *endogenous payoff function*. Indeed, this is the easier interpretation with quantal response functions, because the state-contingent payoffs of individual actions will depend on these endogenous value functions, and then the QRE choice probabilities in turn depend on the state-contingent payoffs of the individual actions. Given any mixed Markov strategy profile, σ, define $V_i^{\sigma}(s)$ for each i, recursively, exactly as in equation (5.5). Then define the expected utility to player i from choosing strategy a in state s, given the Markov strategy profile σ and its implied value function $V_i^{\sigma}(s)$, by

$$U_i^{\sigma}(a, s; V_i^{\sigma}, \sigma_{-i}) = \sum_{a_{-i} \in A} \left\{ \prod_{j \neq i}^{n} \sigma_j(a_j, s) \right\} u_i(a, s) + \delta \sum_{k=1}^{s} \rho(s_k; a_i, \sigma_{-i}, s) V_i^{\sigma}(s_k).$$

A logit Markov quantal response equilibrium (MQRE) is a profile of Markov strategies, σ^*, such that, for all $i \in I$, for all $a_i \in A_i$, and for all $s \in S$,

$$\sigma_i^*(a_i, s) = \frac{e^{\lambda U_i^{\sigma^*}(a_i, s; V_i^{\sigma^*}, \sigma_{-i}^*)}}{\sum_{a' \in A_i} e^{\lambda U_i^{\sigma^*}(a', s; V_i^{\sigma^*}, \sigma_{-i}^*)}},$$

where $V_i^{\sigma^*}$ is given by the unique solution of (5.5).[17]

Breitmoser, Tan, and Zizzo (2010) establish some useful properties of logit MQRE for recursive dynamic games with a finite set of states and strategies. First,

---

[17] In this simple recursive, finite, bounded framework with discounting, existence of a unique value function associated with a Markov strategy σ^* is guaranteed from basic principles of dynamic programming. While each equilibrium strategy profile is associated with a unique value function, this does not rule out multiple MQRE strategy profiles, each of which is associated with its own unique value function.

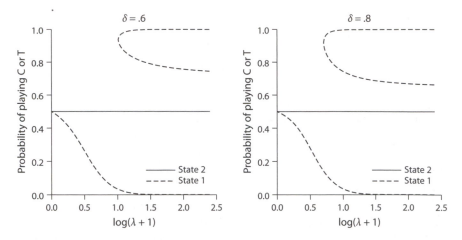

FIGURE 5.2. MQRE for a two-state example.

they show that the existence proof for QRE in normal-form games (McKelvey and Palfrey, 1995) extends with minor modifications to establish the existence of logit MQRE in these more complicated games. They also show how the homotopy method for computing the logit QRE correspondence in normal-form games (Turocy, 2005) can be applied, with only minor changes, to compute the logit MQRE correspondence in dynamic games.[18]

Example 5.4 (Example 5.3 continued: two-state example)*:* Going back to example 5.3, we can actually map the QRE correspondence for different values of δ. This mapping is shown in figure 5.2 for $\delta = 0.6$ and $\delta = 0.8$. That figure displays the MQRE probability of D (or R, as the game is symmetric) in each state as a function of λ. The MQRE in state 2 is always uniform mixing, for all values of δ and λ. For low values of λ, in both cases there is a unique MQRE that converges to the always defect (D, D) equilibrium in state 1. For sufficiently high values of λ, there are two additional MQRE that converge to the always cooperate (C, C) and mixing equilibria in state 1, respectively. As intuition would suggest, these additional equilibria are picked up at a lower value of λ when players are more patient. This is consistent with the general principle that cooperation is easier to support with more-patient players.

---

[18] See Breitmoser, Tan, and Zizzo (2010, online Appendix E) for details.

5.3 EVOLUTIONARY DYNAMICS AND LOGIT QRE

This section provides a unified approach to equilibrium and evolutionary dynamics for a class of models with continuous decisions. The dynamic model is based on an assumption that decisions are changed locally in the direction of increasing payoff, subject to some randomness. In particular, adjustments are based only on current conditions, and in this sense, there is no history. This modeling approach is inspired by the literature on evolutionary game theory, which explores conditions under which strategies with higher payoffs become more widely used.

Such evolution can be driven by increased survival and fitness arguments with direct biological parallels (e.g., Foster and Young, 1990), or by cognitive models in which agents learn to use strategies that have worked better for themselves (e.g., Erev and Roth, 1998), or in which they imitate successful strategies used by others (Vega-Redondo, 1997; Rhode and Stegeman, 2001). An alternative to imitation and adaptation has been to assume that agents move in the direction of best responses to others' decisions. This is the approach we take.

In addition to "survival of the fittest," biological evolution is driven by mutation of existing types, which is the second element that motivates our work. In the economics literature, evolutionary mutation is often specified as a fixed "epsilon" probability of switching to a new decision that is chosen randomly from the entire feasible set (see, e.g., the discussion in Kandori, 1997). Instead of mutation via new types entering a population, we allow existing individuals to make mistakes, with the probability of a mistake being inversely related to its severity (see also Blume, 1993, 1995; Young, 1998; Hofbauer and Sandholm, 2002).

Consider $n \geq 2$ players that make decisions in continuous time. At time t, player $i = 1, \ldots, n$ selects an action $x_i(t) \in [\underline{x}, \overline{x}]$. Since actions will be subject to random shocks, behavior will be characterized by probability distributions. Let $F_i(x, t)$ be the probability that player i chooses an action less than or equal to x at time t. Similarly, let the vector of the $n - 1$ other players' decisions and probability distributions be denoted by $x_{-i}(t)$ and $F_{-i}(x_{-i}, t)$ respectively. The instantaneous expected payoff for player i at time t depends on the action taken and on the distributions of others' decisions:

$$U_i(x_i, t) = \int u_i(x_i, x_{-i}) dF_{-i}(x_{-i}, t). \tag{5.6}$$

We assume that payoffs, and hence expected payoffs, are bounded from above and that expected payoffs are differentiable in $x_i(t)$ when the distribution functions are. The latter condition is ensured when the payoffs $u_i(x_i, x_{-i})$ are continuous.

In a standard evolutionary model with replicator dynamics, the assumption is that strategies that do better than the population average against the distribution of decisions become more frequent in the population. The idea behind such a "population game" is that the usefulness of a strategy is evaluated in terms of how it performs against a distribution of strategies in the population of other players. We use the population game paradigm in a similar manner by assuming that the attractiveness of a pure strategy is based on its expected payoff given the distribution of others' decisions in the population.

To capture the idea of local adjustment to better outcomes, we assume that players move in the direction of increasing expected payoff, with the rate at which players change increasing in the marginal benefit of making that change. This marginal benefit is denoted by $U_i'(x_i(t), t)$, where the prime denotes the partial derivative with respect to $x_i(t)$. However, individuals may make mistakes in expected payoff calculations, or they may be influenced by non-payoff factors. Therefore, we assume that the directional adjustments are subject to error, which we model as an additive disturbance, $\xi_i(t)$, weighted by a variance parameter σ_i:

$$\frac{dx_i(t)}{dt} = U_i'(x_i(t), t) + \sigma_i \xi_i(t). \tag{5.7}$$

Here $\xi_i(t)$ is a standard Wiener (white noise) process that is assumed to be independent across players and time. Essentially, dx_i/dt equals the slope of the individual's expected payoff function plus a normal error with zero mean and unit variance, which captures the idea that adaptation is imperfect and prone to error.

The adjustment rule (5.7) translates into a differential equation for the distribution function of decisions, $F_i(x, t)$. This equation will depend on the density $f_i(x, t)$ corresponding to $F_i(x, t)$, and on the slope, $U_i'(x_i(t), t)$, of the expected payoff function. It is a well-known result from theoretical physics that the stochastic adjustment rule (5.7) yields the Fokker–Planck equation for the distribution function.

Proposition 5.1 (Fokker–Planck characterization of steady state): *The noisy directional adjustment process (5.7) yields the Fokker–Planck equation for the evolution of the distributions of decisions:*

$$\frac{\partial F_i(x, t)}{\partial t} = -U_i'(x, t) f_i(x, t) + \frac{\sigma_i^2}{2} f_i'(x, t) \tag{5.8}$$

for $i = 1, \ldots, n$.

The Fokker–Planck equation has a very intuitive economic interpretation. First, players' decisions tend to move in the direction of greater payoff, and a larger payoff derivative induces faster movement. In particular, when payoff is increasing at some point x, lower decisions become less likely, decreasing $F_i(x, t)$. The rate at which probability mass crosses over at x depends on the density $f_i(x, t)$ at x, which explains the first term on the right-hand side of (5.8). The second term reflects aggregate noise in the system, which causes the density to "flatten out." Locally, if the density has a positive slope at x, then flattening moves mass toward lower values of x, increasing $F_i(x, t)$, and vice versa, as indicated by the second term on the right-hand side of equation (5.8). The variance coefficient σ_i^2 in (5.8) determines the importance of errors relative to payoff-seeking behavior for individual i. Consider the case where $\sigma_i = 0$. If behavior in (5.8) converges, it must be the case that $U_i'(x_i(t), t) f_i(x, t) = 0$, which is the necessary condition for an interior Nash equilibrium: either the necessary condition for payoff maximization is satisfied at x, or else the density of decisions is 0 at x. As σ_i grows large, the noise effect dominates and the Fokker–Planck equation tends to

$$\frac{\partial F_i(x, t)}{\partial t} = \frac{\sigma_i^2}{2} \frac{\partial^2 F_i}{\partial x^2},$$

which is equivalent to the "heat equation" that describes how heat spreads out uniformly in some medium. In this limit, the steady state of (5.8) is a uniform density with $f_i' = 0$.

More generally, in a steady state of the process in (5.8), the right-hand side is identically zero, which yields the equilibrium conditions

$$f_i'(x) = \lambda_i U_i'(x) f_i(x), \tag{5.9}$$

where the t arguments have been dropped since these equations pertain to a steady state and we define $\lambda_i = 2/\sigma_i^2$. These equations can be simplified by dividing both sides by $f_i(x)$ and integrating.

Proposition 5.2 (Evolutionary convergence to logit QRE): *When players adjust their actions in the direction of higher payoff, but are subject to normal error as in (5.7), then any steady state of the Fokker–Planck equation (5.8) constitutes a logit equilibrium*

$$f_i(x) = \frac{e^{\lambda_i U_i(x)}}{\int_{\underline{x}}^{\overline{x}} e^{\lambda_i U_i(y)} dy}. \tag{5.10}$$

This derivation of the logit model is very different from the usual derivations that are static in nature (e.g., the random-utility approach). Here the logit model results from the behavioral assumption of directional adjustment with normal error.

We next consider the dynamics of the system (5.8) and characterize sufficient conditions for a steady state to be attained in the long run. Specifically, we use Liapunov methods to prove stability for a class of games that includes some widely studied special cases. A Liapunov function is nondecreasing over time and has a zero time derivative only when the system has reached an equilibrium steady state. The system is (locally) stable when such a function exists.

Although our primary concern is the effect of endogenous noise, it is instructive to begin with the special case in which there is no decision error and all players use pure strategies. Then it is natural to search for a function of all players' decisions that will be maximized (at least locally) in a Nash equilibrium. In particular, consider a function, $P(x_1, \ldots, x_n)$, with the property $\partial P/\partial x_i = \partial u_i/\partial x_i$ for $i = 1, \ldots, n$. When such a function exists, Nash equilibria can be found by maximizing P. The $P(\cdot)$ function is called the *potential function*, and games for which such a function exists are known as potential games (Monderer and Shapley, 1996).

The usefulness of the potential function is not just that it is (locally) maximized at a Nash equilibrium. It also provides a direct tool to prove equilibrium stability under the directional adjustment hypothesis in (5.7). Indeed, in the absence of noise, the potential function itself is a Liapunov function:

$$\frac{\partial P}{\partial t} = \sum_{i=1}^{n} \frac{\partial P}{\partial x_i} \frac{\partial x_i}{\partial t} = \sum_{i=1}^{n} \frac{\partial u_i}{\partial x_i} \frac{\partial x_i}{\partial t} = \sum_{i=1}^{n} \left(\frac{\partial x_i}{\partial t}\right)^2 \geq 0, \qquad (5.11)$$

where the final equality follows from the directional adjustment rule without noise, that is, $\sigma_i = 0$. Thus the value of the potential function is strictly increasing over time unless all payoff derivatives are 0, which is a necessary condition for an interior Nash equilibrium. The condition that $\partial P/\partial t = 0$ need not generate a Nash equilibrium: the process might come to rest at a local maximum of the potential function that corresponds to a local Nash equilibrium from which large unilateral deviations may still be profitable.

Our primary interest concerns noisy decisions, so we will work with the expected value of the potential function. From (5.6) the partial derivatives of the expected value of the potential function correspond to the partial derivatives of the expected payoff functions:

$$U_i'(x_i, t) = \frac{\partial}{\partial x_i} \int P(x_i, x_{-i}) dF_{-i}(x_{-i}, t). \qquad (5.12)$$

Again, the intuitive idea is to use something that is maximized at a logit equilibrium to construct a Liapunov function, that is, a function whose time derivative is nonnegative and equal to 0 only at a steady state. When λ_i is finite for at least one player i, then the steady state is not generally a Nash equilibrium, and the potential function must be augmented to generate an appropriate Liapunov function. Look again at the Fokker–Planck equation (5.8); the first term on the right-hand side is 0 at an interior maximum of expected payoff, and the $f_i'(x, t)$ term is 0 for a uniform distribution. Therefore, we want to augment the potential function with a term that is maximized by a uniform distribution. Consider the standard measure of noise in a stochastic system, entropy, which is defined as $-\sum_{i=1}^{n} \int f_i \log(f_i)$. It can be shown that this measure is maximized by a uniform distribution, and that entropy is reduced as the distribution becomes more concentrated. The Liapunov function we seek is constructed by adding entropy to the expected value of the potential function, which we call the *stochastic potential*:

$$
P_S = \int_{\underline{x}}^{\overline{x}} \cdots \int_{\underline{x}}^{\overline{x}} P(x_1, \ldots, x_n) f_1(x_1, t) \cdots f_n(x_n, t) dx_1 \cdots dx_n
$$

$$
- \sum_{i=1}^{n} \frac{1}{\lambda_i} \int_{\underline{x}}^{\overline{x}} f_i(x_i, t) \log(f_i(x_i, t)) dx_i. \tag{5.13}
$$

The λ_i parameters determine the relative importance of the entropy terms in (5.13), which is intuitive as λ_i is inversely related to the variance of the Wiener process in player i's directional adjustment rule (5.7). Since entropy is maximized by a uniform distribution (i.e., purely random decision making), it follows that decision distributions that concentrate probability mass on higher-payoff actions will have lower entropy. Therefore, one interpretation of the role of the entropy term in (5.13) is that, if the λ_i parameters are small, entropy places a high "cost" of concentrating probability on high-payoff decisions.

Proposition 5.3 (Evolutionary stability in potential games): *For the class of potential games, behavior converges to a logit equilibrium when players adjust their actions in the direction of higher payoff, subject to normal error as in (5.7).*

The proof follows by using the Fokker–Planck equation to show that the time derivative of the stochastic potential can be expressed in a form analogous to (5.11):

$$
\frac{\partial P_S}{\partial t} = \sum_{i=1}^{n} \int_{\underline{x}}^{\overline{x}} \left(\frac{\partial F_i(x_i, t)/\partial t}{\partial F_i(x_i, t)/\partial x_i} \right)^2 dx_i \geq 0. \tag{5.14}
$$

Since P_S is nondecreasing over time for any potential game, we must have $\partial P_S/\partial t \to 0$ as $t \to \infty$, so $\partial F_i/\partial t$ limits to 0 in this limit. By (5.8) this yields the logit equilibrium conditions in (5.9). The solutions to these equilibrium conditions are the logit equilibria defined by (5.10).

When there are multiple logit equilibria, the equilibrium attained under the dynamical process (5.8) is determined by the initial distributions $F_i(x, 0)$. In other words, the dynamical process is not ergodic. This follows because, with multiple

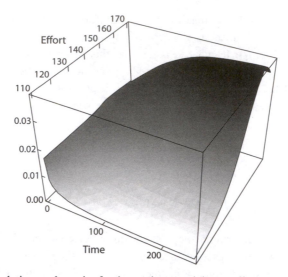

FIGURE 5.3. Evolutionary dynamics for the continuous minimum-effort game of examples 2.9 and 5.5 with two players and $c = \frac{1}{4}$. Starting out from a uniform distribution at $t = 0$, the effort choice density converges to the unique logit equilibrium under the evolutionary process (5.15).

equilibria, the stochastic potential (5.13) has multiple local maxima and minima, and since the stochastic potential cannot decrease over time, any of these extrema are necessarily rest points of the dynamical process.

> **Proposition 5.4** (Stability of QRE in potential games)**:** *A logit equilibrium is locally (asymptotically) stable under the process (5.8) if and only if it corresponds to a strict local maximum of the stochastic potential in (5.13). When the logit equilibrium is unique, it is globally stable.*

> **Example 5.5** (Example 2.9 continued: a continuous minimum-effort coordination game)**:** Consider again the continuous minimum-effort coordination game of section 2.7 for the case of $n = 2$ players. Since the game is symmetric we will drop the player-specific subscripts in (5.8), which becomes

$$\frac{\partial F(x, t)}{\partial t} = -(1 - F(x, t) - c)f(x, t) + \frac{1}{\lambda}f'(x, t), \qquad (5.15)$$

> where we used that $U'(x) = 1 - F(x) - c$ for the two-player minimum-effort game and $\lambda = 2/\sigma^2$. Figure 5.3 describes the time evolution of the effort-choice density when costs are relatively low ($c = \frac{1}{4}$). Initially, the logit

choice is uniform but over time it shifts mass toward higher efforts and converges to the logit equilibrium for this game.

In this example, the stochastic potential has a unique maximizer and the logit equilibrium is unique (see section 9.2 for further details). More generally, it follows that $\partial F_i/\partial t = 0$ when the stochastic potential is (locally) maximized, which, by (5.9) and (5.10), implies that a logit equilibrium is necessarily reached. Recall that, in the absence of noise, a local maximum of the stochastic potential does not necessarily correspond to a Nash equilibrium; the system may come to rest at a point where "large" unilateral deviations are still profitable (see also Friedman and Ostrov, 2010, 2013). In contrast, with noise, local maxima of the stochastic potential always produce a logit equilibrium in which decisions with higher expected payoffs are more likely to be made. In fact, even (local) minima of the stochastic potential correspond to such equilibria, although they are unstable steady states of the dynamical system.

Propositions 5.3 and 5.4 do not preclude the existence of multiple locally stable equilibria. In such cases, the initial conditions determine which equilibrium will be selected. If the initial distributions are "close" to those of a particular logit equilibrium, then that equilibrium will be attained under the assumed dynamic process.

5.4 STOCHASTIC LEARNING EQUILIBRIUM

The previous section's evolutionary model has the property that adjustments depend on the current state (decisions) but not on the observed histories of action profiles. In this section, we consider the effects of histories of action profiles for which the partitions implement a finite memory of m periods. The dynamic properties of the system will be represented by probability distributions of histories. The equilibrium will be defined in terms of a distribution of finite histories, which in turn, determines a distribution of decisions that is stationary in a long-run sense. The specification of the equilibrium in terms of the distribution of histories allows systematic history dependence in the time paths of decisions. Standard arguments will be used to prove existence of this *stochastic learning equilibrium* (SLE), which was introduced in Goeree and Holt (2000, 2003b). The standard analysis of learning in games is to specify a function (e.g., fictitious play) that transforms histories into beliefs. Although these "rules" are admittedly naive, they have been used by theorists to prove convergence to Nash equilibria under general conditions. Others have used fixed learning rules for a different purpose, to explain patterns of adjustment and deviations from Nash predictions in laboratory experiments. Regardless of their theoretical and predictive properties, the mechanical nature of these rules is a concern. For example, a rule that gives recent observations high weight may be inappropriate if such recency effects are not apparent in the data being explained.

5.4.1 Some Alternative Learning Rules

It is useful to begin by discussing a range of simple learning "rules" that have been used by economists and psychologists to evaluate learning and adaptive adjustments in laboratory experiments. First, consider a *reinforcement learning rule* (e.g., Erev and Roth, 1998) that is used to base choice probabilities on earnings received for alternative decisions. For example, let the two possible decisions be denoted by L and R. Each time a decision is selected, the subject receives nonnegative earnings, and let e_L and e_R denote the total earnings amounts received thus far for decisions L and R. The probability that the next choice will be L is specified as $\text{Prob}[L] = (\alpha + e_L)/(2\alpha + e_L + e_R)$, where the denominator ensures that the two choice probabilities sum to 1, and the positive parameter α determines the degree of inertia in the process. In the beginning, with zero cumulative earnings, the initial choice probabilities will be 0.5. If the first decision made yields strictly positive earnings, then its choice probability will rise above 0.5. Reinforcement learning is a psychological theory that does not produce a sharp distinction between beliefs and payoffs; it builds randomness into the combined process.

In contrast, consider a simple model of *belief learning* that depends on observed frequencies (but not earnings) for two events, H and T. The numbers of times that these events have been observed at some point are denoted by N_H and N_T respectively. One commonly used belief learning rule uses relative frequencies to determine beliefs: $\text{Prob}[H] = (\alpha + N_H)/(2\alpha + N_H + N_T)$, where α again determines the degree of inertia in the learning process. There is even a well-known Bayesian foundation for this learning rule, where α corresponds to the strength of the prior distribution on the probability of H. Obviously, the initial prior is 0.5, and the event that is observed more often will have the higher (subjective belief) probability associated with it. These probabilities could be used to determine expected payoffs and choice probabilities via a logit or other quantal response function.

The frequency ratio learning rule just discussed is relevant for discrete random variables. For continuous random variables, the most commonly used rule is an exponentially weighted average of prior observations. For example, suppose that the observation for period t is denoted by x_t and the point forecast for the subsequent period is denoted by F_{t+1}. The fictitious play forecast would be the average of all m observations: $F_{t+1} = \frac{1}{m}(\sum_t x_t)$. If the series has some drift, then it would be better to weight more recent observations more heavily, and a simple rule for doing this is $F_{t+1} = (1 - \rho)(x_t + \rho x_{t-1} + \rho^2 x_{t-2} + \cdots)$, which has geometrically declining weights when $\rho < 1$.[19] In this case, more recent observations carry more weight in the forecast. There is also a Bayesian motivation for this type of rule, based on a conjugate normal prior/sample setup. Such exponentially weighted forecasting rules are sometimes used for the analysis of laboratory data collected in a series of market periods. For example, chapter 9 presents estimates of the weighting parameter for an experiment involving duopoly price choice.

These geometrically weighted learning rules are sometimes referred to as *adaptive learning*, since an equivalent way to express the forecast is as the previous forecast plus an adaptive adjustment for the previous forecast error: $F_{t+1} = F_t + \rho(x_t - F_t)$.[20] So if the most recent observation is higher than the most recent forecast, then that forecast is adjusted upward.

One problem with simple adaptive forecasts is that they can produce persistent, correctable forecast errors when there is a trend in the data. For example, if the time series evolves as $1, 2, 3, 4, \ldots$, then any weighted average of prior observations will be too low. One way to deal with this is to specify a forecasting rule that extrapolates the most recently observed price trend: $F_{t+1} = x_t + \gamma(x_t - x_{t-1})$, where $0 < \gamma < 1$. Haruvy, Lahav, and Noussair (2007) used an extrapolative rule to explain the point forecasts that were elicited in an asset market experiment with strong price trends. The price bubbles observed in these experiments typically rise and then fall, and elicited point price forecasts show a similar pattern, but with a lag. A good forecasting rule would use parameters estimated from a collection of markets with bubbles to fit a model that tracks both the rise and the fall in elicited predictions. A natural way to do this is to specify a forecasting rule that has an adaptive adjustment for errors in the forecast of price levels, $\rho(x_t - F_t)$, and errors in the forecast of price trends, $\gamma\left[(x_t - x_{t-1}) - (F_t - F_{t-1})\right]$. This "double adaptive" rule has been used to analyze elicited forecasts for the rapid price rises and declines that occur during the formation of asset price bubbles in the laboratory.[21]

Any given forecasting rule may be a good procedure for a specific stochastic process, but all of the models discussed thus far are somewhat mechanical in nature. It is natural to think that people will adjust the forecasting process based

---

[19] The $(1 - \rho)$ term in the product is needed to ensure that if there is no variation in the observed data (all x_t values are equal), then the forecast would equal this observed value. A slightly different adjustment would be needed for a finite history of observations.

[20] This equivalence can be verified by recursive substitutions.

[21] See Holt, Porzio, and Song (2015), who estimated parameters for a double adaptive model using laboratory data.

on the data encountered, for example by using the observed time series to estimate a parametric model such as a "least squares learning rule." The approach to be considered in this section is general in that it permits the incorporation of *endogenous* learning rules with forecasting parameters that may depend on the nature of the observed time series. The only requirement is that the learning rule transform histories into belief distributions, which in turn determine the distributions of stochastic decisions and the process through which histories evolve stochastically. The equilibrium will involve a steady-state distribution over finite histories.

It is useful, however, to begin with a simple example, that is, a two-person game with two decisions (1 and 2), and random matching from a population of players of each of the two types. For simplicity, each person is assumed to base decisions on histories that include only the two most recent outcomes, that is, a two-period "memory" model ($m = 2$). In a steady state, each history will have a probability denoted by the ordered subscripts p_{11}, p_{12}, p_{21}, and p_{22}. The goal is to find the vector of probabilities over histories that regenerates itself. That is, if we draw a history randomly from the distribution, use the learning rule to determine beliefs, expected payoffs, and choice probabilities, then it must be the case that the probabilities associated with the new history must equal the probabilities that we used in drawing the original history. In particular, consider how one might end up with a history 12. This could only happen if we start with a history that ends in 1, that is, 11 or 21 and then observe the person choosing a 2. The person making the choice can have any of the four possible histories (11, 21, 12, or 22). Let $\sigma(1|ij)$ denote the resulting quantal response (e.g., logit) probability of choosing decision 1 given the beliefs for history ij. Then with random matching, the probability of ending up with a history 12 would be the product of (a) the probability that the person's most recent observation is 1, which equals $p_{11} + p_{21}$, and (b) the probability that they would choose 2, which is calculated as $p_{11}\sigma(2|11) + p_{12}\sigma(2|12) + p_{21}\sigma(2|21) + p_{22}\sigma(2|22)$. This product is the probability of obtaining a history 12. The steady-state condition is that the probability for each history calculated in this manner turns out to be the probability for that history used in the calculations. In particular, it must be the case that

$$p_{12} = (p_{11} + p_{21})(p_{11}\sigma(2|11) + p_{12}\sigma(2|12) + p_{21}\sigma(2|21) + p_{22}\sigma(2|22)).$$

There would be analogous equations for each of the three histories, 11, 21, 22. Notice that the steady-state probabilities on the right-hand side of the equation appear only as products and do not affect the decision probabilities, which depend only on the beliefs that result from each particular history, not on the probability of encountering that history. Thus the right-hand sides of each the four equations are continuous functions of history probabilities, so the system is a continuous mapping of a convex, compact set of state probabilities into itself, and therefore has a solution.

5.4.2 Beliefs and Probabilistic Choice

The approach taken with the 2×2 example can be generalized in a straightforward manner. Consider an n-person game in which each player has a fixed role. For simplicity, assume that each player has d decisions that can be made in each period, labeled $x_1, \ldots, x_d$, where the subscripts for player types used in earlier sections have been omitted for notational simplicity. There are a large number of people in each role, and the matching involves a random selection of one player from each role, with payoffs determined by the action profile, a, according to a payoff function $u_j(a)$ for player type j. Each person carries a person-specific history that goes back m periods. As before, this history includes own and others' decisions for the previous m periods, so an individual history is a matrix $h \in H$ of $n \times m$ elements, so with d decisions (pure strategies) and n players, there are d^{nm} possible histories. Even though the number of possible histories could be quite large, it is finite, so the probability distribution over possible histories can be represented by a vector p.

The next step is to consider belief formation. For example, Cournot beliefs are that the most recent observation(s) will be repeated with probability 1, that is, there is only one period of memory. Finite fictitious play takes the average for the last m periods as the belief probabilities. A belief learning rule is a continuous mapping from an individual history h (of what all n types have done in the last m periods) to a vector of probabilities $p \in P$ over what other players will do. Notice that the formulation permits some types of endogenous learning rules, since it is possible to use a sufficiently long history of observations to estimate least squares or vector autoregression forecasting functions, which have the property that the estimated rule depends on the history. In any case, the resulting learning rule still maps a history into a belief distribution.

Regardless of how beliefs are determined, the belief distributions determine expected payoffs for each of the possible player types and decisions, $\pi_{ij}(x)$ (player type i, decision j). In this manner, the probabilities associated with each of the possible decisions would be determined by a regular quantal response function with an error parameter λ. Note however, that a sharp distinction between beliefs and actions is not necessarily assumed in all learning models, for example, reinforcement learning in which choice probabilities are assumed to be proportional to some measure of payoffs received from each possible decision. In order to include a wide range of belief, reinforcement, and hybrid models, we will work with a generalized stochastic response function that maps histories directly into the probability associated with an action profile a_j (a vector of decisions, one for each player): $\sigma(a_j|h) : H \rightarrow P$. Note that $\sigma(a_j|h)$ is a product of choice probabilities, for each player, that are associated with the corresponding element in the action profile a_j. This formulation is general enough to include standard belief learning models, as well as learning that has random elements or errors due to faulty perception or recall.

5.4.3 History Formation

The generalized stochastic response function maps histories into choice distributions (probabilities over action profiles), and the actual choices then generate a new action profile and new histories, with the most distant observation being "forgotten" and the most recent observation being appended into the remaining part of the history matrix for each player. Let T denote one of the truncated histories extending for only $m - 1$ periods. Then an initial matrix $[a_{t-m}, T]$ is transformed into the matrix $[T, a_t]$ after the action profile vector a_t is observed and the "old" vector a_{t-m} is dropped. The probabilities for these histories will be denoted by attaching the history as a subscript, so $p_{[T,a_j]}$ represents the probability of one of these histories. These probabilities of new histories are determined uniquely by the history probabilities and the stochastic response functions as the product of having the recent truncated memory T with all possible action profiles that follow (first parenthetical term on the right-hand side below) and the probability of observing a particular action profile a_j given history (second parenthetical term below):

$$p_{[T,a_j]} = \left(\sum_k (p_{[a_k,T]}) \right) \cdot \left(\sum_{h \in H} p_h \sigma(a_j|h) \right) \quad \text{for all histories } [T, a_j].$$

(5.16)

This is a system of equations in the probabilities of the various histories, with σ response functions that are constant with respect to these probabilities (the σ-mapping depends on the history but not on the probability of reaching that history). All joint probabilities are products, as in the previous two-decision example, so the system of equations in (5.16) is a continuous mapping from a closed convex set into itself and contains at least one fixed point by Brouwer's theorem.

5.4.4 Stochastic Learning Equilibrium

The fixed point for (5.16) determines a joint probability distribution over all possible histories with the property that if you draw a vector of histories, one for each of the players, from the joint distribution and use this to calculate beliefs and choice distributions, then the ex ante probability distribution over the new vector of histories is the same as the initial distribution. Here the phrase "ex ante" means prior to the initial draw of a history. We refer to this steady state as a *stochastic learning equilibrium* (SLE) since it permits both errors in belief formation and errors in choice via the quantal response function.

> **Proposition 5.5** (Existence of stochastic learning equilibrium)*: With finite memory and a finite number of decisions, a stochastic learning equilibrium exists for which the vector of probabilities over histories gets mapped into itself.*

> **Example 5.6** (Example 2.1 continued: asymmetric matching pennies game)*:* Here we illustrate the computations underlying stochastic learning equilibria using the asymmetric matching pennies game of chapter 2; see table 2.1. Suppose players recall only the last m of the other's choices, where $m = 1, 2, \ldots$, and then better respond against that sample using the logit choice rule. The logit SLE conditions are

$$
\begin{aligned}
\sigma_{11} &= \sum_{k=0}^{m} \binom{m}{k} \sigma_{21}^{k} (1 - \sigma_{21})^{m-k} \frac{e^{\lambda X k / m}}{e^{\lambda X k / m} + e^{\lambda(1 - k/m)}}, \\
\sigma_{21} &= \sum_{k=0}^{m} \binom{m}{k} \sigma_{11}^{k} (1 - \sigma_{11})^{m-k} \frac{e^{\lambda(1 - k/m)}}{e^{\lambda(1 - k/m)} + e^{\lambda k / m}}.
\end{aligned}
\tag{5.17}
$$

Figure 5.4 shows the solutions to (5.17) for various memory lengths and $\lambda \in [0, \infty)$. The solid lines correspond to SLE correspondences and the dashed line corresponds to the logit QRE correspondence, which arises as the limit case when recall becomes perfect and an infinite sample, $m = \infty$, is used to determine beliefs. This limit property generalizes beyond the asymmetric matching pennies game.

> **Proposition 5.6** (Limit of stochastic learning equilibrium)*: A stochastic learning equilibrium limits to a quantal response equilibrium when memory becomes unbounded ($m = \infty$), that is, when an infinite sample is observed with perfect recall.*

The intuition behind this result is straightforward. When players use an infinite sample and have perfect recall, they know the entire choice distribution. The equilibrium condition that a better response against that choice distribution reproduces itself, is equivalent to the usual QRE condition. With finite memory, however, the two conditions differ as illustrated in figure 5.4.

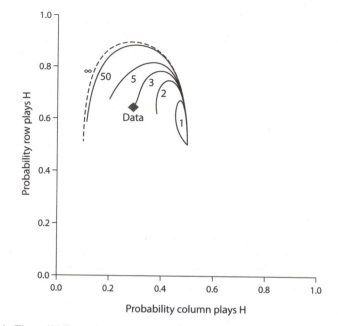

FIGURE 5.4. The solid lines show logit SLE correspondences for $\lambda \in [0, \infty)$ when players recall only the last m periods, where $m \in \{1, 2, 3, 5, 50\}$. The dashed line shows the logit QRE correspondence, which corresponds to the SLE correspondence with perfect recall, i.e., $m = \infty$. The data diamond shows observed averages in the experiment reported in section 2.2.2.

Simple learning models (reinforcement, adaptive and other belief learning rules, and hybrids) have been used extensively to "explain" the patterns of adjustment observed in the early periods of laboratory experiments consisting of sequences of markets or rounds, for example, whether prices tend to converge to a stationary level from above or below.[22] We believe that the stochastic learning equilibrium could be used to model more complex and interesting data patterns, for example, cases where observed data show regular (history-dependent) cycles or patterns of alternating choices.

---

[22] See Camerer (2003) for an extensive discussion of these types of alternative learning models.

6

QRE as a Structural Model
for Estimation

Any noise in the data is, in a superficial sense, sufficient to reject a precise point prediction, for example, a unique Nash pure-strategy equilibrium. In order to construct useful tests of such theories, however, it is necessary to append some kind of error. Even if the researcher is not primarily interested in noise effects per se, it is necessary to add an error specification in order to estimate parameters that are unobserved or are not directly induced in an experiment. The QRE approach is a natural vehicle for such tests, since random elements are fully integrated in an equilibrium framework, and since it includes the Nash equilibrium as a limiting case as the error rate goes to zero, or, equivalently, as the precision parameter goes to infinity.

This chapter discusses the use of maximum-likelihood methods to estimate the precision and other parameters of interest, for example, those that capture risk aversion, inequity aversion, altruism, etc. We will also consider "noisy Nash" and other (nonequilibrium) error specifications. This estimation can be done with any standard statistical package, for example, GAUSS, Mathematica, R, and MATLAB.[1] The sample programs contained in this chapter are specified in a manner that is intended to clarify the underlying structure of the QRE analysis and to illustrate how to use numerical methods to calculate QRE. The chapter is organized around a series of applications, each based on a laboratory experiment that presented interesting estimation issues.

---

[1] The first routines for computing QRE and estimating the logit QRE model were developed by Richard McKelvey in the early 1990s, and this evolved into the much larger Gambit Project in computational game theory. The current and some past versions of Gambit are publicly available both as source code and as compiled executable programs. It is a useful tool for computing equilibrium in simple games. See McKelvey, McLennan, and Turocy (2014) and Turocy (2007).

6.1 THE QRE ESTIMATION APPROACH

The intuitive idea behind maximum-likelihood estimation is to find parameter estimates that maximize the probability of seeing what is observed in the data. The likelihood functions to be considered are based on binomial or multinomial probabilities associated with a discrete number of possible outcomes. For example, consider a coin that yields 2 heads in 6 tosses, and let the probability of heads be denoted by p. Then the probability of obtaining the observed data is $15p^2(1-p)^4$, where the constant represents the number of possible orders in which 2 heads and 4 tails can occur, that is, "6 choose 2" $= 6!/(2!4!) = 15$. In this case, the log-likelihood would be $2\log(p) + 4\log(1-p)$ (plus a constant), which when maximized with respect to p yields $p = \frac{1}{3}$. In a quantal response equilibrium model, the probabilities are the QRE solutions of the model, which will depend on the precision and other parameters to be estimated. Since closed-form solutions typically do not exist, the estimation requires a search over parameter values. If the only parameter to be estimated is the logit precision parameter, for example, then for each given value of λ the model has to be solved for $p(\lambda)$ to evaluate the likelihood function being maximized.

Formally, given some game Γ, let $\sigma^*(\lambda, \beta)$ denote the QRE of Γ, for a given value of λ, and some vector of economic choice parameters β.[2] For example, β might include risk aversion and altruism parameters. As before, the components of σ^* for player i and strategy j will be represented by σ_{ij}^*. For any given data set, let the observed empirical frequencies of strategy choices be denoted by f, so f_{ij} represents the number of observations of player i choosing strategy s_{ij}. Then one can write the log-likelihood function as a function of the parameters (λ, β), given the data, f, as

$$\log L(\lambda, \beta; f) = \sum_{i=1}^{n} \sum_{j=1}^{J_i} f_{ij} \log(\sigma_{ij}^*(\lambda, \beta)).$$

The maximum-likelihood estimates[3] are $(\widehat{\lambda}, \widehat{\beta}) = \arg\max_{\lambda, \beta} \log L(\lambda, \beta; f)$. The estimation is complicated by the fact that the QRE has to be calculated by solving systems of nonlinear equations at each stage in the parameter search process (in order to evaluate the likelihood function at that stage). Therefore, the sample

---

[2] Assume for now there is a unique QRE for all $\lambda \in [0, \infty)$ and for all β in the relevant range. We discuss later in this chapter how to adapt this procedure when Γ has multiple QRE for some values of λ and β.

[3] There are other ways to obtain consistent estimates of λ, β, such as minimizing squared deviations between p and f. For a discussion of the pros and cons of these alternative methods of QRE estimation, see Morgan and Sefton (2002).

TABLE 6.1. Generalized matching pennies with safe and risky choices.

	S	R
S	200, 160	160, 10
R	370, 200	10, 370

programs provided can also be easily modified to calculate quantal response equilibria for exogenously given precision and other parameters.

A simple example can serve to demonstrate how precision and risk-aversion parameters are jointly estimated using logit QRE. The data are taken from a matrix game experiment conducted by Goeree, Holt, and Palfrey (2003) in which each player had a "safe" strategy with similar payoffs and a "risky" strategy, with one payoff being 37 times as large as the other. The payoffs had been previously used in a menu of pairwise lottery choices, in which the probability of the higher payoff in each case was increased in subsequent decision rows. The crossover row in the choice menu, where the probability of the higher payoff is sufficient to induce a person to select the risky option, can be used to infer risk aversion.[4] The game to be considered differed from the lottery choice task in that the probability of obtaining the higher payoff was determined endogenously by the choices of the other player. The equilibrium for the game involves mixed strategies, and intuition suggests that the effect of any risk aversion would be to cause safe strategies to be used somewhat more than otherwise predicted. In contrast, the main qualitative feature of the observed data is an asymmetry: players with one role (Column) select the safe decision twice as often as the risky one, but players with the other role tend to mix equally between their safe and risky decisions. This large asymmetry is not predicted by the Nash equilibrium nor by any QRE with risk neutrality, regardless of the magnitude of the precision parameter. As shown below, the observed asymmetric data pattern arises naturally from a QRE that is calculated using estimated risk aversion and precision parameters. The example provides an opportunity to revisit the discussion about Haile, Hortaçsu, and Kosenok's (2008) concerns about "overfitting" and error specifications. Issues that arise from heterogeneity and multiple QREs are discussed in subsequent sections.

Example 6.1 (Risk aversion in a generalized matching pennies game)*:* The first application to be considered is the Goeree, Holt, and Palfrey (2003) experiment that was based on the game with payoffs (in pennies) shown in table 6.1. This is a game with a unique Nash equilibrium in mixed strategies where Row and Column choose S with probabilities $\sigma_{11}^* = \frac{17}{32}$ and $\sigma_{21}^* = \frac{15}{32}$ respectively.

---

[4] The Goeree, Holt, and Palfrey (2003) choice menu was modeled after the menu used by Holt and Laury (2002) to estimate the effects of increases in payoff scale on risk aversion.

The game was specifically designed to have a strong risk-aversion effect, since there is much less payoff variability for the safe decision. The experiment, which consisted of random matchings from cohorts of 10–12 subjects each, yielded 221 safe decisions out of 340 decisions for Column, and only 160 out of 340 safe decisions for Row. There are some interesting features of the data generated from this game. First, the asymmetry in safe choices cannot be explained by QRE alone. Furthermore, while the increased frequency of Column's safe choices is suggestive of risk aversion, it also the case that the asymmetry cannot be explained by risk aversion alone. What is required is an analysis of the *joint* equilibrium effects of risk aversion *and* noise, which is done using a QRE model with two parameters: the usual precision parameter plus a risk-aversion parameter.

To show that no regular QRE can reproduce the asymmetry observed in the data, one performs the calculations of section 2.4.1 (for an asymmetric matching pennies game) as they apply to the game in table 6.1. Monotonicity dictates that players choose S more often when its expected payoff is higher, which yields the following inequalities:

$$\sigma_{11}^* \leq \frac{1}{2} \quad \text{if } \sigma_{21}^* \geq \frac{15}{32}, \qquad\qquad \sigma_{11}^* \geq \frac{1}{2} \quad \text{if } \sigma_{21}^* \leq \frac{15}{32},$$

$$\sigma_{21}^* \geq \frac{1}{2} \quad \text{if } \sigma_{11}^* \geq \frac{17}{32}, \qquad\qquad \sigma_{21}^* \leq \frac{1}{2} \quad \text{if } \sigma_{11}^* \leq \frac{17}{32}.$$

The region defined by these inequalities allows for only a narrow set of outcomes indicated by the shaded area in figure 6.1. Only outcomes in this set are consistent with *some* risk-neutral regular QRE, which does not include the data.

Next, consider how the Nash equilibrium changes when we transform the monetary payoffs in table 6.1 using a common utility function, for example, for outcome (S, S) Row's utility is $u(200)$ and Column's utility is $u(160)$ etc., where $u(x)$ is some *arbitrary* (but common) nondecreasing function. The mixed-strategy Nash equilibrium can be readily computed by requiring players to be indifferent between their risky and safe options:

$$\sigma_{11}^* = \frac{1}{1+\rho},$$

$$\sigma_{12}^* = \frac{\rho}{1+\rho},$$

where the utility ratio is $\rho = (u(160) - u(10))/(u(370) - u(200))$. Note that $\sigma_{11}^* + \sigma_{12}^* = 1$ and, without any restrictions on the utility function (besides being nondecreasing), ρ can vary between 0 and ∞. For example, for players

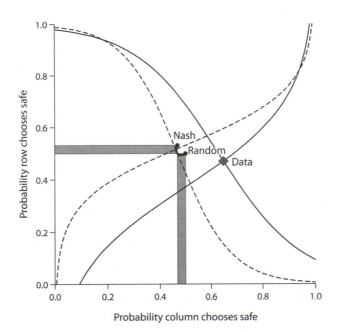

FIGURE 6.1. Logit response functions for the generalized matching pennies game in table 6.1 under risk neutrality (dashed lines) and constant relative risk aversion (solid lines) using the estimated value $r = 0.44$. The shaded area shows possible QRE outcomes, for *any* regular QRE model that assumes risk neutrality, and the dark arc connecting Nash and Random shows the logit equilibrium correspondence under risk neutrality. The diamond marks the data averages.

that are extremely risk loving, $\rho = 0$, while for players who are extremely averse, $\rho = \infty$. This implies that the set of Nash equilibria trace out the entire negative-sloping 45-degree line that starts at $(\sigma_{11}, \sigma_{21}) = (1, 0)$ and ends at $(\sigma_{11}, \sigma_{21}) = (0, 1)$. Note that none of the Nash equilibria looks anything like the asymmetric outcome that is observed in the experiment. So risk aversion by itself cannot explain the experimental findings, and the question remains why Column players choose safe much more frequently than Row players.

To analyze the possible asymmetric equilibrium effects of noise and risk aversion, consider the logit QRE model with risk-averse players. But before applying this model to the data, it is useful to reconsider the question of empirical falsifiability of QRE. In light of the discussion in chapter 2, it is natural to wonder whether a QRE model dressed up with risk aversion is still falsifiable. The answer is affirmative. The combined model allows only for asymmetric outcomes of the type observed in the data, that are "close" to the Nash equilibria on the negative-sloping 45-degree line. Large parts of the outcome space are not consistent with the combined model.

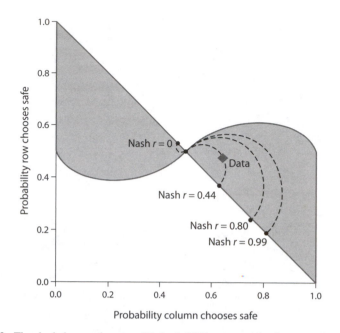

FIGURE 6.2. The shaded areas show possible logit QRE outcomes for the generalized matching pennies game when the payoffs in table 6.1 are transformed into utilities using an *arbitrary* monotonic utility function. The four dashed lines show logit QRE correspondences for the specific case when the utility function is $u(x) = x^{1-r}/(1-r)$, i.e., constant relative risk aversion, for $r = 0$, $r = 0.44$, $r = 0.8$, and $r = 0.99$.

To see this, consider the logit equilibrium equations when the monetary payoffs are transformed using some arbitrary utility function:

$$\sigma_{11}^* = \frac{1}{1 + \exp(\lambda(\sigma_{12}^*(1+\rho) - \rho))},$$

$$\sigma_{12}^* = \frac{1}{1 + \exp(\lambda(1 - \sigma_{11}^*(1+\rho)))},$$

where $\rho = (u(160) - u(10))/(u(370) - u(200))$ can be anything from 0 to ∞ as explained above. The shaded areas in figure 6.2 show the QRE outcomes, $(\sigma_{11}^*, \sigma_{12}^*)$, for all possible values of $\lambda, \rho \in [0, \infty)$. So even when allowing for arbitrary monotonic transformations of the monetary payoffs (including risk aversion, risk lovingness, a mix, etc.), the set of possible QRE outcomes is still far smaller than the entire outcome space; in this example its relative measure is less than a third.[5]

---

[5] In fact, even when we consider *any* regular QRE and *any* nondecreasing utility transformation, outcomes that satisfy $\sigma_{21} \leq \sigma_{11}$ and $\sigma_{21} \leq \frac{1}{2}$, or $\sigma_{21} \geq \sigma_{11}$ and $\sigma_{21} \geq \frac{1}{2}$ are not consistent.

The dashed lines in figure 6.2 show logit equilibrium correspondences for a model of constant relative risk aversion, $u(x) = x^{1-r}/(1-r)$, and different values of the risk parameter: $r = 0$, $r = 0.44$, $r = 0.8$, and $r = 0.99$. In each case, the correspondence starts (for low λ) at random behavior and limits (for high λ) at one of the Nash equilibria on the (negatively sloping) 45-degree line. For $r = 0$, the small arc shows the risk-neutral logit QRE. For $r = 0.44$, the logit equilibrium correspondence intersects the data (diamond) when $\lambda = 0.16$. In figure 6.1, these parameter values are used to draw quantal responses for Row and Column (see the solid lines) to highlight that an increase in safe choices by Column can be consistent with more or less the same number of safe choices by Row. To draw a parallel, when both supply and demand shift outward, the market price can be the same but the quantity sold will go up. In this example, the quantal responses for Row and Column intersect exactly at the data diamond when we use the "optimal" values, $\lambda = 0.16$ and $r = 0.44$. (In contrast, the risk-neutral quantal responses indicated by the dashed lines intersect far from the data.) The next section walks through a detailed estimation program in MATLAB that shows how these optimal parameter values can be found using standard maximum-likelihood techniques.

6.1.1 Estimation Program for Generalized Matching Pennies

The estimation is done by a program, likelihood_2x2_crra, which calculates the value of the likelihood function at each stage of the search process, given the payoff parameters and the data counts. This program, to be explained subsequently, is called from the MATLAB command line using a search routine fminunc that returns the vector of estimates beta along with the value of the likelihood function, some error flags, output, a gradient for the search process, and a Hessian matrix that can be used subsequently to calculate standard errors for the estimates (three dots "..." signal to MATLAB that the following line is a continuation of the current one):

```
>>[beta,fval,flags,output,gradient,hessian]...
    =fminunc(@likelihood_2x2_crra,[1;.2],[ ] )
```

The fminunc command on the right-hand side of the equals sign initiates a search over values of the parameters to be estimated (precision and risk aversion). At each iteration, the program evaluates the likelihood function for the particular precision and constant relative risk-aversion parameters in effect at that iteration, starting with [1; .2], that is, a precision parameter $\lambda = 1$ and a coefficient of relative risk aversion $r = .2$, with the order corresponding to the way the beta parameter vector is set up below. The final "[]" argument of the fminunc command is an empty set of brackets, where options might be specified if needed. Here we used no options. The likelihood function is created in MATLAB as a separate file with name likelihood_2x2 _crra.m that should be stored in the work subdirectory of MATLAB (or some subdirectory created for the purposes of this estimation problem). Before beginning, remember to be sure that the "current folder" box at the top of the MATLAB screen shows the subdirectory where this likelihood file is stored.[6] Please note, even if the appropriate ".m" file shows in the MATLAB editor box, the current folder box at the top must contain the appropriate subdirectory or the fminunc command might generate an alarming array of difficult-to-interpret errors.

The first line in the likelihood_2x2_crra.m file assigns a name logL to the value of the likelihood being returned by the function. The beta vector of parameter estimates consists of a precision parameter lambda in the first position and a coefficient of constant relative risk aversion r in the second position (see lines 2 and 3 below). The next three lines specify the observed data counts, that is, the number of safe decisions for Row (160) and Column (221) and the total number of decisions observed for each role (340). Then the payoff parameters for Row and Column are specified in lines 7 and 8 respectively. Notice that the end of a row in the payoff matrix is indicated by a semicolon, whereas the elements of a row are separated by commas. It is, of course, possible to enter payoffs and data counts directly from the command line, but doing it within the program may reduce data-entry errors. Moreover, data and payoffs for different games can be entered and commented out as needed (using % at the start of the line or using a block comment command).

```
1. function logL = likelihood_2x2_crra(beta)
2. lambda = beta(1);
3. r = beta(2);
4. n_row = 160;
5. n_col = 221;
6. n = 340;
7. row_pay  =  [200, 160; 370, 10];
8. col_pay  =  [160, 10; 200, 370];
9. qre_p = lsqnonlin(@belief_error_2x2_crra,[.5 .5],0,1,[ ],...
          row_pay,col_pay,lambda,r);
10. p_row = qre_p(1);
11. p_col = qre_p(2);
12. logL = n_row * log(p_row) + (n-n_row) * log(1-p_row)...
          + n_col * log(p_col) + (n-n_col) * log(1-p_col);
13. logL = - logL
```

---

[6] To find the current directory, one types pwd at the MATLAB prompt.

Line 9 of the program calls a nonlinear least squares routine, lsqnonlin, which calculates the QRE by minimizing the sum of squared "belief errors," that is, the differences between the belief probabilities and the quantal responses to those beliefs. These differences are zero in equilibrium. The belief_error_2x2_crra function, discussed below, returns a vector qre_p of equilibrium choice probabilities, which are assigned the names p_row and p_col in lines 10 and 11 above. With this notation, the log of the likelihood function is given in line 12, and the negative of the function (line 13) is then minimized. Obviously, a failure to take the negative of the log-likelihood function will produce garbage results or errors that are difficult to interpret and debug. The absence of a semicolon at the end of line 13 will cause the value of logL to be printed on the command line after each iteration.

The final step is to create a file, belief_error_2x2_crra.m, which calculates the difference between the starting belief vector, p, and the vector of quantal responses to those beliefs. The remaining entries on the right-hand side of line 1 of the file assign names to the arguments of this function, that is, the Row and Column payoff matrices and the precision and risk-aversion parameter values used in the current iteration. Lines 2 and 3 assign names to the two components of the p vector, that is, the probabilities with which Row and Column choose safe. For now, it helps to think of these probabilities as beliefs, which will equal the appropriate quantal response probabilities in equilibrium. Then all payoffs are raised to the power 1-r in lines 4 and 5, which expresses them in terms of utility with a constant relative risk aversion. Next, normalize by dividing by 1-r in lines 6 and 7, which ensures that utility is increasing in payoffs in the case of risk preference ($r > 1$). The Row player's expected payoffs for safe and risky are calculated in lines 8 and 9 respectively, and the expected payoffs for Column's two decisions are calculated in lines 10 and 11. The Row player's quantal response for the safe choice is shown in line 13, with an exponential expression of the precision times the expected payoff in the numerator, and a sum of exponentials from line 12 in the denominator. Column's quantal response for the safe choice is calculated similarly in line 15. Finally, the belief errors in lines 16 and 17 are the differences between the belief and response probabilities, and these errors are put into a belief error vector in the final line.

```
1. function [belief_error_vector] = belief_error_2x2_crra...
                                    (p,row_pay,col_pay,lambda,r)
2. p_row = p(1);
3. p_col = p(2);
4. row_pay = row_pay.^(1-r);
5. col_pay = col_pay.^(1-r);
```

6. row_pay = row_pay./(1-r);

7. col_pay = col_pay./(1-r);

8. row_ex_pay_safe = row_pay(1,1)*p_col + row_pay(1,2)*(1-p_col);

9. row_ex_pay_risky = row_pay(2,1)*p_col + row_pay(2,2)*(1-p_col);

10. col_ex_pay_safe = col_pay(1,1)*p_row + col_pay(2,1)*(1-p_row);

11. col_ex_pay_risky = col_pay(1,2)*p_row + col_pay(2,2)*(1-p_row);

12. row_exp_sum = exp(lambda*row_ex_pay_safe)...

+ exp(lambda*row_ex_pay_risky);

13. response_row = exp(lambda*row_ex_pay_safe)/row_exp_sum;

14. col_exp_sum = exp(lambda*col_ex_pay_safe)...

+ exp(lambda*col_ex_pay_risky);

15. response_col = exp(lambda*col_ex_pay_safe)/col_exp_sum;

16. belief_error_row = p_row - response_row;

17. belief_error_col = p_col - response_col;

18. belief_error_vector = [belief_error_row;belief_error_col];

Recall that the belief error vector is then minimized (set to zero) by the nonlinear least squares function in line 9 of likelihood_2x2_crra.m, which produces the quantal response choice probabilities that are used to calculate the likelihood function at each iteration.

The estimation procedure just described yields parameter estimates for this data set of $\lambda = 0.16$ and $r = 0.44$, with a log-likelihood value of -455.21. Note that the magnitude of the precision parameter depends on the units in which payoffs are measured. Expressing payoffs in dollars instead of pennies, for example 3.7 instead of 370, results in estimates of $\lambda = 2.1$ and $r = 0.44$. Note that the estimated risk-aversion parameter did not change (a property of the constant relative risk-aversion model) and that the estimated precision parameter did not increase by a factor of 100 because of the nonlinearity of utility.[7] Finally, the standard errors of the parameter estimates can be obtained from the Hessian matrix by running the appropriate command from the MATLAB command

---

[7] For example, Goeree, Holt, and Palfrey (2003) normalized utility so that the highest payoff (370) was 1 and the utility of the lowest payoff was 0. This normalization would be accomplished by inserting the following commands after line 5 in the belief_error function:

lower = 10^(1-r);

upper = 370^(1-r);

difference = upper-lower;

row_pay = (row_pay-lower)/difference;

col_pay = (col_pay-lower)/difference;

This normalization does not alter the risk aversion, but the change in utility units raises the precision estimate to 6.76, which is the number reported in the paper.

window:

>>[sqrt(inv(hessian))]

The resulting standard errors are 0.056 and 0.073 for the precision and risk-aversion estimates.

Since the lsqnonlin subroutine used to minimize belief errors results in a QRE solution, this function can be embedded in a simple program for calculating quantal response equilibria for any specified precision and risk-aversion values. For $\lambda = 0.6$ and $r = 0.8$, for example, you could enter the command calculate_qre(0.6,0.8) in the MATLAB command window and press enter to activate the program shown below. The order of the arguments (lambda, r) in the first line of program determines the order in which the precision and risk-aversion parameters are entered in the command line. After the payoff matrix is specified in lines 2 and 3, the lsqnonlin function applied to belief errors in line 4 returns the QRE probabilities in the final two lines. Note that these lines do not end in semicolons, thereby ensuring that the equilibrium probabilities are printed out in the MATLAB command box.

```
1. function [qre_p] = calculate_qre(lambda,r)
2. row_pay  =  [200, 160; 370, 10];
3. col_pay  =  [160, 10; 200, 370];
4. qre_p = lsqnonlin(@belief_error_2x2_crra,[.5;.5],0,1,[ ],
          row_pay,...col_pay,lambda,r);
5. p_row = qre_p(1)
6. p_col = qre_p(2)
```

6.2 ESTIMATION AND METHODOLOGICAL ISSUES

This section addresses a number of familiar econometric issues that arise in structural modeling.[8] These issues include multiple equilibria, unobserved heterogeneity, and the method of estimation. Throughout, the logit specification of QRE for games in strategic form is used to illustrate these issues, but the principles

---

[8] This chapter focuses on simultaneous-move games, but similar techniques for extensive-form games are illustrated later in the book. Methodological issues relating to the estimation of QRE models in sequential-move games have been widely studied in the international relations field of political science. See Bas, Signorino, and Walker (2008) for an excellent discussion of these issues, and Signorino (1999) for the first application of QRE to sequential strategic models of international conflict. Chapter 8 also includes a section analyzing a related model of crisis bargaining.

TABLE 6.2. Battle-of-the-sexes game.

	A	B
A	1, 3	0, 0
B	0, 0	3, 1

underlying them apply more generally to other specifications of regular quantal response functions as well as AQRE for games in extensive form.

6.2.1 Multiple Equilibria

Many games have multiple Nash equilibria. Classic examples include the chain-store stage game, the game of chicken, and coordination games. In these cases, the approach often taken in the theoretical analysis of such games is to consider refinements or selections from the set of Nash equilibria, such as perfection, elimination of weakly dominated strategies, risk dominance, symmetry, payoff dominance, stability, and so forth. With QRE, the issue of multiplicity is more subtle for several reasons. First, *in all games*, for low levels of precision where choices have a large random component, the logit QRE is unique. This can be easily seen at the extreme, since if $\lambda = 0$, the unique logit QRE is uniform mixing over the strategy space. Because the quantal response solutions are locally smooth at $\lambda = 0$, the solution to the system of logit QRE equations will necessarily be unique for values of λ sufficiently close to 0. Second, for almost all games, there is a unique continuously differentiable selection from the logit correspondence connecting the centroid of the strategy space to a unique Nash equilibrium point, as λ varies from 0 to ∞, as explained in chapter 2, called *the logit solution* of the game.

These two differences between QRE and Nash equilibrium have two immediate implications. First, for many games with multiple Nash equilibria, the actual behavior in the game may exhibit a sufficient amount of stochastic choice that the multiple Nash equilibrium problem does not pose a practical problem for estimation. Second, the logit solution implies a unique selection for almost all games (including many games studied in the laboratory), and so the estimation can be based on the choice frequencies that are uniquely defined by the logit solution for every value of λ. If there is a unique logit solution to the game, this is the usual approach.

Still, it may be the case that in some games logit QRE has multiple solutions in the relevant portion of the parameter space, or the game being studied may be a nongeneric one that does not have a unique selection because of bifurcations at some critical points of the equilibrium correspondence. A simple example of such a game is the battle-of-the-sexes game where the payoffs off the diagonal are 0 for both players, and the on-diagonal payoffs are (3, 1) and (1, 3); see table 6.2.

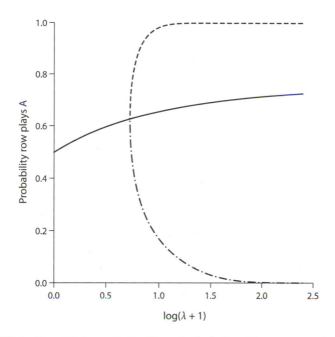

FIGURE 6.3. Logit equilibrium correspondence for the battle-of-the-sexes game in table 6.2.

This game has a symmetric structure and does not have some of the "nice" generic properties of finite games. In particular, the logit equilibrium correspondence is not well behaved; see figure 6.3, which graphs the logit equilibrium correspondence for this game. Note that the upper and lower branches converge to probabilities of 1 and 0 respectively as precision increases. These two branches of the logit equilibrium correspondence approach the two pure-strategy Nash equilibria.

Like in all finite games, the logit equilibrium correspondence is well behaved for low values of λ (high error) in the sense that the logit equilibrium is unique and changes smoothly with respect to λ. However, there is a bifurcation in the graph of the logit equilibrium correspondence, at which point two additional branches of the equilibrium correspondence start simultaneously and are connected to the principal branch that exists for lower values of λ. Hence, there is not a unique continuous selection from the logit correspondence, so this game does not have a "logit solution." For games like this, there are several alternative approaches to estimation.[9] One approach, which works in this particular case, and in most

[9] These games also pose some challenges for computing the logit correspondence. If one uses a path-following algorithm, starting from $\lambda = 0$ and gradually increasing λ, the computation generally

games with a unique symmetric equilibrium, is to select the symmetric branch of
the logit QRE correspondence that converges to the symmetric Nash equilibrium,
and use this for the logit choice frequencies in the estimation.[10]

Another approach, which happens to coincide with the "symmetric solution" in
this particular game, is to select the unique path of the logit QRE correspondence
that defines a continuously differentiable branch as λ varies from 0 to ∞, if
it exists. A third approach is to fit the data to the closest point on the QRE
graph, without imposing assumptions about which branch to use.[11] The idea is
to use observed data and knowledge of the QRE graph to select starting values
for estimation that will yield convergence. In the battle-of-the-sexes game, for
example, if most decisions are A, then starting value probabilities that are close
to this equilibrium could be used, along with a value of λ. All three of the above
approaches implicitly assume that all the data in the experiment come from the
same equilibrium. A fourth approach is to assume that the data come from a
mixture of the points in the logit correspondence, and estimate both λ and the
mixture probabilities.[12]

6.2.2 Estimation Methods

The estimation method for logit QRE described so far is to first compute the
logit QRE correspondence (and apply a selection criterion in the case of multiple
equilibria), and then to search for the value of λ for which the logit QRE is
closest to the observed distribution of choices, where the measure of closeness
is given by the likelihood function. We call this the *equilibrium correspondence
approach* to estimation. This approach has certain advantages, but it also has some
disadvantages from a practical standpoint. Perhaps the biggest disadvantage is that
if the game has more than a few strategies for each player, or has many players,
or many information sets for some players, then the computation of the logit QRE
correspondence (or a selection from it) can be challenging computationally. For
example, in many games of incomplete information with many types and many
strategies, good algorithms for computing QREs are unavailable. This is also

---

runs into problems when the bifurcation value of λ is reached, because the correspondence is not
smooth at that point. In some games (including the example at hand) this problem can be resolved by
following each of the branches corresponding to a Nash equilibrium backward, starting at high values
of λ. These branches will intersect at the bifurcation value of λ. Of course, such an approach is not
fully general and implementation of it requires knowledge about the structure of the Nash equilibrium
of the specific game being studied.

[10] This is the approach applied to the jury voting games analyzed in chapter 8.

[11] This approach can also be applied to well-behaved games that have multiple logit equilibria for
some values of λ.

[12] To our knowledge, this last approach has not been used in applications of QRE to data.

true in some simpler games that have many information sets, such as games of information transmission with rich message spaces.

There are at least two ways to deal with the issue of not being able to compute the logit QRE correspondence. The first approach is to compute the QRE correspondence for a simplified version of the game, by binning strategies and/or types. In the traveler's dilemma experiment to be discussed in chapter 9, subjects could choose any real-valued "claim" between 80 and 200, although almost all actual choices were integers. In the process of estimation, the observed claims were first assigned to bins, for example, 80–84, 85–89, etc., and counts for each bin were used in the estimation. Then the bin size was reduced in width as a clearer idea of the appropriate starting values for the precision parameter emerged. As another example, consider a first-price independent private-value auction, where values are uniform between 0 and 1 and bids can be anywhere between 0 and 1. One could approximate this with a finite game, where bids and values are chopped into 10 categories: $V_1 = B_1 = [0.0, 0.10]$, $V_2 = B_2 = (0.10, 0.20]$, ..., $V_{10} = B_{10} = (0.90, 1.00]$. Then one can approximate all bids in the set B_k by a single bid, say $b_k = 0.05 + (k - 1) \cdot 0.1$, $k = 1, \ldots, 10$ and similarly for values $v_k = 0.05 + (k - 1) \cdot 0.1$, $k = 1, \ldots, 10$. This results in a finite game with a mere 10^{10} strategies if one considers the strategic form of the game. This is still too large for computation of QRE to be feasible, so one could trim the strategy space further by eliminating dominated strategies, such as bidding above one's value. However, this still is a very large set of strategies (10!). Alternatively, one could look at undominated strategies in the AQRE model, which corresponds to behavioral strategies. In this case, a bidder with a value of v_k has only k possible actions, $b_1, \ldots, b_k$. One can now compute the symmetric logit AQRE, viewing the game as a 10-player game, rather than a 2-player game with each player having 10 types, and the types have between 1 and 10 available actions each. Of course, computation of the logit QRE correspondence is still cumbersome as one must search over a fairly high-dimensional space.[13]

The second approach to dealing with the complexity issue is to estimate QRE without actually computing the logit QRE correspondence. This entails a two-stage procedure that we call the *empirical payoff approach*. First, one estimates the expected payoff to every possible strategy, based on the observed *empirical frequency* of the strategies in the data. This gives an *empirical expected payoff function* that assigns an estimated expected payoff for each strategy of a player in the game. In a logit equilibrium, each player's (or each player type's, in the case of a Bayesian game) choice probabilities are logit responses to the expected payoffs

---

[13] Such an equilibrium computation has been carried out for some "simple" auction games with 6 types and 6 possible bids that were studied in the laboratory. The results are discussed in chapter 9.

of the different actions or strategies. In the limit, as the number of observations goes to infinity, and under the maintained hypothesis that all the data are generated by the same logit equilibrium,[14] the empirical frequencies in the data will be exactly equal to the choice probabilities of the logit equilibrium, and the empirical expected payoff function will be exactly the expected payoffs in equilibrium.

For the generalized matching pennies example in table 6.1, the observed counts were 221 safe decisions for Column and 160 safe decisions for Row, so the empirical beliefs would be 221/340 for Column and 160/340 for Row. The quantal responses to these "belief" ratios would be inserted into the likelihood function that is maximized. Hence, this approach provides a method for obtaining a consistent estimate of the precision parameter λ without having to compute the equilibrium correspondence. As in the equilibrium correspondence approach, the empirical payoff approach can be used to simultaneously estimate economic parameters of interest, in addition to estimating the logit precision parameter.

Formally, let f_{-i} denote the empirical choice frequencies of players other than i. Then the empirical expected payoff to player i for using action a_{ij} is equal to

$$\widehat{U}_{ij} = \sum_{a_{-i} \in A_{-i}} f_{-i}(a_{-i}) u_{ij}(a). \tag{6.1}$$

If the logit response parameter is λ, then i's logit response probabilities are given by

$$p_{ij}(\lambda; f_{-i}) = \frac{e^{\lambda \widehat{U}_{ij}}}{\sum_{k=1}^{J_i} e^{\lambda \widehat{U}_{ik}}}. \tag{6.2}$$

Then we can write the log-likelihood function, as a function of the logit precision parameter, λ, given the data, f, as

$$\log L(\lambda; f) = \sum_{i=1}^{n} \sum_{j=1}^{J_i} f_{ij} \log(p_{ij}(\lambda; f_{-i})). \tag{6.3}$$

The maximum-likelihood estimate is simply given by $\widehat{\lambda} = \arg\max_{\lambda} \log L(\lambda; f)$. This provides a consistent estimate of λ because $\widehat{u}_{ij}$ provides a consistent estimate of expected payoffs, under the maintained hypothesis that the data is generated from a QRE for some unknown value of λ. Notice that this is a far simpler approach computationally as it does not require solving a fixed-point problem to compute $p^*(\lambda, \beta)$.

---

[14] This same maintained hypothesis is made in the approach that requires computation of the equilibrium correspondence.

A MATLAB program that uses the empirical payoff approach is given in the likelihood_2x2_empirical.m file shown below, which is called from the MATLAB command line:

>>[beta,fval,flags,output,gradient,hessian]...
   =fminunc(@likelihood_2x2_empirical,[0.25;0.35],[ ])

As with the equilibrium correspondence approach, this program begins with name designations for the parameters to be estimated (lines 2 and 3) and data counts (lines 4–6). The first difference is that the data ratios determine pseudo-equilibrium beliefs (lines 7 and 8). These beliefs are used to construct the utilities (lines 11–14), expected utility payoffs (lines 15–18), and ratios of exponentials that determine quantal responses as was done previously in the belief_error routine. These quantal responses to empirical choice frequencies, which are functions of the stage value of λ, are then used in the likelihood function (instead of the logit QRE probabilities as was done before).

```
1. function logL = likelihood_2x2_empirical(beta)
2. lambda = beta(1);
3. r = beta(2);
% data:
4. n_row = 160;
5. n_col = 221;
6. n = 340;
% empirical belifes:
7. p_row = n_row/n;
8. p_col = n_col/n;
% game payoffs:
9. row_pay = [200, 160; 370, 10];
10. col_pay = [160, 10; 200, 370];
% utilities:
11. row_pay = row_pay.^(1-r);
12. col_pay = col_pay.^(1-r);
13. row_pay = row_pay./(1-r);
14. col_pay = col_pay./(1-r);
% expected payoffs:
15. row_ex_pay_safe = row_pay(1,1)*p_col + row_pay(1,2)*(1-p_col);
16. row_ex_pay_risky = row_pay(2,1)*p_col + row_pay(2,2)*(1-p_col);
17. col_ex_pay_safe = col_pay(1,1)*p_row + col_pay(2,1)*(1-p_row);
18. col_ex_pay_risky = col_pay(1,2)*p_row + col_pay(2,2)*(1-p_row);
% logit quantal responses are ratios of
% exponential expressions:
```

19. row_exp_sum = exp(lambda*row_ex_pay_safe)...
                          +exp(lambda*row_ex_pay_risky);
20. response_row = exp(lambda*row_ex_pay_safe)/row_exp_sum;
21. col_exp_sum = exp(lambda*col_ex_pay_safe)...
                          + exp(lambda*col_ex_pay_risky);
22. response_col = exp(lambda*col_ex_pay_safe)/col_exp_sum;
% quantal responses to be used in the likelihood function:
23. p_row = response_row;
24. p_col = response_col;
25. logL = n_row*log(p_row)+(n-n_row)*log(1-p_row)...
                   + n_col*log(p_col)+(n-n_col)*log(1-p_col);
26. logL = -logL

In this example, the empirical payoff approach yields virtually identical estimates compared with the equilibrium correspondence approach: $\lambda = 0.16$ and $r = 0.45$. In fact, the precision estimate is unchanged, and the risk-aversion estimate is only a bit higher than the 0.44 estimate obtained previously. The estimated standard errors are lower than those obtained previously, but significance levels are comparable. The log-likelihood value and predicted probabilities of choosing safe of 0.65 for Column and 0.47 for Row are also the same as those obtained by the other method. This similarity between the empirical payoff approach and the equilibrium correspondence approach in this example is encouraging, given the computational advantages of the former. One issue that arose is that the program with empirical payoffs did not converge for the previous starting values ($\lambda = 1$ and $r = 0.5$). It did converge when the starting values were set at $\lambda = 0.25$ and $r = 0.35$, which were closer to previously obtained estimates, so application of this approach may require care in selecting starting values. However, when the empirical payoff-based program did converge, it converged quickly and was robust in the sense of yielding the same estimates for a range of starting values.

6.2.3 Unobserved Heterogeneity

There are many different sources of unobserved heterogeneity. The obvious one with respect to QRE is skill heterogeneity, which was explored theoretically in chapter 4. One can estimate a model of individual fixed effects, where each individual is assumed to have a personalized skill level, λ_i, and knows the distribution of other skill levels, as in HQRE. Often the only practical way to do this is to use the empirical payoff estimation approach. For each individual in an experiment, one can compute f_{-i} based on the choice behavior of all individuals other than individual i.[15] This approach makes sense only if one has multiple observations of each subject.[16] Then, for each individual, one would obtain a maximum-likelihood estimate of i's skill level as

$$\widehat{\lambda}_i = \arg \max_{\lambda_i} \log L(\lambda_i; f),$$

where $\log L(\lambda_i; f)$ is given in (6.3). This produces an estimate of the *exact* distribution of λ_i. Thus it is somewhat different from the theoretical development of HQRE, based on the assumption that each λ_i is an independent draw from a distribution of λ_i. There are two alternative interpretations of this method of estimating HQRE. The first interpretation is that we estimate an exact distribution of the individual precision parameters for the subjects in the experiment, and implicitly assume that this exact distribution is common knowledge. An alternative interpretation is that the individual-specific precision parameters are actually independent draws from a distribution of λ_i, and the collection of λ_i's that are estimated represent a nonparametric estimate of that distribution. If the number of subjects were very large, then the distribution of estimated λ_i's would be close to the true distribution. This method of estimating HQRE is illustrated in a common-value auction problem in chapter 9.

An alternative, and more parsimonious, approach to estimating heterogeneous individual effects is to specify a parametric distribution of the λ_i's (e.g., log-normal) and estimate its parameters with the maximum-likelihood techniques discussed earlier. This approach is especially useful in a between-subjects design where each individual is exposed to a single treatment, with multiple decision observations for each individual, and can also be used to estimate heterogeneity with respect to behavioral variables of specific interest. Goeree, Holt, and Laury (2002) used this approach to estimate a truncated normal distribution of altruism parameters for a public goods game (along with a logit precision parameter). More recently, Goeree, Holt, and Smith (2015) used a truncated normal specification to estimate the mean and variance of individual regret parameters (along with a

---

[15] This can be done even if an individual i plays some or all of the player roles. For example, in the asymmetric matching pennies game, one could design an experiment with n subjects, where each subject plays as a column player half the time and a row player half the time. One would compute f_{-i} based on the observed choices (in both Row and Column roles) of the other $n - 1$ subjects.

[16] These multiple observations of a single subject's behavior could be composed of single choices made across multiple player roles, or single choices made across a battery of different games. The estimate of λ_i would normally impose the constraint that it be constant across these different roles, although in principle one could estimate role effects or game effects as well, with sufficient amounts of data.

logit precision) using data from a volunteer's dilemma game, discussed later in chapter 8. In that application, the three-parameter QRE model with a distribution of individual effects provides only a somewhat better fit for data averages, but a dramatically improved fit for individual variability around those averages, which were off by an order of magnitude for homogeneous QRE models.

7
Applications to Game Theory

In this chapter we explore several applications of QRE to specific games in order to illustrate and expand on the wide range of game-theoretic principles and phenomena associated with QRE that have been highlighted in the previous chapters. Probabilistic choice models, both for individual choices and for games with two or more players, are motivated by empirical studies in which observed decisions exhibit some noise. The empirical applications to be considered in this chapter are based on game-theoretic laboratory experiments with players who are motivated by cash payoffs. Anyone who has looked at decisions made by human subjects in such situations will know that, despite strong empirical regularities, there is often a fair amount of unexplained variation across individuals and over time for the same individual. Some of the noise is due to recording errors. For example, we have observed subjects who are shown a signal (e.g., a colored marble drawn from a cup) to occasionally record a signal color that differs from the one they observed. Therefore, it is easy to imagine that decisions are sometimes recorded incorrectly, especially in low payoff situations. Sometimes the variations seem to be due to emotional reactions, for example, to others' prior decisions, or to individual differences in attitudes about relative payoffs or risk. QRE does not rule out the presence of behavioral factors such as social preferences, risk attitudes, or judgment biases. Rather, as illustrated in the previous chapter on estimation, QRE provides a rigorous equilibrium approach to structural estimation of the existence and magnitude of such behavioral factors. Several of the examples in this chapter, as well as the next two chapters (on applications to political science and economics, respectively) use QRE for this kind of estimation and measurement. Of course, there will remain the residual stochastic factors that are unobserved or unmodeled, which constitute the noise effects embodied in QRE. Regardless of the source, systematic bias or residual noise, players in games will respond systematically to the anticipated behavior of other players, which includes stochastic elements to the extent that they are present.

The first application to be considered belongs to the class of continuous games, which were introduced in section 2.7. With a continuum of decisions,

QRE predicts a choice distribution that is not merely a (possibly asymmetric) spread to each side of a Nash equilibrium, since "feedback effects" from deviations by one player alter others' expected payoff profiles, which would induce further changes. The "traveler's dilemma," considered in section 7.1, provides an example in which these feedback effects can cause the QRE choice distribution to differ dramatically from the unique Nash equilibrium.

The second application is a symmetric game with binary actions where players have continuously distributed private information about an unknown state of the world that affects both players' payoffs. The *compromise game* is a model of conflict where each player has a privately known "strength" and can choose to attack the other player or not. If either attacks, a conflict ensues and the player with the higher strength wins a large prize and the other player will have a very low payoff. If neither attacks, then there is peace, and both players earn an intermediate payoff. It will be shown that the unique Nash equilibrium of this game has perpetual conflict. Both parties always attack, regardless of their own strength. However, because of the extreme nature of this solution, the costs of deviating from equilibrium by choosing not to attack are precisely zero for all types, since conflict will occur in any case. QRE immediately implies that one should frequently observe players choosing not to attack. In fact, if they are indifferent, they should attack exactly half the time. But this is only the direct effect of player indifference, and of course it is not an equilibrium for both players to attack half the time, regardless of their own strength. Why? Because if the other player is choosing completely randomly, then players with a relatively low strength are better off not attacking, while players with a higher strength are better off attacking. Thus, noise in the action choices has an asymmetric and systematic effect on the different types of players, leading to the implication that in *any* regular quantal response equilibrium the probability a player chooses to attack will be an increasing function of strength, in stark contrast to the prediction of Nash equilibrium, where all types always choose to attack.

Because the probability of attack is correlated with player types, additional deviations from equilibrium can occur if players do not fully account for this correlation, as in the cursed equilibrium model of Eyster and Rabin (2005).[1] A structural estimation of laboratory data from the compromise game is undertaken to estimate the degree to which a combination of logit QRE and cursed equilibrium can account for the experimental findings. This application also serves two other purposes. First, it illustrates how QRE can be extended to games with continuous types and binary actions, and describes two different approaches for

---

[1] A similar kind of behavioral anomaly can be derived with Jehiel's (2005) analogy-based expectations equilibrium.

doing this. Second, it provides a bridge to the second half of this chapter on extensive-form games, because the experiment explored two versions of the game, one where the play was simultaneous and a second variation where the players choose to attack or not in a fixed sequential order. The QRE predictions (and the data) are different across these two versions of the game, even though they are strategically equivalent in the traditional game-theoretic sense, and have the same Nash equilibrium.

The second half of the chapter looks at three applications to extensive-form games, all three of which are games of incomplete information. The first application is a simple zero-sum signaling game with a unique mixed-strategy Nash equilibrium, called the "simplified poker game." In this game there is a deck of cards with only two cards, one high and one low. Both players ante up a small sum into the pot. A neutral dealer randomly selects one card and gives it to the first player, but the second player is not allowed to see the card. The players then, in sequence, can bet a fixed amount or fold. If they both bet, then the first player wins the pot if the card is high and the second player wins the pot if the card is low. If either player folds, the other player wins the ante. This game has a unique mixed-strategy equilibrium, so in many ways it shares similar strategic properties to the asymmetric matching pennies game. However, the combination of asymmetric information and sequential play adds new and interesting features to the game. In particular, errors by the first player in choosing to bet or not affect the second player's belief about the card held by the first player. This induces some systematic changes in the payoffs for the second player, which feeds back and changes the expected payoffs of the first player. Thus, there are systematic effects in a logit AQRE that imply specific directions of deviation from the Nash equilibrium.

The second application of QRE to extensive-form games is a class of non-zero-sum signaling games that have multiple Nash equilibria. This application explores the refinement and selection properties of logit AQRE, and compares these properties with belief-based refinements, such as the "intuitive criterion," which were developed by game theorists in the 1980 s. Several variations of sender–receiver games have been the subject of laboratory experiments, and this section of the chapter fits data from those games to the logit AQRE solution to each game, and then compares these results to the predictions based on belief-based refinements. The data are also broken down by the experience level of the players and it is shown how AQRE provides insights into the learning dynamics in these games.

The final application, to social learning, focuses on information cascades and herding behavior. Many variations of the social learning model of information cascades lead to the conclusion that in a Nash equilibrium, it will be impossible for social learning to lead to information aggregation, even with an infinite number of pieces of private information, an infinite number of players, and an infinite amount of time. In the simplest model, one for which there is abundant experimental data, there are n players, each of whom has a (binary) piece of information, called their signal, about the state of the world and the players choose, in sequence, one of two actions, each corresponding to one of the states of the world. If a player's action matches the state of the world, they receive a high payoff, otherwise they receive a low payoff. Importantly, players are able to observe the choices of players who were earlier in the sequence but not their signals. Obviously, if every player up to player n chooses according to their own signal and ignores the previous players' choices, and n is large, then player n will have enough information to make an almost perfectly informed decision. But this is not a Bayes–Nash equilibrium. At some point, learning becomes stuck because the accumulated evidence from earlier players' choices swamps the information in the private signals of the later players. However, QRE completely overturns this logic. In *any* regular QRE, there is complete information aggregation in the limit. That is, player beliefs converge with probability 1 to the true state. Put differently, an outside observer will be able to infer the state almost perfectly if there are a large number of observations. This result is very general and does not require strong assumptions about the payoff functions, the number of available actions, or the number and distribution of possible signals (see Goeree, Palfrey, and Rogers, 2006). Section 7.5 provides the theoretical argument for this result and analyzes data from an information cascade experiment.

7.1 THE TRAVELER'S DILEMMA

Basu (1994) first described an interesting social dilemma in which the logic of standard game theory with perfectly rational players seems to be implausible.[2] The game is motivated by a story of two travelers who have purchased identical antiques while on a tropical vacation. When their luggage is lost on the return trip, the airline claims representative informs them,

> We know that the bags have identical contents, and we will entertain any claim between \$2 and \$100, but you will each be reimbursed at an amount that equals the *minimum* of the two claims submitted. If the two claims differ, we will also pay a reward of \$2 to the person making the smaller claim, and we will deduct a penalty of \$2 from the reimbursement to the person making the larger claim.

---

[2] This section includes material from Capra et al. (1999).

Each traveler would presumably know the cost of the lost antique, which would provide a natural focal point. But if one expects the other to claim this cost (or to claim $100 if the cost was above the upper limit), then that person would have an incentive to undercut the anticipated claim for the other person. In fact, each person has an incentive to undercut any common claim that is above $2, so the unique Nash equilibrium for this game is for each to claim the minimum amount, which may seem implausible given the relatively small penalty of $2 for making a higher claim.

The traveler's dilemma is richer than a prisoner's dilemma in the sense that the Nash decision is not a dominant strategy; the best claim to make would be just below the other's claim if it were known. In particular, if the other person is expected to claim $100, then the best response is to claim $99 (if claims are restricted to integer amounts). Obviously, there is no belief about the other's claim that would justify making a claim of $100 in this game. If each player realizes that the other is thinking in this manner, then each will not expect the other to make a claim above $99, but the best response to this upper limit on anticipated claims would then be $98. Reasoning iteratively, one can rule out all claims above the minimum level on the assumption that it is common knowledge that each player is perfectly rational. As Basu (1994) notes, the Nash equilibrium is the unique *rationalizable* equilibrium in this game (with discrete decisions). These iterated-rationality arguments have some appeal, but note that the same arguments would apply if the penalty/reward amount were raised from $2 to $20 or lowered to $1.01. Basu conjectured that claims would not fall to Nash levels for low values of the penalty/reward parameter, but he did not provide a formal theoretical explanation that was sensitive to these payoff parameter changes.[3]

7.1.1 Traveler's Dilemma Experimental Data

The invariance of the Nash equilibrium to variations in the penalty/reward parameter is the motivation behind the laboratory experiment reported in Capra et al. (1999). Subjects were recruited from economics classes at the University of Virginia. There were six sessions, with the number of participants in each session ranging from 9 to 12. Payoffs denominated in US cents were scaled down so that earnings would be appropriate for student subjects. In particular, the instructions stipulated that people would be randomly paired and would have to

---

[3] A different approach might be to consider a model of adjustment, e.g., Cournot best responses to the most recently observed decision of the other player. But the Cournot best response to the other's claim is always to offer a slightly lower claim, regardless of the size of the payoff parameter, so this approach will not provide intuitive predictions, at least not without some modifications to be discussed below.

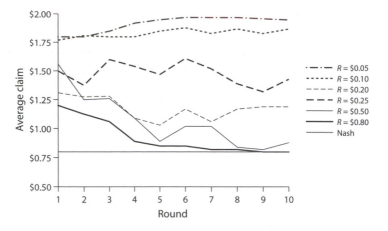

FIGURE 7.1. Average claim data for the traveler's dilemma experiment.

make a claim decision on the interval [80, 200]. Each session was characterized by a single penalty/reward parameter for 10 rounds of play, with random rematching of participants after each round. The penalty/reward parameters used for the six sessions were 5, 10, 20, 25, 50, and 80. As indicated above, the Nash equilibrium for each of these cases is the minimum claim of 80.

The salient feature of the data for this experiment is that the observed claim levels responded sharply to changes in the penalty/reward parameter. The solid lines at the bottom of figure 7.1 show the average claims by round for the two sessions with the highest R values (50 and 80), and the two dot-dashed lines at the top of the figure show the data for the lowest R values (5 and 10 cents). With a high penalty/reward parameter, the average claims started above 120 in the initial round, but fell to levels very near the Nash prediction of 80 in the final round. In contrast, the average claim data for sessions with relatively low penalty/reward parameter values started at about 180 in the initial round, well away from the Nash prediction, and these averages leveled off at levels even farther from this prediction. Two additional sessions, with penalty/reward values of 20 and 25, produced average claims in an intermediate range between those with high values and those with low values, as can be seen from the dashed lines in the figure. Note that there is one "inversion" here in the sense that the average claims for the $R = 20$ session were lower than the average for the $R = 25$ session.

This experiment was somewhat unusual in that a whole range of payoff parameters were used, instead of restricting attention to a couple of parameter values, each with multiple sessions. Nevertheless, it is possible to construct a nonparametric test of the null hypothesis that changes in the payoff parameter

TABLE 7.1. Average claim in data, compared with logit QRE and Nash equilibrium.

Penalty/reward parameter (R)	5	10	20	25	50	80
Average claim in data (last 5 rounds)	196	186	116	146	92	82
Average claim at logit QRE ($\lambda = 0.12$)	183	174	149	133	95	88
Average claim Nash equilibrium	80	80	80	80	80	80

have no effect, as predicted in a Nash equilibrium. The alternative hypothesis is that decreases in the payoff parameter will increase average claims. The most extreme observation would be for the average claims in the second row of table 7.1 to decrease from left to right, but there is one reversal. There are 6! = 720 different ways that the average claims could have been ranked, and of these, only 6 are either as extreme (one reversal in adjacent columns) or more extreme (no reversals). The chances of observing a pattern at least as extreme as the one reported are 6 in 720, and therefore, the null hypothesis can be rejected at the 1% level. To summarize, the Nash equilibrium fails to explain the most salient feature of the data, that is, the responsiveness of average claims to changes in the penalty/reward parameter.

7.1.2 Logit QRE Claim Distributions

The claim averages for the sessions shown in figure 7.1 tend to level off after about five rounds, which suggests that an equilibrium has been approximately reached. Recall the symmetric QRE model for games with a continuum of actions, developed in section 2.7. A player's mixed strategy is a choice probability distribution function, $f(x)$, which is increasing in the expected payoff, $U(x)$, of claim $x \in [80, 200]$. To determine the expected payoff, note that a player's actual payoff depends on the other player's claim, y. This payoff will be $x + R$ if $x < y$ (reward obtained), it will be the claim x if $x = y$ (no penalty or reward), and it will be $y - R$ if $x > y$ (penalty paid). The expected payoff $U(x)$ is therefore

$$U(x) = \int_{80}^{x} (y - R)f(y)dy + (x + R)(1 - F(x)). \tag{7.1}$$

The expected payoffs depend on the probability distribution of claims, which in turn have to satisfy the logit differential equation (2.21) of section 2.7: $f'(x) = \lambda f(x)U'(x)$. For the traveler's dilemma the logit differential equation is given by

$$f'(x) = \lambda f(x)(1 - F(x) - 2Rf(x)) \tag{7.2}$$

with boundary conditions $F(80) = 0$ and $F(200) = 1$.

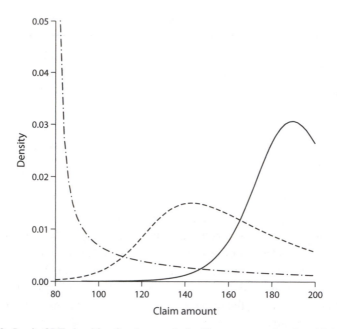

FIGURE 7.2. Logit QRE densities for the traveler's dilemma game for $R = 10$ (solid line), $R = 25$ (dashed line), and $R = 50$ (dot-dashed line) when $\lambda = 0.12$.

The theoretical results of section 2.7 imply that there exists a unique solution (see propositions 2.10 and 2.11), and that increases in the penalty/reward parameter, R, lower claims in the sense of first-degree stochastic dominance (see proposition 2.12). There is no analytical expression for the solution to (7.2), but for given values of λ and R, the solution can be easily determined numerically. Figure 7.2 shows logit QRE claim densities for a low ($R = 10$), medium ($R = 25$), and high ($R = 50$) value of the penalty/reward parameter assuming $\lambda = 0.12$ in all three cases. Notice that an increase in the penalty/reward parameter causes a shift to lower claims as predicted by the comparative statics results of section 2.7.

The logit QRE densities in figure 7.2 were generated using the maximum-likelihood estimate $\lambda = 0.12$ for the data reported in Capra et al. (1999), with payoffs measured in pennies and using data for the final five rounds. The computed logit QRE densities can be used to generate a prediction for the average claim and its standard deviation. For example, predictions for average claims are shown in the third row of table 7.1, just above the Nash prediction, which is $0.80 for all treatments. The logit QRE predictions correctly indicate that claims will converge to near-Nash levels for relatively high values of R and will cluster at the *opposite* end of the set of feasible claims for relatively low values of R.

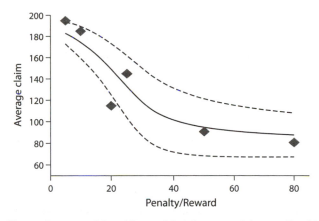

FIGURE 7.3. Observed average claims (diamonds) and average claims predicted by logit QRE (solid line) for $\lambda = 0.12$ and various values of the penalty/reward parameter. The dashed lines show predicted average claims plus or minus one standard deviation.

The accuracy of logit QRE is further illustrated by figure 7.3, which shows predicted average claims (solid line) plus or minus one standard deviation (dashed lines) as well as the data averages (diamonds) for each of the six penalty/reward values used in the experiment.

7.2 THE COMPROMISE GAME

We next turn to an incomplete-information game where adverse selection plays an important role. The game illustrates an application of QRE to games where players have continuous types and binary actions. In addition to showing how observed misbehavior in these adverse-selection games can be explained by QRE, and how QRE predictions change depending on whether the game is played sequentially or simultaneously, the game illustrates how QRE can be used as the underlying error structure for estimation of parameters from several alternative behavioral theories.[4]

In the *compromise game*, each of two opposing players may choose to attack (A) or not to attack (NA) the other player. If an attack occurs, the stronger opponent conquers the weaker opponent, yielding a high positive payoff, W, the win payoff. In the event of an attack by either player, the person with the lower strength earns zero. If an attack does not occur, that is, neither side attacks,

---

[4] This section includes material from Carrillo and Palfrey (2009).

both parties receive an intermediate payoff, $P < W$, the "peace" payoff. Each player knows their own strength but knows only the distribution from which their opponent's strength is drawn. The most obvious application is to the possibility of war between two hostile nations, either one of which would like to conquer the other. One interpretation of the relative payoffs is that if $W < 2P$, then conflict is inefficient and there is a "peace dividend" equal to $2P - W$. On the other hand, it is possible that $W > 2P$, which one might interpret as a case where existing hostilities are so great, or economic benefits from a political merger are so large, that total benefits would be higher were one country to conquer the other.

Formally, let t_i denote player i's strength, which is drawn from a commonly known and continuous density function $f(t)$ with support $T = [0, 1]$. Assume the density is the same for both players and that player types are drawn independently. Further, assume that W and P are common knowledge and lie in the interval $(0, 1)$.

7.2.1 Simultaneous versus Sequential Compromise Games

Carrillo and Palfrey (2009) study this game both theoretically and in the laboratory and find some surprising results, with observed behavior being very different from Nash equilibrium behavior. They show that these findings are consistent with a combination of quantal response equilibrium and "cursed equilibrium" (Eyster and Rabin, 2005). Carrillo and Palfrey (2009) investigate two versions of the game, one with simultaneous play, and the other where one country makes a decision first, followed by a decision of the second country in the case that the first country chooses not to attack. They also vary the size of the compromise, or peace, payoff, P. The motivation for this part of the design was to test the hypothesis implied by QRE, that attack rates will decline with respect to P.

The Bayes–Nash equilibrium is identical for both game versions. In equilibrium, both players always choose A, regardless of strength, and war occurs with probability 1.[5] However, QRE will generally make different predictions in the two game versions, for reasons similar to the example of the chain-store-paradox game in chapter 3. In the sequential game, the second mover's expected utility losses from deviations from equilibrium behavior are greater than the first mover's expected payoff losses and QRE predicts different behavior patterns for the first and second movers. In the simultaneous-move game, the game is symmetric so no such differences are predicted. Also, in equilibrium, the expected payoff losses from deviations from optimal behavior in the simultaneous game are different from the losses for players in either position in the sequential game. Thus, for the

---

[5] In the sequential game, if the first mover attacks, then war occurs regardless, and the second mover's action is irrelevant. If the first mover does not attack, then the second mover attacks.

TABLE 7.2. Aggregate frequencies for A choices by treatment and order of moves. (Number of observations in parentheses.)

Order	Position	$P = 0.39$	$P = 0.50$
Simultaneous	Both	0.66 (560)	0.57 (560)
Sequential	First	0.59 (280)	0.54 (264)
Sequential	Second	0.64 (115)	0.57 (122)

same precision parameter, there should be differences in behavior across the two game versions, and across the two player roles in the sequential game.

7.2.2 The Logic of Equilibrium

Many find the equilibrium reasoning that implies "always A" to be counterintuitive, although the logic is straightforward.[6] Consider the simultaneous version of the game. A strategy is a mapping from one's own strength to a decision on whether to attack the opponent. Given any strategy of the opponent, one's own best reply takes the form of a cutoff rule. That is, choose A if t_i is larger than some cutoff level t^*. Thus it suffices to look for equilibrium in cutoff strategies. Suppose one's opponent uses a cutoff strategy $\hat{t}$. What is the best reply? Suppose one's strength is t. First, observe that one's own choice to attack affects the outcome only in the event that the other player chose NA, so one need consider payoffs conditional on that event only. The expected payoff to A, conditional on one's opponent choosing NA is equal to $W \cdot \min\{1, F(t)/F(\hat{t})\}$ when the loss payoff is 0. The expected payoff to NA, conditional on one's opponent not attacking is simply P. Therefore, if $\hat{t} > 0$, the optimal response cutoff (a point of indifference) is $\tau^*(\hat{t}) < \hat{t}$, and hence the only equilibrium is $t^* = 0$. A similar logic applies to the sequential-move game.[7]

The experimental design uses a uniform distribution of strengths, $W = 1$, and compares behavior for two different values of P (0.39 and 0.50), in addition to comparing behavior in the simultaneous and sequential versions of the game. Table 7.2 displays the attack frequencies across the different games and player roles. There are three main treatment effects on overall attack frequencies. First, attack rates are higher in the $P = 0.39$ treatment than in the $P = 0.50$ treatment, as QRE would predict (see the following section). Second, attack rates are higher

---

[6] Experience assigning this on a problem set for a PhD class shows that even economics PhD students do not understand the logic, until it is carefully explained to them.

[7] If the game were changed slightly so that "it takes two to tango," i.e., war occurs only if both parties choose attack, then the equilibrium has exactly the opposite property: $t^* = 1$, so the outcome is always "peace." That game has not been studied in the lab, but would be interesting and might exhibit anomalies in the opposite direction from the experimental results of the compromise game.

for second movers than first movers, again consistent with QRE. Third, attack rates are higher in the simultaneous-move game than in the sequential-move game.

Looking more closely at the data reveals that the attack probability is also conditional on the strength of the player. For each treatment, empirical conditional attack frequencies as a function of strength can be used to estimate logit QRE parameters, as well as other parameters from alternative behavioral models. Hence the compromise game is a useful example to illustrate how one can estimate these models in a game with continuous types and binary action spaces.

7.2.3 QRE with Continuous Types and Binary Action Spaces

There are a number of ways to specify QRE in incomplete-information games with continuous types and binary actions, and three are considered here.[8] One way is to use the normal form of the game, where the space of pure strategies, S, consists of all measurable functions from the type space, $T = [0, 1]$, to the $\{0, 1\}$ set, that is, a function that maps each type into a decision to either fight or not fight. The set of all mixed strategies is then the set of probability distributions over S. However, this is a cumbersome representation for these games, and for this reason is typically not used for the analysis of QRE in games of incomplete information with continuous types.

A second approach, which is simpler and more standard, is to look at the set of behavioral strategies for each player. A behavioral strategy in this game is a measurable function from the type space, $T = [0, 1]$, to the probability simplex $[0, 1]$, that is, a function that assigns a probability of fighting to each type. This representation corresponds to AQRE.

The third approach is more closely related to the strategic form, but eliminates pure strategies that are weakly dominated in the strategic form version of the game. In continuous games with binary actions, where the payoffs are monotone in type, as in the compromise game, this eliminates all pure strategies that are not cutpoint rules.[9] In this third approach, players are assumed to quantal respond using cutpoint rules.

---

[8] This section uses the compromise game with a uniform distribution of types for illustrative purposes, but it is easy to see how the analysis extends more generally. In the next chapter we will consider an application to voting games that have a similar structure. In the experimental implementation of the compromise game, each player had a finite number (100) of types. The analysis here treats the number of types as infinite.

[9] Monotone in type means that, given any strategy by the opponent, the expected payoff difference between the two strategies is monotone in type. For example, in this particular game, given any strategy by the other player, the expected difference in own payoff between fighting and not fighting is increasing in one's own strength. Thus, if it is optimal for type t to fight, it is also optimal for any type $t' > t$ to fight.

We next go through the characterization of logit QRE for both the behavioral strategy and cutpoint strategy versions of QRE, the cutpoint TQRE model assuming a Poisson distribution of types, and the cursed equilibrium (CE) model hybridized with QRE in behavioral strategies.

THE BEHAVIORAL STRATEGY APPROACH OF AQRE

Consider first the AQRE of the *simultaneous-move* compromise game. Denote a behavioral strategy by $\sigma : [0, 1] \to [0, 1]$, so the probability of fighting, given strength type t, is $\sigma(t)$, and consider the logit AQRE with precision parameter λ. In the logit AQRE, the log odds of the probability of fighting to the probability of not fighting is proportional to the difference between the conditional (on t) expected payoff from fighting and the conditional (on t) expected payoff from not fighting, given the equilibrium behavioral strategy of the other players. To derive the difference in expected utility of using a behavioral strategy σ, conditional on having strength type t, suppose the other player is using some behavioral strategy $\tilde{\sigma}$. Then this difference is equal to

$$\Delta U_1(t; \tilde{\sigma}) = \int_0^1 [1 - \tilde{\sigma}(x)] dx \, (V_A(t) - V_{NA}(t))$$

$$= W \int_0^t [1 - \tilde{\sigma}(x)] dx - P \int_0^1 [1 - \tilde{\sigma}(x)] dx,$$

where $V_{NA}(t) = P$ is the expected utility of not fighting, conditional on the opponent not fighting, and

$$V_A(t) = W \frac{\int_0^t [1 - \tilde{\sigma}(x)] dx}{\int_0^1 [1 - \tilde{\sigma}(x)] dx}$$

is the expected utility of fighting, given the other is not fighting. Hence, in a symmetric logit AQRE with response parameter λ, the equilibrium probability of fighting, conditional on type t is

$$\sigma^*(t) = \frac{e^{\lambda \Delta U_1(t; \sigma^*)}}{1 + e^{\lambda \Delta U_1(t; \sigma^*)}} \quad \forall t \in [0, 1].$$

The sequential game is different because it is an asymmetric game, and the second mover has more information at the time of his move than the first player. The second mover's strategy is simpler because they do not have any choice if the first mover chooses to attack. Denote the first and second mover's strategies by σ_1 and σ_2, respectively, and both map $[0, 1]$ to $[0, 1]$. The form of the first player's payoff is the same, except for the argument of $\Delta U_1(\cdot)$:

$$\sigma_1^*(t_1; \lambda) = \frac{e^{\lambda \Delta U_1(t_1; \sigma_2^*)}}{1 + e^{\lambda \Delta U_1(t_1; \sigma_2^*)}} \quad \forall t_1 \in [0, 1].$$

The second mover's payoff difference, ΔU_2, is now different, because (since 2 observes that 1 has chosen not to attack) the difference is not discounted by $\int_0^1 [1 - \tilde{\sigma}(x)]dx$:

$$\Delta U_2(t_2; \tilde{\sigma}) = W \frac{\int_0^{t_2} [1 - \tilde{\sigma}(x)]dx}{\int_0^1 [1 - \tilde{\sigma}(x)]dx} - P.$$

This "inflation" of player 2's expected payoff differential by a factor of $\int_0^1 [1 - \tilde{\sigma}(x)]dx$ will lead to choice behavior that is more responsive than in the simultaneous-move game, as if he had a higher value of λ. This is the same effect seen before in chapter 3, with the chain-store stage-game example. And in an AQRE, we have

$$\sigma_2^*(t_2; \lambda) = \frac{e^{\lambda \Delta U_2(t_2; \sigma_1^*)}}{1 + e^{\lambda \Delta U_2(t_2; \sigma_1^*)}} \quad \forall t \in [0, 1].$$

CUTPOINT QRE

In a cutpoint QRE, players are modeled as mixing over cutpoint strategies, where a "pure" cutpoint strategy, c, specifies a decision for i to attack if and only if $t_i \geq c$. In a QRE, a player does not always choose the optimal cutpoint, but chooses higher expected payoff cutpoints with higher probability than cutpoints with lower expected payoffs. In the logit QRE, choice probabilities are such that the log odds of choosing one cutpoint rather than another is proportional to the difference in expected payoffs. We derive here the conditions for a symmetric cutpoint QRE in the compromise game, but this generalizes easily to other games, or to asymmetric equilibria. Suppose your opponent is mixing over cutpoints, using the cumulative distribution function, $G(c)$. That is, the probability they use a cutpoint less than or equal to c is equal to $G(c)$. One can then write the expected utility to one player of using cutpoint $\tilde{c}$ if the other player is mixing over cutpoints according to G as

$$U(\tilde{c}) = W \int_{\tilde{c}}^1 t \, dt + \int_0^{\tilde{c}} \left[W \int_0^t (t - c)dG(c) + P \int_0^1 c \, dG(c) \right] dt, \quad (7.3)$$

where the first term corresponds to the probability the player's type, t, is above the cutpoint *and* above the other player's strength. The second part of the expression corresponds to payoffs when the player's type, t, is below the cutpoint, in which case he chooses not to attack, and earns W if the other player attacks with a lower strength, or earns P if the other player does not attack. In a logit equilibrium, the density function over cutpoints is given by

$$q^*(c) = \frac{e^{\lambda U(c)}}{\int_0^1 e^{\lambda U(x)}dx} \quad \forall c \in [0, 1],$$

where $q^*(c) = dQ^*/dc$ is the density function associated with Q^*.

As in the behavior strategy approach, the solution will generally be different in the sequential version of the game, because the second player's choice is made after knowing that the other player did not attack, and will result in an equilibrium where the second mover's choice behavior is more responsive than in the simultaneous-move game.

TQRE

Recall from chapter 4 that the Poisson version of the TQRE model that asymptotically approximates the cognitive hierarchy model has two parameters, γ and τ, where $k\gamma$ specifies the logit response precision of level-k players, and τ is the Poisson precision parameter governing the distribution of level-k types. TQRE is characterized here using the cutpoint strategy approach. If the true distribution of level-k types is $p = (p_0, p_1, \ldots)$, then a level-k player has beliefs about the distribution of types, given by truncating p at type $k - 1$. That is, for $k \geq 1$, the level-k player's beliefs for the event that they face a level-j player ($j < k$) are given by

$$p_j^k = \begin{cases} \dfrac{f(j)}{\sum_{l<k} f(l)} & \text{if } j < k, \\ 0 & \text{if } j \geq k, \end{cases}$$

where $f(j; \tau) = \tau^k e^{-\tau}/k!$ in the Poisson model.

In a TQRE, each level type will have a possibly different distribution over cutpoints, so denote the TQRE cutpoint distribution for level j by Q_j. Then in a Poisson TQRE, the expected utility of using cutpoint $\tilde{c}$ for a level-k player is

$$U_k(\tilde{c}; \tau) = W \int_{\tilde{c}}^1 t\, dt$$

$$+ \sum_{j=0}^{k-1} p_j^k(\tau) \int_0^{\tilde{c}} \left[W \int_0^t (t - c)\, dQ_j(c) + P \int_0^1 c\, dQ_j(c) \right] dt, \quad (7.4)$$

and the logit TQRE cutpoint distribution for k is therefore

$$q_k^*(c; \tau, \gamma) = \frac{e^{k\gamma U_k(c;\tau)}}{\int_0^1 e^{k\gamma U_k(x;\tau)} dx} \quad \forall c \in [0, 1].$$

The TQRE equations for the sequential version of the game can be derived in a similar fashion, but with the second mover's expected payoffs no longer discounted by the probability the other player chooses NA.

α-CURSED QRE

In a cursed equilibrium, players have correct beliefs about the marginal distribution of actions (i.e., the unconditional probability the opponent attacks) but assume this attack probability is independent of the other player's strength. In a partially cursed equilibrium, players assume partial dependence of the attack probabilities on the opponent's strength, where $\alpha \in [0, 1]$ indexes the degree to which players are cursed. If $\alpha = 0$, then players are fully rational and always attack. If $\alpha = 1$, then players are fully cursed, believing that there is no correlation between an opponent's type and their attack choice. While cursed players do not have fully rational beliefs about their opponent's strategy, cutpoint strategies are still optimal given their beliefs. For the α-cursed equilibrium without quantal response choice behavior, the equilibrium cutpoint strategy as a function of P, denoted $c^*_{\alpha P}$, is derived as follows. Formally, in an α-cursed equilibrium each player assumes other players play their α-cursed equilibrium strategy with probability $1 - \alpha$ and make choices that are independent of their type with probability α. Therefore, for a player with strength type t, and assuming the other player is using $c^*_{\alpha P}$, the expected utility of A, conditional on the opponent choosing NA is given by

$$V^\alpha_A(t) = W \left[\alpha \, \text{Prob}[t_j < t] + (1 - \alpha) \, \text{Prob}[t_j < t | a_j = NA; c^*_{\alpha P}] \right]$$

$$= W \left[\alpha t + (1 - \alpha) \min \left\{ 1, \frac{t}{c^*_{\alpha P}} \right\} \right].$$

In an α-cursed equilibrium, a player with strength equal to the α-cursed equilibrium cutpoint must be indifferent between A and NA, conditional on the other player choosing NA, that is, $P/W = \alpha c^*_{\alpha P} + (1 - \alpha)$. Therefore,

$$c^*_{\alpha P} = \begin{cases} 1 - \dfrac{1 - P/W}{\alpha} & \text{if } \alpha > 1 - \dfrac{P}{W}, \\[3mm] 0 & \text{if } \alpha \leq 1 - \dfrac{P}{W}. \end{cases}$$

An obvious difficulty with taking α-cursed equilibrium to the data and trying to estimate α is the zero-likelihood problem: for each α it makes a point prediction for the play of every type, either A or NA. Adding stochastic choice in the form of quantal response avoids this problem.[10] The previous subsection showed how to combine QRE with a cognitive hierarchy model using the cutpoint approach. Here we show how to combine QRE with cursed equilibrium using the behavioral strategy approach. In the simultaneous-move game, a (symmetric) α-cursed QRE

---

[10] Note that since $W > P$, unobserved player heterogeneity with respect to α would not solve the zero-likelihood problem: for any $\alpha \in [0, 1]$, it is always true that $c^*_{\alpha P} \leq P/W$.

is a behavior strategy, or a set of probabilities of choosing A, one for each type $t \in [0, 1]$. As before, we denote such a behavioral strategy evaluated at a specific strength value by $\sigma(t)$. Given λ and α we denote by $\sigma_{\lambda\alpha}^*$ the α-cursed QRE behavioral strategy. If player j is using $\sigma_{\lambda\alpha}^*$ and player i is α-cursed, then the difference in i's expected payoff from choosing A and NA when $t_i = t$ can be shown to be equal to

$$\Delta U(t; \sigma_{\lambda\alpha}^*) = W \left[\alpha t \int_0^1 [1 - \sigma_{\lambda\alpha}^*(x)] dx + (1 - \alpha) \int_0^t [1 - \sigma_{\lambda\alpha}^*(x)] dx \right]$$
$$- P \int_0^1 [1 - \sigma_{\lambda\alpha}^*(x)] dx,$$

and the α-cursed QRE in the simultaneous game is then characterized by

$$\sigma_{\lambda\alpha}^*(t) = \frac{e^{\lambda \Delta U(t; \sigma_{\lambda\alpha}^*)}}{1 + e^{\lambda \Delta U(t; \sigma_{\lambda\alpha}^*)}} \quad \forall\, t \in [0, 1],$$

which can be solved numerically, for any values of α and λ.

The α-cursed QRE in the sequential version of the game can be derived in a similar fashion, but with the second mover's expected payoffs no longer discounted by the probability the other player chooses NA. The derivation is straightforward, but slightly more involved, and details can be found in Carrillo and Palfrey (2009).

7.2.4 Estimating QRE, TQRE, and α-Cursed QRE

Behavioral data for the compromise game were gathered in laboratory experiments conducted at Princeton University in 2005, for both simultaneous and sequential-move games, and for two different values of P. In all games, $W = 1$. An observation is a (type, action) pair. For each of the three models estimated, the theory predicts type-conditional A frequencies between 0 and 1 for any fixed parameters of the models (λ for QRE, (γ, τ) for TQRE, and (α, λ) for α-cursed QRE). These parameter-dependent theoretical predictions are obtained computationally, by solving a fixed-point problem in the case of QRE and α-cursed QRE, and by solving recursively in the case of TQRE.[11] This defines a likelihood function, and table 7.3 reports the parameters for each model that

---

[11] For any parameters, (γ, τ) the Poisson TQRE is solved recursively by first deriving the type-conditional choice frequencies of type 0, then type 1, etc. Note that the type-conditional choice probabilities are not all equal to 0.5 for type-0 players. Because they are uniformly mixing over *cutpoint* strategies, their type-conditional probability of choosing A is increasing in type. Because it is solved recursively, there is no issue of multiple equilibria for the Poisson TQRE model. Based on extensive numerical analysis, there is also not a multiple equilibria issue for the other two models of behavior in the compromise game, at least within the class of symmetric equilibrium.

TABLE 7.3. Model parameter estimates. (Number of observations in parentheses.)

	QRE		TQRE			α-cursed QRE		
	λ	$-logL$	γ	τ	$-logL$	λ	α	$-logL$
Simultaneous all (1120)	16.2	387.8	6.9	2.7	386.3	21.3	0.85	355.6
Sequential all (781)	10.4	265.3	8.0	1.8	263.5	18.4	0.86	248.9
All (1901)	13.0	656.8	10.0	1.8	651.2	20.1	0.85	605.9

maximize the likelihood function, given the type-conditional choice frequencies in our data.[12] The estimates in the first two rows pool the observations for both values of P, reporting estimates separately for the simultaneous and sequential-move treatments. The last row reports the estimates pooling all the data, and shows that the estimates are very similar across the treatments.

Figure 7.4 shows the type A choice frequencies as a function of strength type, for each treatment, and separates the first and second mover data for the sequential version of the game.[13] The fitted/predicted type-conditional choice frequencies for each model are also displayed in the same figure. The predicted values are out-of-sample predictions in the following sense. For the simultaneous-move games, the predicted values are based on the parameter estimates obtained by fitting the sequential-move data, and vice versa. One can see that all three models track the aggregate data very closely, with the α-cursed QRE fitting the data best in terms of log-likelihood.

It is fairly easy to see why the TQRE and α-cursed QRE models fit the data well. In a fully cursed equilibrium the cutpoint is equal to P, leading to unconditional attack rates that are not very different from what is observed in the data, and combining it with QRE smooths it out, which fits the relatively smooth response function in the data. For TQRE, notice that in the standard version of the level-k model, level-0 types choose A about half the time and their actions are independent of type, so the level-1 types act exactly as fully cursed players for $P = 0.5$ and very similarly to fully cursed players for $P = 0.39$. Level-2 types then use a cutpoint of about half the cutpoint of the level-1 players. Given that the estimated distribution of level types in most games is concentrated on levels 0, 1, and 2, these models naturally fit some of the main features of the observed behavior, but also predict "spikes" of behavior at certain strength thresholds. Combining level-k models with QRE (i.e., TQRE) smooths out those spikes.

---

[12] Each observation is treated as independent. One could allow for correlation in the data in various ways, for example by allowing for subject-specific unobserved heterogeneity, as in Goeree, Holt, and Laury (2002a) or Camerer, Nunnari, and Palfrey (2014).

[13] In all cases, we constrain the parameters to be the same for the first and second movers.

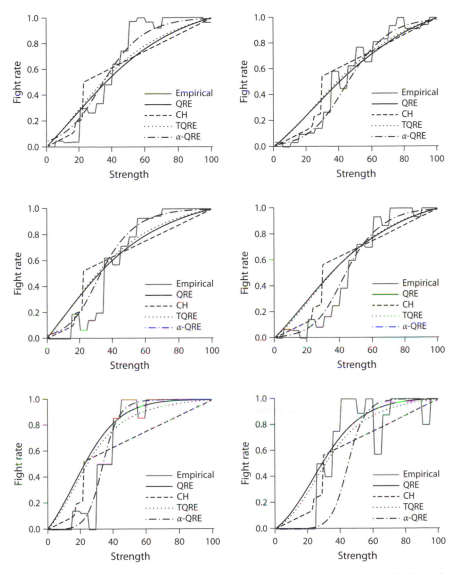

FIGURE 7.4. Conditional attack frequencies. Data and fitted models. (Source: Carillo and Palfrey, 2009.) This figure also includes fitted curves from the cognitive hierarchy model, which is the limiting TQRE, when $\gamma \to \infty$. Left panels: $P = 0.39$. Right panels: $P = 0.50$. Upper panels: simultaneous. Middle panels: sequential (first mover). Bottom panels: sequential (second mover).

An obvious question to ask is, why does QRE fit the data so well? There is nothing "built in" the QRE model that maps neatly into the observed data. However, in the Nash equilibrium players are indifferent between choosing A and NA, because the opponent always chooses A. In QRE, if two strategies yield the same expected payoffs, then they are chosen with equal probability. Thus, a logit best response to Nash equilibrium behavior is to choose randomly. And the best response to the logit best response to Nash equilibrium is to behave like a level-1 player. In games like this, where everything is a best response to a Nash equilibrium, logit QRE predictions can be very far away from Nash equilibrium, even for relatively high values of the response parameter, λ. This theme will appear again in chapter 9, in the context of a common-value auction game.

7.3 SIMPLIFIED POKER

A classic example in game theory to illustrate sequential equilibrium in extensive-form games with incomplete information is the simplified poker game. There are two players, which we call S (for *sender*) and R (for *receiver*). Before the poker game begins, each player is required to make an initial bet (the "ante") equal to some fixed amount $A > 0$. There are two cards, a high card, H, and a low card, L. Nature selects one of the two cards at random, with equal probability and that card is given to S. After S observes the card (R does not observe the card), S can either bet an exogenously fixed amount, $B > 0$, or can fold. If S folds, the game is over and R wins the initial bet of A, and the payoff to S is $-A$. If S chooses to bet, then R has a choice to either match the B bet (call) or fold. If R folds, then S wins the initial bet of A, and the payoff to R is $-A$. If R calls, then S wins $A + B$ if the card was H and R wins $A + B$ if the card was L.

The sequential Nash equilibrium of this zero-sum game is in mixed strategies. In equilibrium S will always bet with probability $\sigma_H = 1$ with H and bet with probability $\sigma_L = B/(2A + B) < 1$ with L (i.e., "bluff"). This makes R indifferent between calling and folding, following S's bet. In equilibrium, R must call with a probability that makes S indifferent between betting B and folding at the first move, if the card is L. This probability of calling is $\sigma_R = 2A/(2A + B) < 1$.

In the AQRE, there are several effects. First, S does not necessarily always bet with H, so the AQRE consists of three *interior* probabilities, $(\sigma_H, \sigma_L, \sigma_R)$.[14]

---

[14] As in the chain-store stage game and the centipede game (see chapter 3), the normal-form QRE analysis is different and will lead to different predictions about the probabilities of betting at each information set. The strategic form of the game is a 4×2 game, because player S has four strategies: BB (always bet); FF (always fold); BF (bet only with H); and FB (bet only with L). The last two of S's strategies are dominated, but they still play a role in the normal-form QRE analysis.

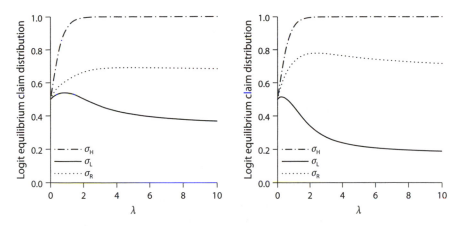

FIGURE 7.5. Logit AQRE correspondence for a simplified poker game with $A = B = 1$ and $\text{Prob}[H] = \frac{1}{2}$ (left panel) and $\text{Prob}[H] = \frac{1}{3}$ (right panel).

Second (and related to the first observation), S can bluff either more or less than in the Nash equilibrium, which will depend on the relative values of A and B. Similarly, R may call less or more than in a Nash equilibrium. Fixing λ, the logit AQRE must satisfy the following equations:

$$\sigma_R = \frac{e^{\lambda U_R(\text{bet}|S \text{ bet},\sigma_L,\sigma_H)}}{e^{\lambda U_R(\text{bet}|S \text{ bet},\sigma_L,\sigma_H)} + e^{\lambda U_R(\text{fold}|S \text{ bet})}},$$

where

$$U_R(\text{bet}|S \text{ bet}, \sigma_L, \sigma_H) = \frac{\sigma_L - \sigma_H}{\sigma_L + \sigma_H}(A + B)$$

and $U_R(\text{fold}|S \text{ bet}) = -A$, since, by Bayes' rule, the conditional probability of R winning the bet when he calls is equal to $\sigma_L/(\sigma_L + \sigma_H)$. The conditions for S's mixing probabilities, conditional on H and L, respectively, are given by

$$\sigma_H = \frac{e^{\lambda U_S(\text{bet}|H,\sigma_R)}}{e^{\lambda U_S(\text{bet}|H,\sigma_R)} + e^{\lambda U_S(\text{fold}|H)}} = \frac{e^{\lambda(A+\sigma_R B)}}{e^{\lambda(A+\sigma_R B)} + e^{-\lambda A}}$$

and

$$\sigma_L = \frac{e^{\lambda U_S(\text{bet}|L,\sigma_R)}}{e^{\lambda U_S(\text{bet}|L,\sigma_R)} + e^{\lambda U_S(\text{fold}|L)}} = \frac{e^{\lambda(A-\sigma_R(2A+B))}}{e^{\lambda(A-\sigma_R(2A+B))} + e^{-\lambda A}}.$$

The left panel of figure 7.5 illustrates the logit AQRE correspondence for $A = B = 1$, for which the sequential equilibrium behavioral strategy profile is $(\sigma_H^*, \sigma_L^*, \sigma_R^*) = (1, \frac{1}{3}, \frac{2}{3})$. Note that for these parameters, the AQRE bluffing probability is always higher than in the sequential equilibrium, and the QRE calling strategy is always below the sequential equilibrium. Also, as λ increases,

TABLE 7.4. Data and fitted AQRE probabilities for simplified poker game (Popova, 2006).

Treatment	Strategy	Data	AQRE	Nash
$\text{Prob}[H] = \frac{1}{2}$	σ_L	0.438	0.528	0.333
	σ_R	0.603	0.649	0.667
$\text{Prob}[H] = \frac{1}{3}$	σ_L	0.409	0.403	0.167
	σ_R	0.723	0.729	0.667

the logit equilibrium probability of calling converges to the sequential equilibrium faster than the logit equilibrium probability of bluffing.[15]

Popova (2006) reports the results from a laboratory experiment, for which one of the treatments was the above game with parameters $A = B = 1$. Subjects in the experiment played the game 20 times, with random rematching between games. There was significant learning in the first 10 matches, and with experience, subjects moved in the direction of the sequential equilibrium over time. Qualitatively, the results are consistent with the logit AQRE. Averaging all the data in the last 10 matches, bluffing probabilities were too high, the calling probability was too low but closer to the sequential equilibrium than the bluffing probability, and the probability of betting with H was very close to 1. That paper reported a second treatment, which also used the parameters $A = B = 1$, but where the two cards were not drawn with equal probability: the probability of H was $\frac{1}{3}$ rather than $\frac{1}{2}$. This changes both the sequential equilibrium and the logit AQRE correspondence, although the latter has similar qualitative properties to the first game. In this variation, the sequential equilibrium profile is $(\sigma_H^*, \sigma_L^*, \sigma_R^*) = (1, \frac{1}{6}, \frac{2}{3})$, so the only difference is that the equilibrium bluffing probability is much lower. The logit AQRE is displayed in the right panel of figure 7.5.

As in the first treatment, the deviations from sequential equilibrium, even after considerable experience, are consistent with logit AQRE. Table 7.4 reports the data averages for the last 10 matches for both treatments, as well as the sequential equilibrium probabilities, and the fitted logit AQRE probabilities.

7.4 SIGNALING GAMES

A natural application of the refinements implied by QRE is to simple sender–receiver signaling games. In the mid-1980s such games were intensely studied to better understand sequential equilibrium and belief-based refinements of

---

[15] The probability of betting with H, p_H, converges very quickly to 1, as this is a strictly dominant strategy.

sequential equilibrium. Belief-based refinements were developed to rule out equilibria in extensive-form games with incomplete information that were considered improbable because they were supported by off-the-equilibrium-path beliefs that seemed starkly at odds with intuition. The fact that these issues manifested themselves in many games of significant economic relevance (such as the Spence job-market signaling model) justified the efforts to develop plausible refinements to overcome a severe multiple equilibria problem.

The multiple equilibria problem arises for a simple reason. It is often the case that there are some information sets that are not reached in a sequential equilibrium. If information sets are not reached, then Bayes' rule does not apply and beliefs about player types at those information sets are (nearly) unrestricted. This extra degree of freedom (to assign beliefs arbitrarily) allows a theorist to construct implausible beliefs to keep the play of the game away from certain information sets, which in turn supports an equilibrium that avoids those information sets. Players avoid the information sets because they are implicitly "threatened" by behavior, which is apparently rational if one assigns implausible beliefs to the decision maker at the unreached information set.

The QRE approach to extensive-form games avoids the issue of assigning arbitrary beliefs and supporting multiple equilibria in two distinct ways. First, and most important, there are no unreached information sets. Thus, beliefs about player types at every information set are always derived from Bayes' rule, based on the equilibrium (totally mixed) strategies of the players. This leads to a standard and highly intuitive model of belief formation. In particular, player types who have more to gain from reaching an information set compared to other types of the same player are assigned higher probability if that information set is reached. Second, the games typically have multiple sequential equilibria, and often more than one of these is "plausible" by various belief-based refinement criteria. In contrast, the logit AQRE, generically selects a unique equilibrium as the limit of the unique connected path in the QRE correspondence as the error rate goes to zero, the logit solution.

Figure 3.4, showing the logit AQRE correspondence for the chain-store stage-game experiments, illustrates this selection. Notice that in the strategy version of that game there are multiple logit AQRE for high values of λ. However, there is a unique connected path starting at the unique equilibrium at $\lambda = 0$. The unique connected path selection is the darker curve in that figure.

In this section, we focus on games of incomplete information where belief-based refinements have been proposed to winnow down the set of sequential equilibria.[16] As we show below, in some cases, the logit solution of the game is

[16] This section includes material from McKelvey and Palfrey (1998).

TABLE 7.5. Signaling games considered in Banks, Camerer, and Porter (1994).

BCP # 2 (Nash versus sequential)

m_1	a_1	a_2	a_3	m_2	a_1	a_2	a_3	m_3	a_1	a_2	a_3
t_1	1, 2	2, 2	0, 3	t_1	1, 2	1, 1	2, 1	t_1	3, 1	0, 0	2, 1
t_2	2, 2	1, 4	3, 2	t_2	2, 2	0, 4	3, 1	t_2	2, 2	0, 0	2, 1

Nash: (m_1, a_2, a_2, a_2) sequential: (m_3, a_2, a_2, a_1)

BCP # 3 (sequential versus intuitive)

m_1	a_1	a_2	a_3	m_2	a_1	a_2	a_3	m_3	a_1	a_2	a_3
t_1	0, 3	2, 2	2, 1	t_1	1, 2	2, 1	3, 0	t_1	1, 6	4, 1	2, 0
t_2	1, 0	3, 2	2, 1	t_2	0, 1	3, 1	2, 6	t_2	0, 0	4, 1	0, 6

sequential: (m_2, a_1, a_3, a_1) intuitive: (m_1, a_2, a_1, a_1)

BCP # 4 (intuitive versus divine)

m_1	a_1	a_2	a_3	m_2	a_1	a_2	a_3	m_3	a_1	a_2	a_3
t_1	4, 0	0, 3	0, 4	t_1	2, 0	0, 3	3, 2	t_1	2, 3	1, 0	1, 2
t_2	3, 4	3, 3	1, 0	t_2	0, 3	0, 0	2, 2	t_2	4, 3	0, 4	3, 0

intuitive: (m_2, a_3, a_3, a_2) divine: (m_3, a_2, a_2, a_1)

the same as what is selected by standard refinement theory, but this is not always the case. In one of the signaling games below, the unique connected path does not correspond to *any* of the belief-based refinements of sequential equilibrium proposed in that extensive literature. For reasons of space, we do not reexamine the results of all the experimental signaling games that have been conducted to date. We focus on a subset of the experimental games reported by Banks, Camerer, and Porter (1994) and Brandts and Holt (1992, 1993).

7.4.1 Banks, Camerer, and Porter (1994)

The Banks, Camerer, and Porter (1994) experiment consisted of a series of two-person signaling games, with two sender types, three messages, and three responses. Each game was designed to have two Nash equilibria, one of which was always further up the chain of deductive refinements than the other. The experiments were intended to give data on the conditions under which subjects play the more refined equilibrium. We analyze games 2, 3, and 4 reported by Banks, Camerer, and Porter (1994). These games are given in table 7.5. Each of

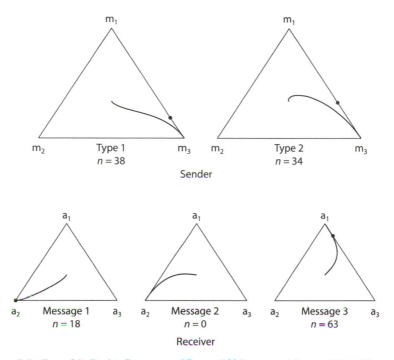

FIGURE 7.6. Game 2 in Banks, Camerer, and Porter (1994): sequential versus Nash. The curved lines are the logit AQRE and the dots indicate data averages.

these games has two Nash equilibrium outcomes, with one being a more refined equilibrium than the other.[17]

Experimental procedures were standard. There were 13 sessions with 6 subjects in each session. In each session two or three different payoff configurations were used. Each payoff configuration was repeated 10 times, with random pairing and role assignment between each game.[18]

The logit AQRE graphs for games 2, 3, and 4 are displayed in figures 7.6–7.8. The estimation of λ is reported in table 7.6 for each game. We see that in each of the three cases, the QRE selects the more refined equilibrium, which corresponds to what was observed in the data.

---

[17] In game 2, one equilibrium is sequential and the other is not. In game 3, both are sequential, but only one is "intuitive." In game 4, both are intuitive, but only one is "divine." In the graphs that follow, the QRE correspondence is displayed in the probability simplices (one for each information set for the tree) as a curved line beginning at the centroid (for $\lambda = 0$) and approaching the logit solution (as $\lambda \to \infty$). The aggregate data from the experiment are represented by a black dot in each simplex.

[18] A total of seven different payoff configurations were investigated in the study. The analysis here focuses on three of them.

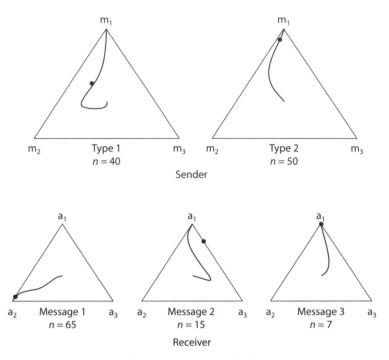

FIGURE 7.7. Game 3 in Banks, Camerer, and Porter (1994): intuitive versus sequential. The curved lines are the logit AQRE and the dots indicate data averages.

Moreover, the QRE predictions pick up quite well which strategies are relatively more overplayed (or underplayed) compared to the refined equilibrium predictions. In game 2, the predicted equilibrium has both types sending message 3, with the receiver choosing action 1 in response to message 3, and choosing action 2 in response to messages 1 and 2. For both types of senders, AQRE predicts that nearly all message errors should involve sending message 1. Indeed in all 18 deviations from this equilibrium, senders use message 1, and never use message 2.[19] Response errors to the equilibrium message are predicted by AQRE to be more than 10 times as likely to be action 3 than to be action 2, and indeed action 3 errors are observed in all 10 of the responders' deviations from equilibrium. Also, AQRE predicts that type 1 senders are less likely to deviate from the sequential equilibrium than type 2 senders, which was also observed in the data.

In game 3 the predicted (i.e., unique intuitive) equilibrium has both types sending message 1, with the receiver choosing action 2 in response to

---

[19] Neither m_1 nor m_2 are dominated strategies.

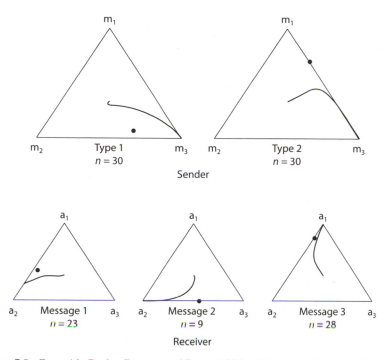

FIGURE 7.8. Game 4 in Banks, Camerer, and Porter (1994): divine versus intuitive. The curved lines are the logit AQRE and the dots indicate data averages.

message 1, and choosing action 1 in response to both nonequilibrium messages. Of the 33 deviations from the intuitive equilibrium predicted strategies, most were deviations by type 1 senders, who are predicted by the AQRE model to send the intuitive message approximately half the time. (As it turns out, type 1 senders were observed to send the intuitive message exactly half the time, 20/40.)

In game 4, the predicted (i.e., divine) equilibrium has both agents sending message 3, with the receiver choosing action 1 in response to the equilibrium message 3, and action 2 otherwise. In AQRE, both sender types are predicted to send the divine equilibrium message only about half the time. Aggregating both types of senders, this is what we observe in the data, where message 3 is sent 28 times out of 60 chances; message 1 is predicted to be sent slightly more than a third of the time, and we observe it sent 23 out of 60 times. The frequency of actions in response to each message is predicted even more accurately (see table 7.6).

TABLE 7.6. Experimental results and logit AQRE fit for the games considered in Banks, Camerer, and Porter (1994).

		BCP game 2			BCP game 3			BCP game 4		
		n	f_i	QRE	n	f_i	QRE	n	f_i	QRE
	m_1	7	0.184	0.103	20	0.500	0.497	2	0.067	0.274
t_1	m_2	0	0.000	0.021	14	0.350	0.339	9	0.300	0.221
	m_3	31	0.816	0.875	6	0.150	0.165	19	0.633	0.504
	m_1	11	0.324	0.176	45	0.900	0.927	21	0.700	0.422
t_2	m_2	0	0.000	0.026	4	0.080	0.057	0	0.000	0.032
	m_3	23	0.677	0.798	1	0.020	0.016	9	0.300	0.546
	a_1	0	0.000	0.049	4	0.062	0.153	9	0.391	0.298
m_1	a_2	18	1.000	0.837	61	0.939	0.704	13	0.565	0.593
	a_3	0	0.000	0.113	0	0.000	0.142	1	0.044	0.109
	a_1	0		0.184	14	0.778	0.686	0	0.000	0.044
m_2	a_2	0		0.797	0	0.000	0.174	4	0.444	0.647
	a_3	0		0.019	4	0.222	0.140	5	0.556	0.308
	a_1	53	0.841	0.726	7	1.000	0.999	23	0.821	0.704
m_3	a_2	0	0.000	0.026	0	0.000	0.001	5	0.179	0.235
	a_3	10	0.159	0.248	0	0.000	0.000	0	0.000	0.062
	λ			2.249			1.598			1.193

7.4.2 Brandts and Holt (1992, 1993)

As a follow-up to an earlier, unpublished version of Banks, Camerer, and Porter (1994), Brandts and Holt (1992) proposed two alternatives to game 3 (which differentiated the intuitive equilibrium from a nonintuitive sequential equilibrium), which they ran using similar [20] procedures. The games, which differ slightly from those studied by Banks, Camerer, and Porter (1994) because the sender has only two available messages, are given in table 7.7. The two games each have a sequential and an intuitive equilibrium at the same strategies. In each of the Brandts and Holt games, 4 trials were conducted, each with 12 subjects, matched in pairs over a sequence of 6 periods. Each subject participated in 2 such sequences, once as a sender, and once as a receiver. For details see Brandts and Holt (1992).

The point of the Brandts and Holt experiment was to attempt to construct a game in which play might conceivably converge to the *less* refined equilibrium.

---

[20] There were some minor procedural differences. Brandts and Holt successfully replicated the results of game 3 in Banks, Camerer, and Porter (1994).

TABLE 7.7. Signaling games considered in Brandts and Holt (1993).

BH # 3 (sequential versus intuitive)

$m = I$	C	D^n	E	$m = S$	C	D	E
A	45, 30	15, 0	30, 15	A	30, 90	0, 15	45, 15
B	30, 30	0, 45	30, 15	B	45, 0	15, 30	30, 15

sequential: (S, D, C) intuitive: (I, C, D)

BH # 4 (sequential versus intuitive)

$m = I$	C	D^n	E	$m = S$	C	D	E
A	30, 30	0, 0	50, 35	A	45, 90	15, 15	100, 30
B	30, 30	30, 45	30, 0	B	45, 0	0, 30	0, 15

sequential: (S, D, C) intuitive: (I, C, D)

Brandts and Holt proposed a descriptive story of how a particular dynamic of play in early rounds could potentially lead to the nonintuitive equilibrium if the payoffs in game 3 of Banks, Camerer, and Porter (1994) were changed slightly. Their learning dynamic is related to the intuition behind the logit AQRE selection; a natural intuition for the logit solution is that if subjects begin an experiment at a QRE with relatively high error rates (i.e., low value of λ) and "follow" the equilibrium selection as λ increases with experience, then they should converge along the principal branch of the equilibrium correspondence.

The data, along with logit AQRE estimates, are given in table 7.8 and figures 7.9 and 7.10. Table 7.8 presents results of a maximum-likelihood estimation of the logit AQRE model, and the observed frequencies for the later periods in the experiment (periods 4–6). Figures 7.9 and 7.10 present graphically the results of estimation with a breakdown by period. These figures are similar to those of the Banks, Camerer, and Porter (1994) experiments for the receiver. However, the sender in the Brandts and Holt experiments has just 2 messages. Therefore we graph the sender data in a unit square, with type A's frequency of choosing I on the horizontal axis and type B's on the vertical axis. The curves represent the predicted move frequencies for the logit AQRE selection, as a function of λ. As in figures 7.6–7.8, the QRE correspondence starts at the centroid (the center of the square and the center of the simplices) for low values of λ, and converges to the logit solution, which is the intuitive equilibrium for game 3 and the (nonintuitive) sequential equilibrium for game 4, as $\lambda \to \infty$. There is an overall trend in the expected direction of increasing λ, as indicated by comparing data from the first three periods (open circles) to the last three periods (solid circles).

TABLE 7.8. Data and estimates for the Brandts and Holt experiments. Periods 4–6.

| | | BH game 3 | | | BH game 4 | | |
		n	f_i	QRE	n	f_i	QRE
$t = A$	$m = I$	32	0.970	0.946	0	0.000	0.026
	$m = S$	1	0.030	0.054	38	1.000	0.974
$t = B$	$m = I$	18	0.462	0.601	15	0.441	0.280
	$m = S$	21	0.539	0.399	19	0.559	0.720
	$a = C$	48	0.960	0.686	1	0.067	0.252
$m = I$	$a = D$	2	0.040	0.178	13	0.867	0.729
	$a = E$	0	0.000	0.136	1	0.067	0.019
	$a = C$	4	0.182	0.109	52	0.912	0.888
$m = S$	$a = D$	18	0.818	0.719	2	0.035	0.050
	$a = E$	0	0.000	0.173	3	0.053	0.062
	λ			0.108			0.095

7.5 SOCIAL LEARNING

In social learning games, players learn about an unknown state of the world through a combination of observing the actions of others and observing private signals that are informative about the true state of the world. The simple canonical version of the game is due to Bikhchandani, Hirschleifer, and Welch (1992): there are two states of the world, private signals are binary, actions are binary, and payoffs are either 1 or 0, depending on whether or not an individual chooses the best action for the true state. Each individual makes a single action choice, in sequence, and observes a private signal as well as all previously chosen actions prior to choosing an action.[21]

The Bikhchandani, Hirschleifer, and Welch (1992) model has been used to understand fads, fashions, rumors, and other phenomena that seem to be driven by individuals who copy the behavior of other people. *Herding phenomena* like this are, in the context of social learning models, understood as resulting from an *information cascade*, whereby, over time, the cascade of information from the accumulated observation of many others' actions trump any single individual's own private information, leading individuals to eventually just "follow the crowd."

---

[21] This section contains material from Goeree, Palfrey, and Rogers (2006) and Goeree et al. (2007).

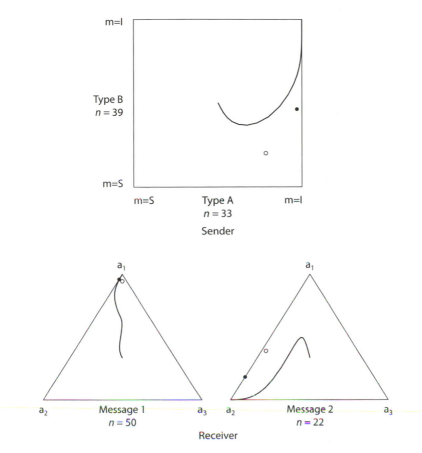

FIGURE 7.9. Game 3 in Brandts and Holt (1993): intuitive versus sequential. The curved lines are the logit AQRE and the dots indicate data averages: first three periods (open circles) and last three periods (solid circles).

7.5.1 The Basic Model

There is a set $I = \{1, \ldots, T\}$ of agents who sequentially choose between one of two alternatives, A and B. Agent $t \in I$ chooses at time t, and $c_t \in \{\alpha, \beta\}$ denotes agent t's choice. There are two states of the world denoted by $\omega \in \{A, B\}$, and the state is unknown to the agents, who have common prior beliefs that the two states are equally likely. Further, each individual t receives one conditionally independent private signal $s_t \in \{a, b\}$ that is informative about the state of the world. If $\omega = A$ then $s_t = a$ with probability $q \in (\frac{1}{2}, 1)$ and $s_t = b$ with probability $1 - q$. Likewise, when $\omega = B$, then $s_t = b$ with probability q and $s_t = a$ with probability $1 - q$. Payoffs are such that an agent receives a payoff

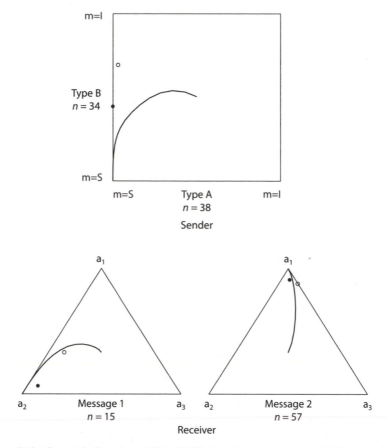

FIGURE 7.10. Game 4 in Brandts and Holt (1993): intuitive versus sequential. The curved lines are the logit AQRE and the dots indicate data averages: first three periods (open circles) and last three periods (solid circles).

of 1 if their choice is α in state A, or their choice is β in state B. Otherwise the agent's payoff is 0.[22]

The interest in social learning models is to understand how social learning affects the evolution of agents' beliefs, and how these beliefs coevolve with choices. Agent t observes the choices of all her predecessors, $\{c_1, \ldots, c_{t-1}\}$ but not their signals. Thus a history H_t for agent t is simply a sequence $\{c_1, \ldots, c_{t-1}\}$ of choices by agents $1, \ldots, t-1$, with $H_1 = \emptyset$. Agents care about the history

---

[22] There are many variations on this simple model. For example, there can be more than two states, more than two actions, the signal structure can be more complicated, payoffs can depend on other agents' action choices, and communication can be added.

only to the extent that it is informative about which alternative is correct. So let $p_t \equiv \text{Prob}[\omega = A|H_t]$ denote the (common knowledge) posterior belief that the state is A given the choice history H_t, with $p_1 \equiv \frac{1}{2}$, the initial prior. Agent t's private posterior beliefs given the public beliefs p_t and given her signal s_t is obtained by applying Bayes' rule. If $s_t = a$, agent t believes that alternative A is correct with probability

$$\pi_t^a(p_t) \equiv \text{Prob}[\omega = A|H_t, s_t = a] = \frac{q\, p_t}{q\, p_t + (1-q)(1-p_t)}. \tag{7.5}$$

Likewise,

$$\pi_t^b(p_t) \equiv \text{Prob}[\omega = A|H_t, s_t = b] = \frac{(1-q)p_t}{(1-q)p_t + q(1-p_t)} \tag{7.6}$$

is the probability with which agent t believes that A is the true state if her private signal is $s_t = b$. Notice that $\pi_t^a(p_t) > p_t > \pi_t^b(p_t)$ for all $0 < p_t < 1$.

7.5.2 Nash Equilibrium

The unique trembling-hand perfect equilibrium of the game, identified by Bikhchandani, Hirshleifer, and Welch (1992), involves rapid convergence to an information cascade, as follows. The first agent chooses α if $s_1 = a$, and chooses β if $s_1 = b$, so that her choice perfectly reveals her signal. If the second agent's signal agrees with the first agent's choice, the second agent chooses the same alternative, which is strictly optimal. On the other hand, if the second agent's signal disagrees with the first agent's choice, the second agent is indifferent, as she effectively has a sample of one a and one b. We assume all agents follow their private signal when indifferent.[23] The third agent faces two possible situations: (i) the choices of the first two agents coincide, or (ii) the first two choices differ. In case (i), it is strictly optimal for the third agent to make the same choice as her predecessors, even if her signal is contrary. Thus her choice imparts no information to her successors. The fourth agent is then in the same situation as the third, and so also makes the same choice, a process which continues indefinitely, so that all agents herd on the action chosen by the first two players. In case (ii), however, the choices of the first two agents reveal that they have received one a signal and one b signal, leaving the third agent in effectively the same position as the first. Her posterior (before considering her private information) is $p_3 = \frac{1}{2}$, so that her signal completely determines her choice. The fourth agent would then be in the same situation as the second agent described above, and so forth. Thus a cascade begins after some even number of agents have chosen and

---

[23] There are other Nash equilibria where players randomize with different probabilities when indifferent, all eventually resulting in information cascades.

$|\#\alpha - \#\beta| = 2$, where $\#\alpha$ (respectively, $\#\beta$) is the number of agents who have chosen α (respectively, β).

One quantity of interest is the probability that "correct" and "incorrect" cascades will form in equilibrium. First, the probability of being in neither cascade vanishes rapidly as t grows. The probability of eventually reaching a correct cascade for large t is

$$\frac{q(1+q)}{2 - 2q(1-q)},$$

and the complementary probability of eventually reaching an incorrect cascade is[24]

$$\frac{(q-2)(q-1)}{2 - 2q(1-q)}.$$

Once a cascade forms, all choices occur independently of private information, so public beliefs that the true state is A remain unchanged at either $p_t = 1/(1 + (1/q - 1)^2)$ if $\#\alpha > \#\beta + 1$ or $p_t = 1/(1 + (1/(1-q) - 1)^2)$ if $\#\beta > \#\alpha + 1$.

7.5.3 AQRE in the Information Cascade Game

In a Nash equilibrium, beliefs get "stuck" because after some point each individual's action is completely uninformative about that agent's private information, so information aggregation is never achieved. This is no longer the case in a quantal response equilibrium, because following *any* history, a player is *always* more likely to choose action α after an a signal than after a b signal, and furthermore, both actions occur with positive probability at any history. Thus, beliefs never get stuck and $p_t \neq p_{t+1}$ for all t. The logit AQRE for this game is very simple to characterize. The expected payoff of choosing α at history H_t if agent t observes private signal s_t is $\pi_t^{s_t}(p_t)$ and the expected payoff of choosing β is $1 - \pi_t^{s_t}(p_t)$. Thus given agent t's signal, the probability of choosing α is[25]

$$\sigma(H_t, s_t) = \frac{1}{1 + \exp(\lambda(1 - 2\pi_t^{s_t}(p_t)))}, \tag{7.7}$$

and β is chosen with the complementary probability $1 - \sigma(H_t, s_t)$. It is easy to show that as $\lambda \to \infty$ the logit QRE converges to the pure cascade Nash

---

[24] After the first two choices, the probabilities of a correct cascade, no cascade, or incorrect cascade are $\frac{1}{2}q(1+q)$, $q(1-q)$, and $\frac{1}{2}(q-2)(q-1)$, respectively. More generally, after $2t$ choices, these probabilities are $\frac{1}{2}q(1+q)(1 - (q(1-q))^t)/(1 - q(1-q))$, $(q(1-q))^t$, and $\frac{1}{2}(q-2)(q-1)(1 - (q(1-q))^t)/(1 - q(1-q))$. Taking limits as t approaches ∞ yields the long-run probabilities of the three regimes. Thus as q increases from 0.5 to 1, the probability of landing in a good cascade grows from 0.5 to 1.

[25] Indifference occurs with probability zero under the logit specification, and, hence, plays no role.

equilibrium in which indifferent subjects randomize uniformly.[26] On the other hand, as λ approaches 0 choices are independent of beliefs and become purely random.

To derive the evolution of the public belief that A is the true state, note that given p_t there are exactly two values that $p_{t+1} = \text{Prob}[\omega = A | H_t, c_t]$ can take depending on whether c_t is α or β. These are denoted p_t^+ and p_t^- respectively. The computation of the posterior probabilities p_t^+ and p_t^- given p_t is carried out by agents who do not know the true state, and so cannot condition their beliefs on that event. In contrast, the transition probabilities of going from p_t to p_t^+ or p_t^- (i.e., of a choice for α or β) depend on the objective probabilities of a and b signals as dictated by the true state. Thus when computing these transition probabilities, it is necessary to condition on the true state. Conditional on $\omega = A$, the transition probabilities are

$$T_t^{\omega=A} = \sigma(H_t, \omega = A)$$

$$= \sigma(H_t, s_t = a)\,\text{Prob}[s_t = a | \omega = A]$$

$$+ \sigma(H_t, s_t = b)\,\text{Prob}[s_t = b | \omega = A]$$

$$= \frac{q}{1 + \exp(\lambda(1 - 2\pi_t^a(p_t)))} + \frac{1-q}{1 + \exp(\lambda(1 - 2\pi_t^b(p_t)))}, \qquad (7.8)$$

with the probability of a β choice given by $1 - T_t^{\omega=A}$. Similarly, conditional on $\omega = B$, the probability agent t chooses α is

$$T_t^{\omega=B} = \frac{1-q}{1 + \exp(\lambda(1 - 2\pi_t^a(p_t)))} + \frac{q}{1 + \exp(\lambda(1 - 2\pi_t^b(p_t)))}. \qquad (7.9)$$

Using Bayes' rule, we now obtain the two values that p_{t+1} may take as

$$p_t^+ \equiv \text{Prob}[\omega = A | H_t, c_t = \alpha] = \frac{p_t T_t^{\omega=A}}{p_t T_t^{\omega=A} + (1 - p_t) T_t^{\omega=B}}, \qquad (7.10)$$

and

$$p_t^- \equiv \text{Prob}[\omega = A | H_t, c_t = \beta]$$

$$= \frac{p_t(1 - T_t^{\omega=A})}{p_t(1 - T_t^{\omega=A}) + (1 - p_t)(1 - T_t^{\omega=B})}. \qquad (7.11)$$

These expressions can be used to derive the following properties of the belief dynamics, where without loss of generality we assume the true state is $\omega = A$.

---

[26] This is because for any $\lambda \in (0, \infty)$, an agent chooses equiprobably when indifferent.

The key result is that for *every* value of λ, p_t converges to the truth. That is, if the true state is A, then the public belief that the state is A converges to 1. If the true state is B, then the public belief that the state is A converges to 0.

> **Proposition 7.1:** *For all $\lambda > 0$ there is a unique logit QRE with the following properties:*
>
> (i) *Beliefs are interior:* $p_t \in (0, 1)$ *for all* $1 \leq t \leq T$.
> (ii) *Actions are informative:* $p_t^- < p_t < p_t^+$ *for all* $1 \leq t \leq T$.
> (iii) *Beliefs about the true state rise on average:* $E(p_{t+1}|p_t, \omega = A) > p_t$ *for all* $1 \leq t, t + 1 \leq T$.
> (iv) *Beliefs converge to the truth: conditional on* $\omega = A$, $\lim_{t\to\infty} p_t = 1$ *almost surely.*

The intuition for the full information aggregation result (iv) is as follows. First, for all possible histories and signal realizations, both options are chosen with strictly positive probability. This implies there are no herds and learning never stops. Second, choice probabilities are increasing in expected payoffs so actions reveal some (private) information. Hence, for any belief, a decision maker is more likely to choose α with an a signal than with a b signal. As a result, more and more information gets revealed over time leading to full information aggregation in the limit.[27]

Finally, note that, in QRE, beliefs converge to the true state, but actions do not converge to the optimal action for the true state because action choice is still stochastic. Thus, while QRE unambiguously leads to full informational efficiency in the limit, it is not true that it leads to first best outcomes in the limit. Of course, for sufficiently high values of λ, action choices will be approximately optimal.

7.5.4 Belief Dynamics Implied by QRE

The evolution of beliefs, and the effect on it of the exact sequencing of action choices, has some interesting features in the logit QRE model. One interesting feature is that information cascades are "self-correcting," in the sense that beliefs can be very wrong for long periods of time, then these wrong beliefs at some

---

[27] This result generalizes to any finite set of states, signals, and actions, and more general preferences, in the following way. Denote the set of states by $K = \{1, \ldots, K\}$, the set of actions by $A = \{1, \ldots, A\}$, and the set of signals by $S = \{1, \ldots, S\}$. Signals are conditionally independent, and $q(s|k)$ denotes the probability an agent receives signal s in state k, so this defines an $S \times K$ matrix Q that fully specifies the information structure of the game. Common-value payoffs are given by an $A \times K$ matrix Π, with elements $\pi_{ak} \in (0, 1)$ specifying the payoff of choosing action a in state k. It is assumed that signals and payoffs are *admissible*, that is, signals are informative and nonredundant, and the payoff change from switching between two actions differs across states. Proposition 7.1 generalizes to such environments; see Goeree, Palfrey, and Rogers (2006) for details.

point become corrected via a surprisingly short sequence of actions that would be suboptimal in the state for which public beliefs are (incorrectly) close to 1. In fact, when λ is very large, then the length of time that these wrong beliefs can persist is much longer. Because following the cascade is optimal, players who nearly optimize (λ large) will follow the cascade with very high probability, regardless of the signal they receive, while players whose choice behavior is subject to more randomness (low λ) are relatively likely to choose actions against the herd when they receive signals that are contrary to the current public belief. The fact that false cascades that have persisted for hundreds of periods can be reversed by a relatively short sequence of actions is due to the strong updating that occurs following highly unexpected action choices. The information contained in such unexpected actions is equivalent to a large number of the more recent very uninformative action choices that resulted mainly from agents following the herd.

As an illustrative example, the upper panel of figure 7.11 displays the belief evolution from 10 specific sequences of signal draws over 2000 periods, where there are two ex ante equally likely states, A and B, and the signal precision is $q = \frac{3}{4}$. Agents receive 1 if they choose α in state A and β in state B and they receive an *additional* 2 if they choose β. In other words, choosing β results in a payoff of 2 when A is the true state and it results in a payoff of 3 when B is the true state. Hence, irrespective of the true state, the payoff from choosing β is strictly higher than that of choosing α.

Nevertheless, if the true state is $\omega = A$, beliefs will converge with probability 1 to the true state, that is, $\lim_{t \to \infty} P_t^{\omega = A} = 1$. The lower panel of figure 7.11 indicates that the long-run frequency of α choices is less than $\frac{1}{2}$ because in state A, the payoff of choosing β is 2 while the payoff of choosing α is 1. Hence, the long-run average will converge to $\lim_{t \to \infty} f_t^\alpha = \frac{1}{1+e} \approx 0.27$. However, the many repeated action choices of β provide very little information about the state, since β is the optimal choice in both states (and there is a cascade), while action choices of α are highly informative because they are much more likely to be the result of an agent receiving an a signal than a b signal. As a result, beliefs converge to the true state, $\omega = A$, as shown in the upper panel of figure 7.11. Notice that convergence to full information aggregation is nonmonotonic, and that the corrections from incorrect to correct beliefs often happen in a narrow time window.

7.5.5 An Experiment on Self-Correcting Cascades

There have been a number of experiments that explore information cascades and herding behavior, mostly based on the Bikhchandani, Hirshleifer, and Welch (1992) model, with binary states, actions, and signals, and a symmetric information and payoff structure. Most experiments use relatively short sequences, for example, $T = 6$, which precludes a full analysis of cascades forming, breaking,

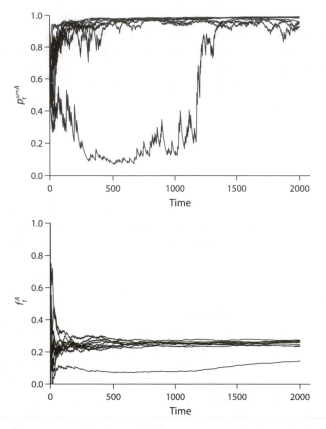

FIGURE 7.11. The asymmetry in private valuations causes most choices to be for alternative 2 (lower panel). The few choices for alternative 1, however, are sufficient to eventually tilt beliefs to the correct alternative (upper panel).

forming again, etc.[28] To test the logit AQRE prediction that beliefs never get stuck but eventually converge to the true state, Goeree et al. (2007) report results from an experiment that uses much longer sequences. The Goeree et al. (2007) experiment is based on a simple two-by-two design (for the standard setup of Bikhchandani, Hirshleifer, and Welch, 1992) that varies the sequence length ($T = 20, 40$) as well as the signal quality ($q = \frac{5}{9}, q = \frac{6}{9}$).[29]

[28] Anderson and Holt (1997) report the first such experiment. They also provide estimates of a logit error parameter for a quantal response equilibrium, based on results from the Anderson (1994) dissertation. See also Anderson (2001) for an application of logit AQRE to the analysis of data from a six-move information cascade experiment.

[29] A total of 206 games were run for $T = 20$ and 116 games were run for $T = 40$. See Goeree et al. (2007) for details.

The logit AQRE belief dynamics shown in figure 7.11 exhibit several proper-ties, which reflect two important characteristics: cascades do not last forever, and over time, choice behavior can herd on different actions. To describe these, it is useful to identify different kinds of information cascades that are observed in the experiment. First, using Nash equilibrium as a benchmark, we define a cascade as *forming* at period t, if the number of α actions up to t exceeds the number of β actions by at least 2 (or vice versa). A *pure cascade* is one that forms initially and is persistent to the end of the experiment. A *temporary cascade* is one that forms at some point, but is broken before the end of the experiment (i.e., some subject chooses an action opposite to the herd). The *length* of a cascade is the number of periods before it is broken. A *correct (respectively, incorrect) cascade* is one where players herd on the correct (respectively, incorrect) action. A cascade is said to *reverse* if it is followed by a cascade on the opposite action. An incorrect cascade is said to *self-correct* if it is followed by a correct cascade.

Logit AQRE makes the following predictions with regards to the *frequency* and *length* of cascades:[30]

(C1) For any q and sufficiently large T, the probability of observing a pure cascade is decreasing in T, converging to 0 in the limit. For any q and $T > 2$, the probability of observing a temporary cascade is increasing in T, converging to 1 in the limit.

(C2) For any T, the probability of a pure cascade is increasing in q.

(C3) The expected number of cascades is increasing in T.

(C4) The expected number of cascades is decreasing in q.

(C5) The probability that a cascade, which has already lasted τ periods, will break in the next period is decreasing in τ.

(C6) The average length of cascades is increasing in T and q.

Logit AQRE also predicts that cascades *self-correct*, which is essential for information aggregation to occur:

(SC1) Incorrect cascades are shorter on average than correct cascades.

(SC2) Incorrect cascades are more likely to reverse than correct cascades.

(SC3) Correct cascades are more likely to repeat than incorrect cascades.

(SC4) Later cascades are more likely to be correct than earlier ones.

The first finding in Goeree et al. (2007) is that pure cascades almost never happen in the longer sequences; see table 7.9. Pure cascades were observed in only 17

---

[30] Several of these hypotheses are sensible only if T is sufficiently large. At least 2 periods are required for any cascade to form, and at least 6 periods are required to observe a cascade and its reversal. For example, $\{\alpha, \alpha, \beta, \beta, \beta, \beta\}$ is the shortest possible sequence for a reverse from an A cascade to a B cascade, and $\{\alpha, \alpha, \beta, \alpha\}$ is the shortest possible sequence for a repeat A cascade.

TABLE 7.9. Frequency of pure and temporary cascades. (Source: Goeree et al., 2007.)

	$q = \frac{5}{9}$	$q = \frac{5}{9}$	$q = \frac{6}{9}$	$q = \frac{6}{9}$
	$T = 20$	$T = 40$	$T = 20$	$T = 40$
# sequences	116	56	90	60
# sequences with pure cascades	5	0	12	8
# sequences without cascades	0	0	0	0
# sequences with temporary cascades	111	56	78	52

TABLE 7.10. The top panel shows the frequency of pure and temporary cascades, and the bottom panel shows the frequency of repeat and reverse cascades.
(Source: Goeree et al., 2007.)

		$q = \frac{5}{9}$		$q = \frac{6}{9}$	
		$T = 20$	$T = 40$	$T = 20$	$T = 40$
		$M = 116$	$M = 56$	$M = 90$	$M = 60$
average number cascades	Data	3.47	7.54	2.99	3.73
	AQRE-BRF	3.85	7.31	2.81	4.19
	Nash	1.00	1.00	1.00	1.00
average length cascades	Data	2.43	2.00	3.27	7.83
	AQRE-BRF	1.59	2.04	3.79	6.50
	Nash	17.32	37.25	17.43	37.44
average number repeat cascades	Data	2.14	5.98	1.69	2.60
	AQRE-BRF	2.39	5.37	1.62	2.91
	Nash	0.00	0.00	0.00	0.00
average number reverse cascades	Data	0.34	0.55	0.30	0.13
	AQRE-BRF	0.50	0.93	0.26	0.30
	Nash	0.00	0.00	0.00	0.00

out of 206 sequences with $T = 20$ decision makers, and only 8 of 116 sequences with $T = 40$ decision makers. The Nash equilibrium probability of a pure cascade with T decision makers is $1 - (q(1 - q))^{T/2}$, so the data clearly contradict the standard Nash predictions. Also, the frequency of pure cascades in the data is sharply increasing in q, while the Nash equilibrium predicts essentially no effect. In contrast, temporary cascades are the norm in all treatments. Table 7.9 also shows the frequency of temporary cascades in our data: essentially all cascades observed were temporary.

Multiple temporary cascades were observed in most games. The top panel of table 7.10 displays the average number of cascades in each treatment.

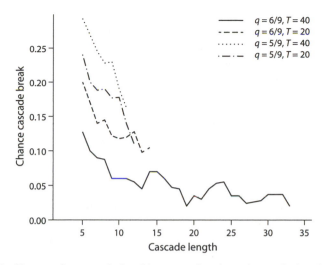

FIGURE 7.12. Chance of a cascade breaking as a function of cascade length. The lines show five-period moving averages of the probability of a break in each of the treatments. (Source: Goeree et al., 2007.)

The number of temporary cascades rises with the sequence length, T, and falls with the signal precision, q, not only on average. The table also shows the contrasting Nash prediction of exactly 1 cascade per sequence, independent of q and T, and predictions generated by the AQRE-BRF model that is discussed below.[31] The top panel also displays the average length of cascades by treatment, which increases with q, as predicted by logit AQRE.

The bottom panel of table 7.10 shows the average number of repeat and reverse cascades per sequence, by treatment, and also gives theoretical expectations according to the Nash and out-of-sample AQRE-BRF predictions. The number of repeat cascades is increasing in T and decreasing in q, which is consistent with the logit AQRE model.

Figure 7.12 graphs, separately for each treatment, the empirical probability of collapse as a function of the duration of the cascade, that is, the probability of a collapse in period $t + s$, given the cascade started in period t. This probability is sharply decreasing in s. In other words, longer cascades are more stable (see also Kübler and Weizsäcker, 2005), which is predicted by logit AQRE but is not true in the Nash equilibrium.

---

[31] The AQRE-BRF (base-rate fallacy) model is a two-parameter model: one parameter, λ, is the logit response parameter, and the other parameter, ρ, measures the degree to which individuals are prone to base-rate neglect.

TABLE 7.11. Transitions between correct and incorrect cascades.
(Source: Goeree et al., 2007.)

		$T = 20$			$T = 40$	
	From/To	Correct	Incorrect	From/To	Correct	Incorrect
$q = \frac{5}{9}$	Correct	92.7%	7.3%	Correct	93.6%	6.4%
	Incorrect	22.7%	77.3%	Incorrect	11.0%	89.0%
# Correct	Total = 252	Start = 65	Final = 78	Total = 237	Start = 29	Final = 34
# Incorrect	Total = 151	Start = 51	Final = 38	Total = 185	Start = 27	Final = 22

	From/To	Correct	Incorrect	From/To	Correct	Incorrect
$q = \frac{6}{9}$	Correct	91.4%	8.6%	Correct	98.7%	1.3%
	Incorrect	30.5%	69.5%	Incorrect	20.0%	80.0%
# Correct	Total = 197	Start = 66	Final = 71	Total = 186	Start = 48	Final = 52
# Incorrect	Total = 72	Start = 24	Final = 19	Total = 38	Start = 12	Final = 8

Table 7.11 shows how frequently correct and incorrect cascades repeat or reverse themselves. Averaging over the four treatments shows that when a correct cascade breaks, it reverses to an incorrect one only 6% of the time, while an incorrect cascade that breaks leads to a self-corrected cascade more than 21% of the time.

To summarize, all of the logit AQRE predictions regarding features of temporary cascades, (C1)–(C6), and their self-correcting nature, (SC1)–(SC4), are confirmed by the Goeree et al. (2007) data. We next explore how well logit AQRE performs at the individual choice level and compare it with a model that allows for non-Bayesian updating.

7.5.6 Estimation

We start by describing the estimation procedure for the basic logit AQRE model, where the only parameter is the precision λ. Since this precision parameter affects choices, it also affects public beliefs. The evolution of public beliefs can be solved recursively—see (7.10) and (7.11)—so we can write the public belief as $p_t(\lambda, H_t)$. Given $\{\lambda, s_t, H_t\}$, the probability of observing agent t choosing α is given in (7.7):

$$\text{Prob}[c_t = \alpha | \lambda, s_t, H_t] = \frac{1}{1 + \exp(\lambda(1 - 2\pi_t^{s_t}(p_t(\lambda, H_t))))},$$

and $\text{Prob}[c_t = \beta | \lambda, s_t, H_t] = 1 - \text{Prob}[c_t = \alpha | \lambda, s_t, H_t]$. Therefore, the likelihood of a particular sequence of choices, $c = (c_1, \ldots, c_T)$, given the sequence of signals is simply

$$l(\lambda, c) = \prod_{t=1}^{T} \text{Prob}[c_t | \lambda, s_t, H_t].$$

Finally, assuming independence across sequences, the likelihood of observing a set of M sequences $\{c^1, \ldots, c^M\}$ is just

$$L(\lambda, c^1, \ldots, c^M) = \prod_{m=1}^{M} l(\lambda, c^m).$$

We include a detailed MATLAB estimation program using the notation of chapter 6.

```
1.  function logL = loglik(lambda,data,T)
2.  M = length(data)/T; p=1/2; q = 5/9; logL = 0;
3.  for m = 1:M
4.  for t = 1:T
5.  pi_a = q*p/(q*p+(1-q)*(1-p));
6.  pi_b = (1-q)*p/((1-q)*p+q*(1-p));
7.  P_A_a = 1/(1+exp(lambda*(1-2*pi_a))); P_B_a = 1-P_A_a;
8.  P_A_b = 1/(1+exp(lambda*(1-2*pi_b))); P_B_b = 1-P_A_b;
9.  p_plus = (p*q*P_A_a+p*(1-q)*P_A_b)/...
       ((p*q+(1-p)*(1-q))*P_A_a +(p*(1-q)+(1-p)*q)*P_A_b);
10. p_minus = (p*q*P_B_a+p*(1-q)*P_B_b)/...
       ((p*q+(1-p)*(1-q))*P_B_a +(p*(1-q)+(1-p)*q)*P_B_b);
11. signal = data((m-1)*T+t,1);
12. choice = data((m-1)*T+t,2);
13. if signal==1 && choice==1
14. p = p_plus; logL = logL+log(P_A_a);
15. elseif signal==0 && choice==1
16. p = p_plus; logL = logL+log(P_A_b);
17. elseif signal==1 && choice==0;
18. p = p_minus; logL = logL+log(P_B_a);
19. elseif signal==0 && choice==0;
20. p = p_minus; logL = logL+log(P_B_b);
21. end
22. end
23. end
```

The program is simple because the information cascade game is essentially an individual decision-making problem (with informational externalities) and no fixed-point equation needs to be solved to derive QRE probabilities. The MATLAB command to call the program is similar to the one used in chapter 6, except that the function to be minimized has more inputs now so the variable over which it is to be minimized has to be specified (with @(lambda)).

TABLE 7.12. Estimation results for the logit AQRE model with and without including the base-rate fallacy (BRF).

		$q = \frac{5}{9}$		$q = \frac{6}{9}$		*Pooled*
		$T = 20$	$T = 40$	$T = 20$	$T = 40$	
		(2320)	(2240)	(1800)	(2400)	(8760)
Logit AQRE	λ	11.36	7.19	4.38	4.69	6.12
	logL	-981.0	-1181.4	-682.0	-634.0	-3650.3
Logit AQRE-BRF	λ	7.07	3.68	3.47	4.09	4.23
	ρ	2.33	2.97	2.01	1.67	2.46
	logL	-947.7	-1156.3	-660.8	-627.9	-3466.0

```
>>[beta, fval, flags, output, gradient, hessian]...
= fminunc(@(lambda) -loglik(lambda,data,T),6)
```

where the final entry, "6" is a starting value. The results of this program applied to the Goeree et al. (2007) data are shown in the top panel of table 7.12.

Since comparison with Nash equilibrium does not provide a particularly informative benchmark for logit AQRE, we also consider an alternative to the basic model. This allows us to access the extent to which choice behavior in the data is explained by quantal-response decision errors as opposed to non-Bayesian updating, since there is considerable evidence in the literature on cascade experiments that players are non-Bayesians. A particularly prevalent judgment bias is the *base-rate fallacy* (BRF). In the context of our social learning model, the base-rate fallacy would imply that agents weight the public prior too little relative to their own signal. Because past experiments have been suggestive of these effects, we construct an analytical model of this and estimate it using the error structure of logit AQRE.

We formalize this idea as a non-Bayesian updating process in which the private signal is counted by the decision maker as ρ signals, where $\rho \in (0, \infty)$.[32] Rational agents correspond to $\rho = 1$, while agents have progressively more severe base-rate fallacies as ρ increases above 1.[33] While agents overweight their private signals we retain the assumption that they have rational expectations about others' behavior. This implicitly assumes that ρ is common knowledge (as well as λ).

---

[32] This could also be loosely interpreted as a parametric model of "overconfidence" bias.

[33] Values of $\rho < 1$ correspond to underweighting the signal, or "conservatism" bias. Although this latter kind of bias has less support in the experimental literature, it is sufficiently plausible that we choose not to assume it away.

The updating rules in (7.5) and (7.6) now become

$$\pi_t^a(p_t|\rho) = \frac{q^\rho p_t}{q^\rho p_t + (1-q)^\rho(1-p_t)} \tag{7.12}$$

and

$$\pi_t^b(p_t|\rho) = \frac{(1-q)^\rho p_t}{(1-q)^\rho p_t + q^\rho(1-p_t)}. \tag{7.13}$$

Here, the public belief, p_t, is derived recursively using (7.10) and (7.11). In particular, this means that subjects not only overweight signals, but also take into account that other subjects overweight signals, and the public belief is updated accordingly. Thus, for $\rho > 1$, the public belief is updated more quickly than in the pure Bayesian model.

There is good reason to think this model may better describe some features of the data. First, when $\rho = 1$, AQRE predicts that indifferent agents randomize uniformly. But in the data, 85% of indifferent subjects follow their signals, which is consistent with $\rho > 1$. Second, when $\rho > 1$, cascades take longer to start.[34] The base-rate fallacy therefore provides one possible explanation for the prevalence of length-zero temporary cascades in our data set.

The estimation results for the AQRE-BRF model are reported in the bottom panel of table 7.12. For all treatments, the base-rate fallacy parameter, ρ, is significantly greater than 1. To test for significance we can simply compare the log-likelihood of the AQRE-BRF model to that of the constrained model (with $\rho = 1$) in the top panel. Obviously, the base-rate fallacy parameter is highly significant.[35] Furthermore, the constrained model yields a significantly (at the 0.01 level) higher estimate of λ for all treatments.

To summarize, the data from the Goeree et al. (2007) experiment show clear evidence for noisy decision making and overweighting of own information. Logit AQRE provides a convenient structural econometric model to estimate these two factors. As shown by the "AQRE-BRF" rows in table 7.10, the estimated model captures all the essential features of the temporary cascades observed in the experiment.

---

[34] For example, after two α choices the third decision maker need not choose α if she sufficiently overweights her b signal.

[35] For the pooled data the difference in log-likelihoods is nearly 200. A simple t-test also rejects the hypothesis that $\rho = 1$, with a t-statistic of 14.6.

8

Applications to Political Science

In many strategic situations, players must choose between two decisions, for example, whether or not to vote, volunteer, enter a market, or contribute to a public good. These binary-choice games are typically modeled in a very stylized way, often with sharp and surprising predictions. Kahneman (1988) reported an early laboratory experiment in which subjects had to choose whether or not to enter a market, where the observed number of entrants was on average approximately equal to a market capacity parameter. He remarked, "To a psychologist, it looks like magic."[1] This tendency for aggregate behavior to nearly equate expected payoff levels for entry and exit, as predicted in a Nash equilibrium, has been documented by others (e.g., Ochs, 1990; Meyer et al., 1992). Some subsequent experimental studies that put these games to a tougher test, however, report overentry and others report underentry, and some even find overentry relative to Nash predictions in some treatments and underentry in others. The pattern of over- and underentry varies systematically with the asymmetric losses associated with deviations from Nash equilibrium exactly in the direction implied by regular QRE. The addition of "noise" in these game-theoretic models of symmetric binary-choice games flattens the best-response functions in the relevant range, which generates equilibrium explanations of behavioral anomalies such as the divergence of results on whether there is excess entry or not.

This chapter considers a set of closely related binary-choice games that have been applied to model questions in political science. We start with an analysis of "participation games," which are n-player games where each player has only two possible actions: to participate or not. The payoff for either decision depends on the number of other players who make that decision. In some cases, a threshold level of participation is required for the group benefit to be obtained. The first example studied below, the "volunteer's dilemma," pertains to the special case where the threshold is 1, that is, only a single volunteer is needed. For example, it takes only one person to cast a politically costly veto or to sponsor a bill authorizing a pay raise for all members of a legislative body. The dilemma here is

---

[1] To a game theorist or economist, it looks like equilibrium.

that the act of volunteering is costly, and therefore, each person would prefer that someone else bears the cost.

In the volunteer's dilemma, one person's decision to volunteer imposes a positive externality on others. But in other situations the decision to participate imposes a negative externality on others. In a market-entry game, for example, the payoff for each entrant will generally be a decreasing function of the number of entrants. The general model of participation games is formulated in a manner that pertains to both types of externalities.

It is also possible that the decision to participate benefits some and harms others. In two-party elections, for instance, the decision to vote for one candidate benefits others who favor that candidate but not the supporters for the rival candidate. QRE explains some of the anomalous findings about turnout in "voting games," both when voting costs are commonly known and when there is incomplete information in the form of privately known voting costs.

A major question is whether elections aggregate all private information and result in welfare-maximizing outcomes. This question applies to elections where privately informed voters have opposing interests (e.g., voters have private values for either candidate winning), but also to voting in committees where everyone benefits from choosing the correct alternative, for example, acquitting an innocent defendant. The key result is that QRE overturns some of the stark and counterintuitive predictions that the Bayes–Nash equilibrium makes for such "jury" games.

The chapter ends with an analysis of bargaining situations, including an application of AQRE to bilateral "crisis bargaining" and of Markov QRE to the Baron–Ferejohn model of multilateral "legislative bargaining." Laboratory experiments that study these bargaining protocols often produce data patterns that differ systematically from standard Nash predictions, but are reasonably well explained by QRE, both in qualitative and quantitative terms.

8.1 PARTICIPATION GAMES

Example 8.1 (The volunteer's dilemma)*:* The volunteer's dilemma, introduced by Diekmann (1986), is a game in which each of the n players in a group receives a benefit, B, if at least one of them incurs a cost, $C < B$.[2] The expected payoff from volunteering is $B - C$, since a decision to volunteer ensures that the threshold is met. Next, consider the expected payoff from not

---

[2] This section includes material from Goeree and Holt (2005b) and Goeree, Holt and Smith (2015).

TABLE 8.1. Frequencies of volunteer decisions and no-volunteer outcomes
(Franzen, 1995).

Group size n	2	3	5	7	9	21	51	101
Frequency of volunteer decisions	0.65	0.58	0.43	0.25	0.35	0.30	0.20	0.35
Frequency of "no-volunteer" cases	0.12	0.07	0.06	0.13	0.02	0.00	0.00	0.00

volunteering in a symmetric equilibrium in which each of the $n - 1$ others volunteers with probability p. In this case, there is no cost, and there is a benefit B if at least one of the others volunteers, which is calculated as 1 minus the probability that none of them volunteer. Thus the expected payoff for not volunteering is $B(1 - (1 - p)^{n-1})$, and the expected payoff difference is $B(1 - p)^{n-1} - C$.

The symmetric Nash equilibrium is in mixed strategies and the Nash volunteer probability, p_N, can be found by equating the expected payoff difference to 0:

$$p_N = 1 - \left(\frac{C}{B}\right)^{1/(n-1)}. \tag{8.1}$$

This probability is decreasing in C and n, and increasing in B. These results are intuitive. In contrast, the probability that there will be no volunteer is $(1 - p_N)^n$, and it follows from (8.1) that the Nash equilibrium probability of no volunteer is $(C/B)^{n/(n-1)}$, which is an *increasing* function of the number of potential volunteers that limits to $C/B > 0$ as the number of potential volunteers goes to infinity. Intuitively, one might expect that with more potential volunteers it would be more likely to have *at least one* volunteer.

Table 8.1 shows the results of a one-shot volunteer's dilemma experiment, with $B = 100$ and $C = 50$ (Franzen, 1995). The volunteer rates in the middle row generally conform to the intuitive Nash predictions that an increase in n decreases the chances that any single person decides to volunteer. In contrast with the Nash prediction, however, the individual propensity to volunteer does not fall to zero even with as many as 101 possible volunteers. As a result, the frequency of a no-volunteer outcome, shown in the bottom row, declines to zero as n grows large, which is the opposite of the Nash prediction that this frequency should be increasing and converge to $C/B = 0.5$.

The Franzen results are qualitatively similar to data reported by Goeree, Holt, and Smith (2015) for a different parameterization of the volunteer's dilemma, with random matching, in a between-subjects design with groups of various sizes. The Nash volunteer rate predictions reported in the latter paper for groups of size 2, 3, 6, 9, and 12 are 0.75, 0.50, 0.24, 0.16, and 0.12, respectively. In contrast, the data show a sequence of volunteer rates that decline more slowly: 0.52, 0.42, 0.28, 0.19, and 0.19, respectively, which are pulled down toward 0.5 for group size 2 and are pulled up toward 0.5 for larger groups. This "flatness" causes the rates of no-volunteer outcomes to decline for large groups, again contradicting the nonintuitive Nash predictions.

The QRE model considered here is based on the distribution-function approach of section 2.2.2. The individual volunteer probability is a smoothly increasing function of the payoff difference between volunteering and not volunteering, that is,

$$p = F(\lambda(B(1-p)^{n-1} - C)), \tag{8.2}$$

where $F(\cdot)$ is some cumulative distribution function that is symmetric around 0, so $F(0) = 0.5$. The left-hand side of (8.2) is increasing in p while the right is decreasing, so the QRE volunteer probability is unique. Moreover, it is readily verified that the QRE volunteer probability falls with n, C, and rises with B,[3] as was the case for the Nash equilibrium. Interestingly, the QRE volunteer probability does *not* tend to 0 as the number of potential volunteers, n, grows large. Indeed, from (8.2) it follows that $\lim_{n\to\infty} p = 1 - F(\lambda C) > 0$. As a result, the probability that there will be no volunteer tends to 0. These results are intuitive: even when there are (many) other volunteers and the benefit of volunteering is 0, the QRE propensity to volunteer is $p = F(\lambda(0 - C)) = 1 - F(\lambda C)$. To summarize, QRE explains the main experimental findings of Franzen (1995); see table 8.1.

Proposition 8.1 (QRE properties for the volunteer's dilemma game): *In the unique QRE for the volunteer's dilemma, individual volunteer probabilities remain strictly positive in the limit when the number of potential volunteers grows large and, as a result, the probability that there will be no volunteer vanishes.*

8.1.1 General Participation Games

A mixed strategy in a participation game will be characterized by a probability, p, of the active participation strategy (e.g., entry, voting, volunteering, contributing), and $1 - p$ will denote the probability of the "exit-out" strategy. In a symmetric equilibrium in which all others participate with the same probability p, a player's expected payoff from participation in general can depend on both p and the number of others $(n - 1)$ and is assumed to be finite. When participation choices

---

[3] For example, $(\partial P/\partial B)(1 + (n-1)(1-p)^{n-2} Bf(B(1-p)^{n-1} - C)) = (1-p)^{n-1} f(B(1-p)^{n-1} - C)$, from which it follows that $\partial P/\partial B > 0$.

are independent, the expected payoff for participating can be calculated from a binomial distribution with parameters $n - 1$ and p, and will be denoted $U_1(n - 1, p)$. Note that this is the expected payoff for a player who participates (*with probability* 1) when all $n - 1$ others participate with probability p. Similarly, the expected payoff from exit, denoted $U_2(n - 1, p)$, represents the payoff from entry with probability 0 when the others enter with probability p.

As with the volunteer's dilemma game, it is convenient to use the distribution-function approach of section 2.2.2 to model the QRE probability of participation,

$$p = F\big(\lambda(U_1(n - 1, p) - U_2(n - 1, p))\big), \tag{8.3}$$

where $F(\cdot)$ is some cumulative distribution function that is symmetric around 0, so $F(0) = 0.5$. Equation (8.3) characterizes QRE as a fixed-point equation, since the probability used to calculate expected payoffs on the right matches the response probability on the left.

By applying the inverse of the distribution function to both sides of equation (8.3), one obtains an expression that separates the factors affecting the noise terms, on the left, from the expected payoff difference on the right:

$$\frac{F^{(-1)}(p)}{\lambda} = U_1(n - 1, p) - U_2(n - 1, p), \tag{8.4}$$

where the inverse distribution function, $F^{(-1)}(\cdot)$, is increasing on (0,1) with $F^{(-1)}(0.5) = 0$, as shown by the curved *inverse distribution line* in figure 8.1.

Equation (8.4) implies that the quantal response equilibrium occurs at the point where the expected payoff difference line crosses $F^{(-1)}(p)/\lambda$, the inverse distribution line, even though it includes the effects of both λ and the inverse of $F(\cdot)$. An increase in λ makes the inverse distribution line flatter in the center of figure 8.1, and conversely, a reduction in precision (more noise) makes the line steeper in the center.

8.1.2 Participation Games with Negative Externalities

There are two expected payoff difference lines in figure 8.1: a low payoff difference line on the left, and a higher payoff difference line, the relevant part of which is shown on the right. Both lines are continuously decreasing in p, as would be the case for games with congestion effects, that is, in which participation by one person decreases the participation payoffs for others. For the game with the lower payoff difference line, the expected payoff for participation is higher than for exit if the $n - 1$ others participate with a probability that is less than 0.2, and the expected payoff for exit is higher to the right of 0.2 in the figure. When $p = 0.2$, the expected payoffs are equal, and this point represents a Nash equilibrium in mixed strategies. The Nash equilibrium for the game with the

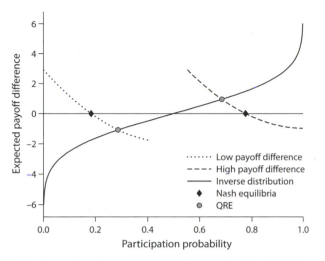

FIGURE 8.1. Equilibrium points for an entry game with negative externalities. (Key: Nash equilibria are diamonds, QREs are circles.)

higher payoff difference line is located at 0.8, as shown by the diamond mark on the right-hand side of the figure. The QRE for each case is marked with a large dot.

Since the inverse distribution line is continuously increasing with a range from $-\infty$ to ∞, and the expected payoff difference line is decreasing for games with negative externalities, a symmetric QRE will exist and it will be unique. Moreover, any factor that shifts the expected payoff difference line downward will lower the QRE participation probability. For example, in a market entry game, an increase in the number of potential entrants will raise the expected number of entrants for each given level of p, which will reduce the expected payoff from entry as a function of p. If the cost of entry is not affected, then it follows that an increase in the number of potential entrants will lower the QRE entry probability. The same logic can also be used to derive a similar prediction for the Nash probability of entry. One way in which the Nash and QRE predictions differ when the expected payoff difference is strictly decreasing in p is that the QRE prediction is always closer to 0.5, which reflects the fact that noise pulls the probabilities closer to the center, and this difference will be accentuated when there is more noise (lower λ), which makes the inverse distribution line steeper around 0.5.

The qualitative nature of the QRE prediction, that entry rates will be pulled toward 0.5 relative to the Nash prediction, is generally borne out in market entry experiments. First, note that if the expected payoff line crosses the zero point at $p = 0.5$, then it will also intersect the inverse distribution line at that point, so the Nash and quantal response equilibria coincide. Meyer et al. (1992) ran an experiment with groups of size 6 and parameters in their baseline treatment which caused the equilibrium number of entrants to be 3, so the Nash equilibrium probability of entry was 0.5. The average number of entrants was not statistically different from 3 for the baseline sessions, even when the game was repeated for 60 periods (see their tables 3 and 5). In contrast, results that deviate from the Nash prediction are reported in Camerer and Lovallo (1999). They ran an experiment in which subjects had to choose whether or not to enter a market with an integer-valued "capacity" of c. Entrants were randomly ranked and the top c entrants were given shares of a $50 pie, based on their rank. The exit payoff was set to 0. The capacities and group sizes (14–16) were selected so that the Nash entry probability was greater than or equal to 0.5 in all cases. Underentry was observed in all of the 8 baseline sessions, which resulted in positive expected payoffs from entry (see their table 4).[4] Some of the most striking evidence in support of the QRE prediction is found in Sundali, Rapoport, and Seale (1995). Their design used an entry game with 10 different capacity parameters, which resulted in Nash equilibrium entry rates at about 10 more or less equally spaced probabilities on the interval from 0 to 1. On average, there was overentry in all 5 cases with Nash predictions below 0.5, and there was underentry in 4 of the 5 cases with Nash predictions above 0.5, with the only exception being a treatment in which the Nash prediction was close to 0.5.

8.1.3 Participation Games with Positive Externalities

In many games of interest, participation by one player may tend to increase, not decrease, the participation payoff of others. For example, consider a production process in which individuals decide whether or not to help in a group production activity in which each person's productivity is an increasing function of the number of others who participate. The game can be structured so that it does not pay to participate unless enough others do so, and therefore, the expected payoff difference is negative for low p, positive for high p, and increasing in p. Figure 8.2 illustrates this case of positive externalities. Since both the expected payoff difference and the inverse distribution lines are increasing, it is possible to have multiple crossings, that is, multiple QRE. In this case, any parameter change that causes the expected payoff difference line to shift upward will raise

---

[4] Camerer and Lovallo (1999) also report a second treatment in which subjects were told that entrants would be ranked on the basis of performance on a sports trivia quiz. They observed overentry in this case, which they attributed to "overconfidence" and "reference group neglect," i.e., neglecting the fact that other participants might think in the same manner.

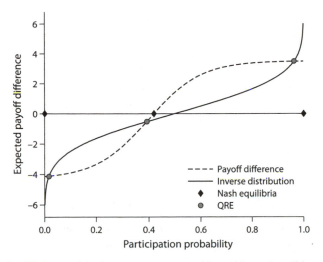

FIGURE 8.2. Equilibrium points for an entry game with positive externalities. (Key: Nash equilibria are diamonds, QREs are circles.)

the equilibrium participation probability for some of the equilibria and lower it for others.

The indeterminacy of comparative statics effects for participation games with positive externalities can be dealt with using an analysis of dynamic stability. Suppose that the participation probabilities tend to increase when the current participation probability, p, is less than the stochastic best response to p, that is, when $p < F(\lambda(U_1(n - 1, p) - U_2(n - 1, p)))$, or, equivalently, when $F^{(-1)}(p)/\lambda < U_1(n - 1, p) - U_2(n - 1, p)$. Thus the participation probability would tend to increase (respectively, decrease) when the inverse distribution line in figure 8.2 is below (respectively, above) the expected payoff difference line. For example, consider the QRE at about $p = 0.4$. At probabilities a little above 0.4, there is an upward pressure on the participation probability, and there is a downward pressure at probabilities a little below 0.4. Therefore this equilibrium is unstable. Conversely, the two more extreme QRE, at probabilities of about 0.05 and 0.95, are stable. Note that the stable equilibria are those for which the inverse distribution line intersects the expected payoff difference line from below as one moves from left to right. For these stable equilibria, any parameter change that shifts the payoff difference line upward will raise the QRE probability of participation. One lesson to be drawn from this analysis is that in cases where comparative statics effects are indeterminate or nonintuitive, it may help to consider adjustment dynamics. The results of this section can be summarized as follows.

Proposition 8.2 (QRE comparative statics for symmetric participation games)*: There is at least one symmetric quantal response equilibrium in a symmetric binary-choice participation game. If the expected payoff difference line is a decreasing function of the probability of participation, then the QRE is unique and it lies between the Nash equilibrium probability and the probability of 0.5. In this negative-externality case, any exogenous factor that increases the participation payoff or lowers the exit payoff will raise the QRE participation probability. The same comparative statics result holds when there are multiple equilibria and attention is restricted to the stable equilibria.*

8.1.4 Positive and Negative Externalities in Voting Games

Strategic voting is a key component of an interesting and considerably more complicated class of participation games: voter turnout games. Such games are more complicated for two reasons. First, the games are not generally symmetric. Second, payoffs are not monotone functions of the number of players who choose to participate. Thus, they do not fall neatly into the two categories of participation games analyzed in the previous two sections. Rather, they are games with a mixture of positive and negative externalities from participation.

This section introduces the application of QRE methods to voter turnout games, where the payoff calculations must take into account the (typically small) effects of vote decisions on election outcomes. Of course, small differences can have large effects in a QRE model where choice probabilities based on small payoff differences are particularly sensitive to random shocks.

The voting game to be considered is similar to that of Palfrey and Rosenthal (1983). In that model, there is an election with two candidates, or parties, A and B. There are $N_A + N_B$ voters, with $N_A > 0$ voters preferring candidate A and $N_B \geq N_A$ voters preferring candidate B. Hence, we refer to candidate A as the minority party candidate (or underdog) and candidate B as the majority party candidate (or frontrunner). The outcome is determined by majority rule, with ties determined by the toss of a fair coin. The payoff to a voter is normalized to be 1 if the player's preferred outcome is selected, and 0 otherwise. Each individual who votes incurs a cost, $c \in [0, 0.5]$, irrespective of the outcome of the election.

Voting is assumed to be simultaneous, so each voter has a choice to either vote for A, vote for B, or abstain. As is standard, we assume throughout this section that dominated actions are not taken, so voters never vote for their less preferred candidate. Thus each player faces a binary choice: vote or abstain. In the absence of a cost of voting ($c = 0$), each person has a weakly dominant strategy to vote, and the frontrunner always wins, except in the special case where $N_B = N_A$ in which case the outcome is a tie. If the cost were greater than 0.5 then each voter has a strictly dominant strategy to abstain, and the outcome would also be a tie. Otherwise, except for the special case of $N_B = N_A$, where it is a Nash equilibrium for everyone to vote, there is no pure-strategy equilibrium of the game.

EQUAL-SIZED PARTIES

Considering again the special case where $N_B = N_A = N$, there may also be symmetric mixed-strategy equilibria in which each individual votes with a probability p, irrespective of their group identity.[5] To analyze these equilibria, we need to derive formulas for the probabilities of specific vote sums, using binomial probability calculations. For example, with N_B voters, the probability that there will be exactly n_B votes from this group is

$$\binom{N_B}{n_B} p^{n_B} (1-p)^{N_B - n_B}.$$

From the point of view of an A voter, the relevant events also depend on the number of votes cast by the $N_A - 1$ other A voters, since this number, along with n_B, determines whether the voter's decision to vote will either *create a tie* (raising the value of the outcome from 0 to 0.5) or *break a tie* (raising the expected value from 0.5 to 1). A decision to vote will create a tie whenever there are j votes on the other side and $j - 1$ votes from the others on one's own side, so the probability of an A voter creating a tie, $\text{Prob}_T[p]$, if all other voters are voting with probability p, is given by

$$\text{Prob}_T[p] = \sum_{n_A=0}^{N_A - 1} \binom{N_B}{n_A + 1} p^{n_A + 1} (1-p)^{N_B - 1 - n_A} \binom{N_A - 1}{n_A} p^{n_A} (1-p)^{N_A - 1 - n_A}$$

$$= \sum_{n_A=0}^{N-1} \binom{N}{n_A + 1} \binom{N-1}{n_A} p^{2n_A + 1} (1-p)^{2(N-1-n_A)},$$

where the second equation uses the symmetry condition that $N_A = N_B = N$. Similarly, a decision to vote will *break a tie* whenever there are j votes on the other side and j votes on one's own side, so the probability of an A voter breaking a tie, $\text{Prob}_W[p]$, if all other voters are voting with probability p is given by

$$\text{Prob}_W[p] = \sum_{n_A=0}^{N-1} \binom{N-1}{n_A} \binom{N}{n_A} p^{2n_A} (1-p)^{2(N-n_A)-1}.$$

---

[5] This kind of equilibrium generalizes nicely when $N_B \neq N_A$, where quasi-symmetric mixed equilibria exist. In a quasi-symmetric mixed-strategy equilibrium, there are two mixing probabilities, p_A for voters in party A and p_B for voters in party B.

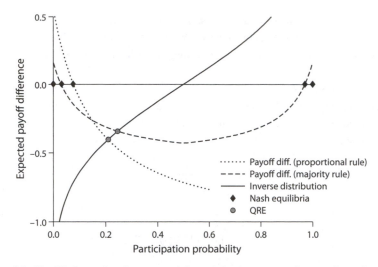

FIGURE 8.3. Equilibrium points for a symmetric participation game under two alternative voting rules (Key: Nash equilibria are diamonds, QREs are circles.)

Note that both "pivot probabilities" depend on N and p. It follows that the gain in expected payoff from voting compared to not voting, as a function of p, equals $\frac{1}{2}(\text{Prob}_T[p] + \text{Prob}_W[p]) - c$.

Figure 8.3 shows this expected payoff gain from voting under majority rule as a function of the common participation probability, as shown by the U-shaped dashed line. This was calculated for the parameters used in an experiment reported by Schram and Sonnemans (1996), with $c = 0.28$ and $N = 6$. The intuition behind this U-shape is that the expected payoff difference is highest at each endpoint, that is, when none of the others vote ($p = 0$), in which case a vote will result in a sure win, or when all of the others vote ($p = 1$) and a vote will create a tie. As before, any point where this payoff difference curve is exactly equal to 0, that is, $\frac{1}{2}(\text{Prob}_T[p] + \text{Prob}_W[p]) = c$, corresponds to a mixed-strategy Nash equilibrium. As in this example, there is typically one with a low probability of voting, and one with a high probability of turnout. This participation game has two interesting strategic dimensions. First, each person in a winning group has an incentive to "free ride" on the turnout of the other members of its own party, just as players in a public goods game have an incentive to free ride on the contributions of other members of their group. On the other hand, there is a competitive force between parties that creates a "rat-race" effect and works in the opposite direction of the free rider problem. Because the groups are equal in size, from a welfare standpoint it makes no difference which party wins, so in this game (in contrast to public goods games) the rat-race effect, as opposed

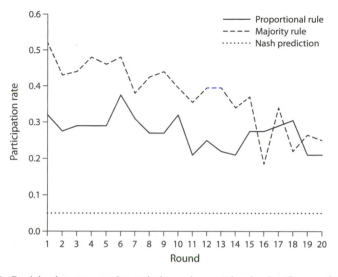

FIGURE 8.4. Participation rates under majority and proportional rule. (Source: Schram and Sonnemans, 1996.)

to the free rider problem, creates inefficiencies. The symmetric high-participation equilibrium is Pareto inferior to the low-turnout equilibrium, as voting is costly, so turnout produces only deadweight losses in this game.

Turning to the experiment, Schram and Sonnemans (1996) used a design in which subjects were randomly rematched into competing groups of size 6 in a series of 20 voting participation games. The average participation rates under this majority rule treatment are shown in figure 8.4 as the dashed line. Note that the participation rates start at about 0.5 and fall to the range from 0.2 to 0.3 in the final rounds. These rates are well above the Nash prediction of 0.05 for this treatment. To illustrate the QRE prediction, the inverse distribution function line is graphed in figure 8.3. This line is plotted for a logit model with $\lambda = 0.9$, resulting in an intersection with the expected payoff difference line at a participation rate in the observed range, between 0.2 and 0.3.[6]

A second treatment was implemented to study proportional voting rules, where a voter's payoff is proportional to its party's share of the total votes cast. As before, voting entails a cost c that is the same for all, and there are N voters

---

[6] Even though the subjects in the experiments all faced the same, deterministic cost of voting, c, the theoretical analysis of the QRE would coincide with a Bayesian Nash equilibrium in which each voter faces a randomly determined, i.i.d. voting cost with mean c. In other words, the epsilon shocks in the QRE model could be interpreted as cost shocks, which corresponds exactly with the incomplete information model developed in Palfrey and Rosenthal (1985).

in each group. Consider the decision of a voter in group A. If n_A votes are cast by the other members of group A and n_B votes are cast by those in group B, then the change in group A's vote share caused by an additional vote is $(n_A + 1)/(n_A + n_B + 1) - n_A/(n_A + n_B)$. The expected change in vote share, denoted $\Delta(N, p)$, is calculated as

$$
\Delta(N, p) = \sum_{n_B=0}^{N} \sum_{n_A=0}^{N-1} \binom{N}{n_B} \binom{N-1}{n_A} \left(\frac{n_A + 1}{n_A + n_B + 1} - \frac{n_A}{n_A + n_B} \right)
$$
$$
\times p^{n_A + n_B} (1 - p)^{2N - n_A - n_B - 1},
$$

where the outer sum is taken with respect to the number of votes from the other group and the inner sum is taken with respect to the number of votes in one's own group. With this notation, the gain in expected payoff from voting can be expressed as a function of N, p, and c: $\Delta(N, p) - c$. The dotted curve in figure 8.3 shows this expected payoff difference for the parameters used in this treatment: $c = 0.32$ and $N = 6$. The line is a smoothly decreasing function of p because each vote has less effect on the proportion when the others are more likely to vote, following a similar intuition to the participation games with negative externalities in section 9.1.1. The (unique) Nash equilibrium for these parameters is $p^* = 0.09$, as indicated by the diamond in the figure where the dotted curve representing the expected payoff difference crosses zero on the vertical axis, and the QRE for $\lambda = 0.9$ is at essentially the same point as was the case for the majority voting rule treatment. The data from the experiment are consistent with these predictions. Under proportional voting, the participation rates (shown by the solid curve in figure 8.4) start lower than was the case under majority voting, but the rates for the two treatments end up being approximately the same, in the 0.2–0.3 range.

The declining participation rates observed in both treatments could be due to a dynamic process in which subjects are responding to the relatively low impact of their vote decision on the payoff outcome. For the middle range of participation rates (in the center of figure 8.3), the expected payoff difference lies below the inverse distribution line, so the quantal responses to these low incentives to vote would reduce participation rates. Thus the QRE analysis explains both the falling participation rates and the tendency for these rates to stay well above the Nash predictions.

UNEQUAL-SIZED PARTIES

This analysis can be generalized to the case of two groups of differing sizes, by studying the quasi-symmetric equilibrium involving two participation rates,

p_A and p_B, which are determined by solving simultaneous equations, one for each group. Cason and Mui (2005) used the asymmetric participation game model to analyze strategic interactions between majority and minority groups in a laboratory experiment of turnout in winner-take-all elections. In the "certain roles" treatment, all subjects knew their own roles with certainty. A majority (3 of 5 voters) were given payoffs that caused them to favor a "reform" option over a "status quo" option, and the minority had opposing preferences. All voters were given the same voting cost, regardless of group. Subjects were randomly rematched in groups of size 5 in a series of rounds, with the reform decision being determined by a majority-rule vote in each round. The cost of voting was varied across treatments to test the Nash equilibrium prediction that participation rates for both rates would be inversely related to the cost of voting. This negative relationship between voting costs and participation rates was observed for the majority group, but not for the minority group. Cason and Mui (2005) use numerical calculations to show that these conflicting patterns are both consistent with a QRE with a moderate amount of noise. QRE also predicts the directions of deviations from Nash participation rate predictions (extreme predictions were pulled in the direction of a fifty-fifty split). Finally, any incidence of incorrect votes for the less preferred alternative cannot be explained by a theory that does not incorporate random elements.

The main focus of Cason and Mui (2005) was on a comparison of the "certain roles" treatment (just discussed) with one in which some voters faced uncertainty about whether or not they would benefit from the reform. Specifically, two voters knew that they preferred the reform, but the other three only knew that one of them (selected at random ex post) would benefit from the reform. The payoff parameters were selected so that a majority would always favor reform ex post, but the uncertainty would cause a majority to vote against the reform if voting were costless. With costly voting, the Nash (and QRE) equilibrium probabilities of reform are lower with role uncertainty. This reduction in the incidence of reform under uncertainty was observed in the data.

The analysis of asymmetric participation games suggests a wider range of applications. For example, political scientists are interested in studying other factors that may increase the chances that a revolution or other radical reform may occur, but such events are often decided in contest settings that do not correspond to majority voting. It may be reasonable to model the probability of a successful revolution as an increasing function of the number who incur a cost to support the revolution, but this function may not have a discrete jump in value from 0 to 1 as that number switches from a minority to a majority. In particular, a proportional voting rule would implement the assumption that the chance of success is equal to the proportion of support among activists (those who incur the cost of supporting their preferred alternative).[7] Cost differences or asymmetries in the probability-of-success function could be introduced to induce a bias in favor of (or against) the status quo. In this case, the revolution could be modeled as an asymmetric participation game with a modified proportional voting rule. As noted above, the interesting strategic features of this game are that it incorporates elements of both public goods and coordination games, since each person would prefer to free ride on the costly activities of others in their group, but all people in a group might be better off if they could coordinate on high activity levels.

8.2 INCOMPLETE INFORMATION AND VOTER TURNOUT

QRE has also been applied to voter turnout games with incomplete information about voting costs.[8] The approach we describe in this section has its roots in the Palfrey and Rosenthal (1985) model, which is a partial equilibrium version of the original Ledyard (1984) general equilibrium model of elections.[9]

As in the previous section, there are two candidates, and voters receive a utility of 1 if their preferred candidate wins and a utility of 0 otherwise, with two party sizes, N_A and N_B. The difference is that voting costs are heterogeneous and private information. Each voter i's cost, c_i, is an independent draw from a distribution of costs with a known cumulative distribution function F. Each voter, after observing her own cost, decides to vote for their preferred candidate or abstain. Payoffs are exactly as described above, and the formulas for Prob_T and Prob_W are as described in the previous section. Because this is a game with a continuum of types and binary choices, with payoffs that are monotone in types, the structure of equilibrium is similar to the compromise game described in the previous chapter of this book. Equilibrium takes the form of a cutpoint rule, one for each party, denoted c_A^* and c_B^*, respectively, and in the Bayes–Nash equilibrium, a voter in party k turns out to vote if and only if $c_i \leq c_k^*$. By continuity of payoffs, a voter in party k with a cost exactly equal to c_k^* is exactly indifferent between voting and abstaining. This gives two equations that must hold in an equilibrium:

$$\frac{1}{2}\left(\text{Prob}_T^A[p_A, p_B] + \text{Prob}_W^A[p_A, p_B]\right) = c_A^*,$$

$$\frac{1}{2}\left(\text{Prob}_T^B[p_A, p_B] + \text{Prob}_W^B[p_A, p_B]\right) = c_B^*.$$

---

[7] This proportional success probability function is widely used in the rent-seeking and contest literatures, which will be discussed in the final section of this chapter.

[8] See, e.g., McKelvey and Patty (2006), Levine and Palfrey (2007), Herrera, Morelli, and Palfrey (2014), Kartal (2015a, 2015b), and Tulman (2015).

[9] This section includes material from Levine and Palfrey (2007).

There are two additional equilibrium conditions because the turnout rate for party j is equal to the probability a j voter has a voting cost less than or equal to c_j^*, which depends on the distribution of voting costs:

$$p_A = F(c_A^*), \qquad p_B = F(c_B^*).$$

8.2.1 QRE Model of Turnout with Private Information

In this voter participation game, subjects with costs close to the equilibrium cutpoint are nearly indifferent between voting and abstaining. Consequently, very small errors in their judgment about the pivot probabilities, or other factors, could lead them to make a suboptimal decision. Quantal response equilibrium in this game can be modeled in the same way as in the compromise game, either from a behavioral strategy approach, or the cutpoint strategy approach. The former approach was adopted in Levine and Palfrey (2007) to analyze data from a voter turnout experiment based on this model. The computation is somewhat more complicated, but the basic idea is the same as described in the previous section on participation games with complete information. Logit QRE turnout rates will generally be closer to 50% than in the Bayesian Nash equilibrium of the game.

In the logit QRE a voter's turnout probability is a *continuous* strictly decreasing function of voting cost, $\sigma(\cdot)$, which is equal to 0.5 precisely at the voting cost at which the voter is exactly indifferent between voting and abstaining. If π_j is the probability a voter in party j is pivotal (which will generally be different for minority and majority voters), then in a logit equilibrium such a voter's turnout probability, if his voting cost is c, follows the formula

$$\sigma_j(c; \lambda) = \frac{1}{1 + e^{\lambda(c - \pi_j)}},$$

where λ is the logit response parameter. Integrating over all possible voting costs, we obtain the voter's ex ante turnout probability:

$$p_j^*(\lambda) = \int_{-\infty}^{\infty} \sigma_j(c; \lambda) f(c) dc$$

for $j \in \{A, B\}$. These equations are analogous to those defining the Nash equilibrium.

8.2.2 Experimental Data

The Levine and Palfrey (2007) experiment controlled all the parameters of the voter turnout game, including the distribution of voting costs which was given by the uniform distribution on the interval $[0, 0.55]$. The experiment varied N_A and N_B and conducted elections with $N_A + N_B = \{3, 9, 27, 51\}$. For each of these

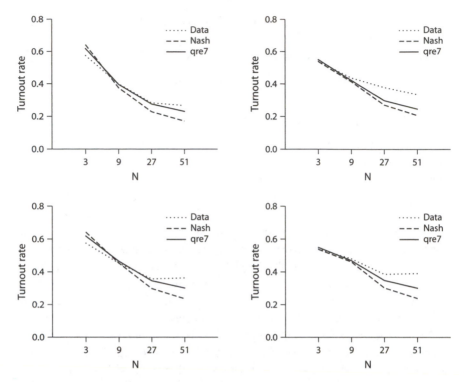

FIGURE 8.5. Turnout data and theoretical predictions. The top panels pertain to the landslide treatment, and the bottom panels pertain to the toss-up treatment. In each case, the majority party turnout rate is shown in the left panel. (Source: Levine and Palfrey, 2007.)

electorate sizes, there were two subtreatments,[10] one where $N_B = 2N_A$ and one where $N_B = N_A + 1$, called the *landslide* and *toss-up* treatment, respectively. In each session, $N_A + N_B$ was held fixed throughout the entire session. For the $N_A + N_B > 3$ sessions, there were two subsessions of 50 rounds each; one subsession was the toss-up treatment and the other subsession was the landslide treatment.[11] Overall, nearly 300 subjects participated in a total of more than 2300 elections.

The data was fitted to the logit QRE described above, estimating the logit response parameter, λ, by maximum likelihood. To avoid overfitting, λ was constrained to be identical across all treatments and produced a maximum-likelihood estimate equal to $\widehat{\lambda} = 7$. The estimated value, $\widehat{\lambda} = 7$, along with the actual draws of voting costs, was then used to compute for each treatment and each party, theoretical QRE turnout rates. The Nash equilibrium, QRE ($\widehat{\lambda} = 7$), and observed turnout rates are graphed for each treatment in figure 8.5.

---

[10] When $N_A + N_B = 3$ these two treatments collapse into one treatment.
[11] The $N_A + N_B = 3$ treatments were conducted somewhat differently.

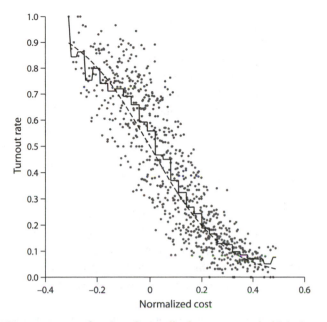

FIGURE 8.6. Turnout rates as a function of normalized cost, compared with logit response curve for $\widehat{\lambda} = 7$. (Source: Levine and Palfrey, 2007.)

The QRE model provides the right qualitative correction to the Nash model. As one can see in the figure, the Nash model underpredicts turnout in larger elections and is very close for small and intermediate-sized elections. The QRE predicts less turnout than Nash equilibrium in the smallest election, and more turnout than Nash equilibrium in the large elections. QRE also predicts even greater amounts of overvoting (relative to Nash), as the size of the election becomes larger and larger, which is also seen in the data.

As an alternative way to see how well the logit equilibrium model is capturing voter behavior, figure 8.6 graphs turnout rates by "normalized" voting cost. For each treatment and each party, we define a normalized voting cost as the difference between a voter's actual voting cost and the logit equilibrium cutpoint ($\lambda = 7$). Thus, for example, if the QRE cutpoint for an A voter in some treatment were, say, 15, and their actual cost were 25, their normalized voting cost would be 10. Thus our normalization allows us to display all the voting behavior in a single graph. According to the logit QRE, the voting probabilities should follow a logit curve, which is the smooth decreasing curve in the figure (for $\lambda = 7$). The decreasing step-function curve averages the data across normalized cost intervals of 0.03.

8.2.3 Asymptotic Voter Turnout Predicted by QRE

There are also interesting implications for QRE in large elections. Even though the equilibrium pivot probabilities converge to 0, turnout rates will not converge to 0 and in fact can be very large. This is quite intuitive, especially if the mean of the distribution of voting costs is a lot less than the difference in expected utility from one's candidate winning or losing the election (which is normalized to 1 in the model). That difference can be interpreted as the willingness to pay for a voter to unilaterally dictate the outcome of the election one way or the other. In important elections, it is easy to imagine that, for an average voter, this value is orders of magnitude higher than the cost of voting.

Examining first Nash equilibrium, it is well known that as the population increases to infinity, the participation rate falls to zero. Suppose that for one of the parties it does not. Then the probability of being pivotal must fall to zero because there is an infinite number of voters. This in turn implies that the optimum threshold for participation is falling to zero, leading to a participation rate that falls to zero, contradicting the original hypothesis. Indeed, this is a version of the "paradox of voter participation."

In contrast, the quantal response model has more intuitive implications about turnout in large elections. For a given value of λ we can compute the participation rate (using the distribution in the experiment):

$$p_j^*(\lambda) = \int_0^{0.55} \frac{1}{1 + e^{\lambda(c - \pi_j)}} \frac{1}{0.55} dc,$$

where the QRE pivot probability, π_j, is endogenous. However, just as in the Nash equilibrium, in a large population, the QRE probability of being pivotal is approximately 0, so we have the following large population approximation for p_j^*:

$$p_j^*(\lambda) \approx \int_0^{0.55} \frac{1}{1 + e^{\lambda c}} \frac{1}{0.55} dc = \frac{1}{0.55} \left(0.55 - \frac{\log(1 + e^{\lambda 0.55})}{\lambda} + \frac{\log(2)}{\lambda} \right).$$
$$(8.5)$$

We may then compute the asymptotic participation rate for various values of λ in (8.5). In particular, for the value of λ estimated with the data, the asymptotic participation rate is 17%. For the participation rate to fall to 5%, for example, would require $\lambda = 28$, which is an order of magnitude higher than we estimate.

To compare with elections in the field, it is more reasonable to use a lower distribution of voting costs than in the experiment. The average voting cost in the experiment is equal to one-quarter the value of being a "dictator" over the outcome of the election, which is probably several orders of magnitude too big for important elections where voters have relatively strong preferences over the alternatives or candidates. To consider lower distributions of voting costs, suppose that the maximum voting cost is C and the distribution is uniform. Then the asymptotic expected turnout in the logit QRE is equal to

$$T(C) = \frac{1}{C}\left(C - \frac{\log(1 + e^{C\lambda})}{\lambda} + \frac{\log(2)}{\lambda}\right).$$

For $\lambda = 7$ this expression implies 48% turnout (42% if $\lambda = 28$) when $C = 0.02$.

More broadly if the distribution has an arbitrary shape, and the fraction of voters with costs less than or equal to C is F then we can compute voter turnout by multiplying by F. So, for example, if only 80% of voters have costs below 0.02 and $\lambda = 7$, the corresponding asymptotic turnout is $0.8 \times 0.48 = 0.384$. In fact, for *all* values of λ, the turnout percentage converges to 50% as C converges to 0 in the uniform case, and as the random variable representing the cost distribution converges to 0 in probability in the general case. In other words, even with incredibly rational voters ($\lambda > 100$, for example), very high asymptotic turnout rates can be supported if the distribution of costs is low (or the stakes are very high). It is *only* in the knife-edge case of *perfectly* rational voters ($\lambda = \infty$) that the turnout paradox arises.

8.3 INFORMATION AGGREGATION BY VOTING

Elections are typically thought of as mechanisms for settling differences among individuals who do not agree on particular issues. There is, however, a second function of voting procedures: the aggregation of information across individuals who may agree on policy matters but who have different sources of information. The idea that elections aggregate information dates to the eighteenth century writings of Condorcet (1785), and has generated a large body of research since then. The main insight of this literature can be illustrated with a simple example. Suppose that N members of a jury observe a trial and form independent opinions about the guilt of the defendant. If the defendant is actually guilty and the trial is informative, then it is more likely that jurors will vote to convict. The way the trial information is interpreted is noisy, so some jurors may vote either way, but there will be a greater tendency to vote for a conviction if the defendant is actually guilty. Let $\text{Prob}[V_C|G]$ be the probability that a juror will vote to convict, V_C, if the defendant is guilty, and let $\text{Prob}[V_C|I]$ be the probability that a juror will vote to convict if the defendant is innocent. In a large jury with independent decisions, the fraction who vote to convict will be approximately $\text{Prob}[V_C|G]$ if the defendant is guilty, and this fraction will be approximately $\text{Prob}[V_C|I]$ otherwise, with $\text{Prob}[V_C|I] \leq \text{Prob}[V_C|G]$. Now let ρ be the fraction of the jury that is required to vote for conviction in order to obtain a conviction, for example, $\rho = 0.5$ under majority rule and $\rho = 1$ under unanimity. If ρ is set too low, below Prob[$V_C|I$], then innocent individuals would tend to get convicted. Obviously, one would like to set the required fraction to be between Prob[$V_C|I$] and Prob[$V_C|G$], but these cutoffs may not be known with precision and may differ from trial to trial. In any case, the chances of convicting an innocent person can be reduced by using a higher value of the cutoff fraction ρ, *if these probabilities are not affected by the voting rule cutoff itself.*

This intuition, that a higher standard of conviction will tend to protect the innocent, is implicitly based on a naive model of voting, that is, jurors form opinions based on their interpretation of the trial evidence and vote without any strategic consideration for how others might vote. Naive voting, however, may not be a rational way to use one's information if unanimity is required (Austin-Smith and Banks, 1996; Feddersen and Pesendorfer, 1996, 1999). To see this, suppose that each person receives one of two informative signals about the state, either a guilty signal or an innocent signal, with a guilty signal being more likely if the defendant is guilty, and an innocent signal being more likely if the defendant is innocent.[12] If each juror votes naively (to convict if they see a guilty signal and to acquit if they see an innocent signal), then consider the dilemma of a juror who sees an innocent signal. Under unanimity, with $\rho = 1$, a vote to acquit will preclude conviction, regardless of how the others vote. Even a vote to convict will not result in conviction if at least one of the others voted to acquit. The only way that a person's vote will matter, then, is if *all* of the others voted to convict.[13] Under naive voting, this means that all of the $n - 1$ others saw a guilty signal, so a juror who really believes that the others are voting naively, might want to vote to convict, contrary to the signal. In this case, naive voting is not a best response to naive voting, so voting naively would not be a Nash equilibrium.[14] Moreover, the incentive for a juror with an innocent signal to deviate from naive behavior by voting to convict is stronger for a larger jury. In the Nash equilibrium to be

[12] Notice that the information structure is similar to the information cascade model of chapter 7.

[13] In many contexts, there may be three possible outcomes: conviction, acquittal, and mistrial, which may occur if there is not total agreement among the jurors. Following Feddersen and Pesendorfer (1998), we will not consider the possibility of mistrials.

[14] The key insight here is that being a "swing voter" is an informative event, and a voter should take this into account in making a decision, just as a bidder in a first-price, common-value auction should realize that a bid will be relevant only if it is the high bid, which may mean that the person overestimated the value of the prize. Failure to adjust for this implicit information may cause the winning bidder to lose money due to the "winner's curse." Similarly, failure to properly condition the voting decision on pivotal events may lead to a "swing-voter's curse" (Feddersen and Pesendorfer, 1996). See Battaglini, Morton, and Palfrey (2008, 2010) for laboratory studies of the swing-voter's curse in a variety of settings that include larger electorates.

explained next, the probability of voting to convict given an innocent signal will be an increasing function of n, the jury size.[15]

8.3.1 Rational Strategic Voting in Juries

Consider a game in which the defendant is guilty (state G) with probability 0.5 and innocent (I) with probability 0.5. Each of the n jurors receives an independent Bernoulli signal that is γ with probability p if the defendant is guilty, and is ι with probability p if the defendant is innocent, where $p > 0.5$. After observing their own private signals, jurors vote to convict or acquit. The outcome under unanimity is a conviction, C, if and only if all n jurors vote V_C. Jurors have identical utilities that depend on the state and outcome, and on a parameter q that determines the relative cost of making a mistake in either direction; for a false conviction, $u(C, I) = -q$, and for a false acquittal, $u(A, G) = -(1 - q)$. The utilities for the correct decision are normalized to 0: $u(A, I) = u(C, G) = 0$. Let $\widehat{\text{Prob}}[G]$ be the voter's posterior probability of G, given relevant information (as will be explained below). Then the expected payoffs for a jury decision to convict or acquit, denoted by U_C and U_A respectively, are

$$U_C = -q(1 - \widehat{\text{Prob}}[G]), \qquad U_A = -(1 - q)\widehat{\text{Prob}}[G],$$

so

$$U_C - U_A = \widehat{\text{Prob}}[G] - q. \tag{8.6}$$

Thus the preferred outcome is conviction when the posterior probability of guilt is above q, which represents a "threshold of reasonable doubt." In what follows, we assume that $p > q$, which implies a jury of one would vote to convict with a guilty signal and to acquit otherwise.

In this game, a strategy is a probability of voting for conviction, conditional on the private information $\sigma : \{\gamma, \iota\} \to [0, 1]$. There are two ways that a vote for conviction can be generated, depending on whether the signal is "correct" or not. Given the probability, p, of a correct signal, the conditional probabilities of a vote to convict are

$$\text{Prob}[V_C | G] = p\sigma(\gamma) + (1 - p)\sigma(\iota), \tag{8.7}$$

$$\text{Prob}[V_C | I] = (1 - p)\sigma(\gamma) + p\sigma(\iota). \tag{8.8}$$

The symmetric equilibrium to be constructed will have the property that those with γ signals always vote to convict ($\sigma(\gamma) = 1$), and those with ι signals vote

---

[15] This section includes material from Guarnaschelli, McKelvey, and Palfrey (2000).

to convict with a probability $\sigma(\iota) < 1$, so the equations in (8.7) and (8.8) can be expressed as

$$\text{Prob}[V_C|G] = p + (1-p)\sigma(\iota),$$

$$\text{Prob}[V_C|I] = (1-p) + p\sigma(\iota).$$

$$(8.9)$$

Next consider the decision for a juror who receives the ι signal. Under unanimity, the only case in which a vote will matter is when the $n-1$ others vote to convict, so the juror should evaluate the probability of guilt conditional on both an ι signal and on $n-1$ other V_C votes. Given the simplifying assumption that each state (G or I) is equally likely a priori, the 0.5 terms cancel out of the numerator and denominator of Bayes' rule, and the conditional probability of G is

$$\widehat{\text{Prob}}[G|\iota, n-1 \text{ other } V_C \text{ votes}] = \frac{(1-p)\,\text{Prob}[V_C|G]^{n-1}}{(1-p)\,\text{Prob}[V_C|G]^{n-1} + p\,\text{Prob}[V_C|I]^{n-1}}$$

$$= q,$$

$$(8.10)$$

where the $(1-p)$ terms represent the probability of getting an ι signal when the defendant is guilty, and the p term in the denominator is the probability of an ι signal when the defendant is innocent. The final equality in (8.10) follows from (8.6) and the requirement that the voter must be indifferent between the two outcomes in a mixed-strategy equilibrium. (More precisely, indifference between the two outcomes implies indifference between the two vote decisions that might affect these outcomes.) Equation (8.10) can be solved for the probability of a vote to convict, conditional on the defendant being innocent:

$$\text{Prob}[V_C|I] = R_n \,\text{Prob}[V_C|G],$$

$$(8.11)$$

where the ratio of conviction vote probabilities is

$$R_n = \left(\frac{(1-q)(1-p)}{pq}\right)^{1/(n-1)}.$$

$$(8.12)$$

Then (8.12) can be used with (8.9) to solve for the equilibrium probability:

$$\sigma(\iota) = \frac{R_n p - (1-p)}{p - R_n(1-p)},$$

$$(8.13)$$

where R_n is expressed in terms of the exogenous parameters in (8.12).[16]

---

[16] By construction, a juror with an ι signal is indifferent and can do no better than to randomize. The final step is to show that a juror with a γ signal would vote to convict with probability 1 if jurors with the ι signal are randomizing. Suppose (in contradiction) that those with γ signals were to randomize in equilibrium, so they must be indifferent between voting to convict or acquit. But then those with ι

Consider what happens to the probability of convicting an innocent defendant as the jury size grows. As n tends to ∞, R_n tends to 1 and $\sigma(\iota)$ tends to 1 in (8.13). In this case, it follows from (8.7) and (8.8) that the probabilities of a vote to convict in either state converge to 1. These tendencies suggest that unanimity rules may not protect innocent defendants as the jury size increases. More precisely, the probability of convicting an innocent defendant is the probability that all n jurors vote to convict, $\text{Prob}[V_C|I]^n$, and this probability can be calculated by substituting the solution for $\sigma(\iota)$ into the right equation in (8.9) and raising the result to the power n. If $q = 0.5$ and $p = 0.7$, for example, the Nash equilibrium probability of convicting the innocent under unanimity is 0.14 with a jury of size 3, and it *increases* to 0.19 with a jury of size 6. Moreover, Feddersen and Pesendorfer (1998) show that the limiting value of this error probability is 0.23 as n goes to ∞ for this example. In fact, they show that this limit is strictly positive for all parameter values under consideration, that is, for $1 - p < q < p$.[17]

In contrast, consider what happens under majority rule as the jury size increases. If mistakes in either direction are equally costly ($q = 0.5$), then there is an equilibrium in which jurors vote naively, with $\sigma(\gamma) = 1$ and $\sigma(\iota) = 0$. The intuition is that under the symmetric environment with a symmetric voting rule, since being pivotal means that the others are evenly split or nearly so, the event is not informative. Thus a naive vote would result from the appropriate use of private information. With naive (and in this case rational) voting under majority rule, the intuition for the Condorcet information aggregation result holds, and large juries would tend to make no mistakes. From a game-theoretic perspective, what was wrong with the original intuition drawn from the Condorcet jury voting literature was the assumption that voting behavior is unaffected by the voting rule itself. Whether actual behavior is naive or not in specific settings, however, is a question that can be, and has been, addressed with laboratory experiments, which is the next topic.

8.3.2 Experimental Data

Guarnaschelli, McKelvey, and Palfrey (2000) report a laboratory experiment done with $p = 0.7, q = 0.5$, and juries of size 3 and 6. Here we focus on the results for the unanimity rule (they also ran some trials with a majority rule). Subjects were recruited in groups of 12 and were randomly matched into juries in a sequence of jury meetings, with payoffs that were contingent on the jury decision and the

---

signals would not be indifferent; they would always vote to acquit, so the only votes to convict would be cast by those with γ signals. Then a person with an ι signal would be a swing voter only if all others have γ signals, and hence this person would prefer to vote to convict, which is a contradiction.

[17] The formula for this limit is $((1 - q)(1 - p)/qp)^{p/(2p-1)}$.

TABLE 8.2. Nash equilibrium mistake frequencies when $q = 0.5$ and $p = 0.7$.
(Source: Guarnaschelli, McKelvey, and Palfrey, 2000.)

	$n = 3$	$n = 6$	$n = 12$	$n = 24$
Prob[$C|I$] under unanimity	0.14	0.19	0.21	0.22
Prob[$C|I$] under majority	0.22	0.07	0.04	0.01
Prob[$A|G$] under unanimity	0.50	0.48	0.48	0.47
Prob[$A|G$] under majority	0.22	0.26	0.12	0.03

TABLE 8.3. Proportion of incorrect convictions (Source: Guarnaschelli, McKelvey, and Palfrey, 2000.)

	Data	Nash	QRE
Smaller jury ($n = 3$)	0.19	0.14	0.19
Larger jury ($n = 6$)	0.03	0.19	0.07

state, which was announced after each round. With these parameters, the Nash equilibrium mistake probabilities are shown in table 8.2. The prediction is that innocent defendants will not be protected under unanimity for juries of size 6 and above.

One finding in the experiment is that subjects tend to vote strategically (not naively) when the Nash prediction is that they will vote strategically. Under unanimity, the fraction of votes to convict after receiving an ι signal goes from 0.36 for juries of size 3 to 0.48 for juries of size 6. These proportions are roughly consistent with the corresponding Nash predictions of 0.31 and 0.65 respectively, which are derived from equation (8.13). Thus the data are qualitatively consistent with the game-theoretic intuition that jurors with innocent signals will be more likely to vote to convict in larger juries, where the swing-voter's curse has more of an effect. The results contrast sharply with the naive voting rule, where the probability of voting to convict with an ι signal is 0 for all jury sizes.[18]

Next consider the nonintuitive Nash prediction, that an increase in the size of the jury will *increase* the probability of conviction under unanimity. This did not happen. In fact, the proportion of incorrect convictions of innocent defendants went from 0.19 down to 0.03, instead of rising as predicted in a Nash equilibrium. See the first two columns of table 8.3 for a comparison of the data and Nash predictions for the rate of incorrect convictions under unanimity voting.

---

[18] Note that all voting was simultaneous in this experiment. Ali et al. (2008) compare simultaneous and sequential (public) voting (under unanimity) and find similar patterns of strategic voting in each case.

8.3.3 QRE Voting in Juries

For a given private signal, voting is a binary decision; with QRE the probability of a vote to convict is an increasing function of the expected payoff difference for the two decisions, conditional on the private signal, $s \in \{\iota, \gamma\}$, and on the strategies used by the $n - 1$ others. Let these two expected payoffs be denoted by $U_{V_C}(s, n - 1, \sigma(\iota), \sigma(\gamma))$ and $U_{V_A}(s, n - 1, \sigma(\iota), \sigma(\gamma))$. Then, again following the distribution function approach to QRE, the equations analogous to (8.4) are

$$\frac{F^{(-1)}(\sigma(\iota))}{\lambda} = U_{V_C}(\iota, n - 1, \sigma(\iota), \sigma(\gamma)) - U_{V_A}(\iota, n - 1, \sigma(\iota), \sigma(\gamma)),$$

$$(8.14)$$

$$\frac{F^{(-1)}(\sigma(\gamma))}{\lambda} = U_{V_C}(\gamma, n - 1, \sigma(\iota), \sigma(\gamma)) - U_{V_A}(\gamma, n - 1, \sigma(\iota), \sigma(\gamma)).$$

$$(8.15)$$

Solving these equations requires calculation of the probability that a change in one's vote from V_A to V_C will alter the outcome from A to C. Then the expected utilities for these outcomes can be used to calculate the expected payoff differences on the right-hand sides of (8.14) and (8.15). Even without detailed calculations, we can obtain a useful limit result, which generalizes Guarnaschelli, McKelvey, and Palfrey (2000, Theorem 1), that assumed the logit specification.

Proposition 8.3 (QRE asymptotic acquittal for unanimous juries): *Fix $\lambda < \infty$. Under unanimity, for every $\delta > 0$, there exists a number $N(\delta, \lambda)$ such that for all $n > N(\delta, \lambda)$, the probability of conviction in any regular QRE is less than δ regardless of whether the defendant is innocent or guilty.*

Proof: Since all outcome-contingent payoffs are less than 1, it must be the case that

$$U_{V_C}(\iota, n - 1, \sigma(\iota), \sigma(\gamma)) - U_{V_A}(\iota, n - 1, \sigma(\iota), \sigma(\gamma)) < 1.$$

Then the equilibrium conditions, (8.14) and (8.15), imply that $F^{(-1)}(\sigma(\iota)) < \lambda$ and $F^{(-1)}(\sigma(g)) < \lambda$. It follows that both $\sigma(\iota)$ and $\sigma(\gamma)$ are less than $F(\lambda)$, which is strictly positive when the preference shocks have positive density on the real line. (In a logit equilibrium, this bound is $1/(1 + e^\lambda)$.) Thus the probability of voting for *acquittal* is greater than $F(\lambda)$ for both signals. Since this lower bound on the acquittal probabilities is independent of n it follows that the probability of obtaining at least one vote to acquit tends to 1 as n tends to ∞. ∎

To summarize, the Nash equilibrium conviction probability for an innocent defendant under unanimity is bounded away from 0 for any jury size, which is inconsistent with the tendency for the incidence of false convictions to decrease as the jury size is increased in the experiment. In contrast, QRE has the property that the probability of conviction goes to 0 under unanimity as the jury size increases, due to noise effects. The QRE predictions for the parameters and estimated precisions from the Guarnaschelli, McKelvey, and Palfrey (2000) experiment, shown in the far right column of table 8.3, exhibit this pattern, which is consistent with the trend observed in the "Data" column of the table.

8.4 MARKOV QRE AND DYNAMIC LEGISLATIVE BARGAINING

Much of the business conducted by committees and legislatures involves distributional politics.[19] That is, a more or less fixed budget of resources is to be divided among the members or the committee (or the polities represented by those members). A classic example is the process of allocating a fixed budget of funds for public infrastructure improvements across states and congressional districts by the US Congress, often loosely referred to as "pork-barrel politics." A standard theoretical approach to analyzing these allocation processes is the Baron–Ferejohn (1989) model of legislative bargaining. In that stylized model, committee members have the opportunity to offer proposals for an allocation according to some formal agenda rules, and these proposals are then voted on according to some voting rule. Once a proposal has passed, that proposed allocation is implemented. If no proposal succeeds, then some default outcome occurs.

This section applies MQRE to one version of this game (Kalandrakis, 2004), where this allocation process takes place repeatedly over an infinite number of periods, with the default outcome in any given period being the allocation decided on by the committee in the immediately preceding period. In this stylized model, there are an odd number ($n \geq 3$) of members of the committee, and the budget, normalized at $B = 1$, is perfectly divisible. An allocation in period t is a profile $x^t = (x_1^t, \ldots, x_n^t)$, where $x_i^t \geq 0$ for all i and $\sum_{i=1}^{n} x_i^t = 1$. Member i's utility for the allocation in period t is denoted $u(x_i^t)$, and all members have the same utility function. Each individual i values the infinite sequence of allocations, $x^\infty = (x^1, \ldots, x^t, \ldots)$ by discounting at a constant rate δ, so $W_i(x^\infty) = \sum_{t=1}^{\infty} \delta^{t-1} u(x_i^t)$.

The agenda rule is very simple. In each period t, one of the n members is selected at random to propose an allocation, $\hat{x}$, for period t. If a majority of

---

[19] This section includes material from Battaglini and Palfrey (2012).

members vote in favor of the proposed allocation it passes and is implemented, so $x^t = \hat{x}$. If the proposal fails, then last period's allocation (the status quo) is implemented instead, that is, $x^t = x^{t-1}$. The initial status quo is some exogenously fixed allocation x^0. This defines a stochastic game, which has been studied both theoretically and experimentally, in terms of the Markov perfect equilibrium of the game.

The special case of linear utility, where $u(x_i^t) = x_i^t$, has been the main focus of analysis, because simple closed-form solutions for the Markov perfect equilibria (MPE) have been obtained by construction. These equilibria typically involve extremely unfair allocations in each period, except possibly the first few, with the current proposer receiving a 100% share of the pie. In each period the proposer proposes this allocation, so that voters end up voting between the current proposer receiving 100% or some other member receiving 100% (last period's proposer). Every member except the current proposer and the previous proposer is indifferent and votes in favor of the proposal. The current proposer also votes in favor, and therefore it passes. The state space for MPE is the set of possible status quo points, which is just the simplex of allocations, $X = \{x = (x_1, \ldots, x_n) | x_i^t \geq 0 \ \forall i, \sum_{i=1}^{n} x_i^t = 1\}$. A full description of the equilibrium is therefore a specification of proposal and voting behavior for each status quo point, not just the points on the vertex of the simplex. The details of the MPE at nonvertex points are a bit cumbersome. Briefly, if one starts at a point off the vertices, then within a few periods the equilibrium path leads to one of the vertices, and from then on the status quo is always one of the vertices.[20]

The properties of equilibria for the case of concave utility, that is, $u''(x_i^t) < 0$, have not been established analytically. Battaglini and Palfrey (2012) compute logit MQRE for a finite approximation of the game, and for very high values of λ.[21] This computational approach allows an approximation of the MPE of the game as limiting logit MQRE. Because utility functions are strictly concave, members would prefer to have some consumption smoothing and would be willing to give up something in the short run in exchange for smoother allocations in the future. In particular, the sequence of allocations involving an equal split in every period Pareto-dominates the "randomly rotating dictatorship" characteristic of the MPE with linear utilities. The equilibrium computations show that the limiting logit MQRE incorporates some consumption smoothing, and the amount of consumption smoothing increases with the concavity of the utility functions.

---

[20] The equilibrium is established by construction.

[21] See chapter 5 for the formal definition of logit MQRE. The state space for this application is the simplex of feasible one-shot allocations, and the finite approximation represents the state space as a finite grid on that simplex.

8.4.1 Approximating MPE by Logit MQRE

This limiting MQRE approach is used to compute an approximate MPE for the family of utility functions with constant relative risk aversion:

$$u(x_i) = \frac{x_i^{1-r}}{1-r},$$

where x_i is the allocation to i. The coefficient of relative risk aversion, r, measures the curvature of the utility function: the higher is r, the more concave is utility. It is interesting to compare two polar cases: the linear case, $r = 0$, and a strictly concave case, $r = 0.95$. A discount factor of $\delta = \frac{5}{6}$ is assumed, and the game is solved using discrete approximation of a unit simplex where allocations are in increments of $\frac{1}{12}$. This reduces the set of states to 91. For any discrete approximation, existence of a symmetric Markov equilibrium follows from standard fixed-point arguments, so the limiting logit MQRE is in fact an approximation to an MPE of this finite version of the game.

Given the smooth properties of this Markov logit equilibrium (MQRE) path, there is a simple and (relatively) fast path-following algorithm which will find this solution, similar to the techniques described in chapter 6. First, the solution at $\lambda = 0$ is known: all behavioral strategies are chosen with equal probability, and this implies a unique value function. Hence, one starts with this known solution at $\lambda = 0$ and then uses that solution as the starting value to find the MQRE for an incrementally larger value, say $\lambda = \epsilon$. Because it is guaranteed that for small enough ϵ the starting value obtained from $\lambda = 0$ is very close to the solution at $\lambda = \epsilon$, the fixed-point algorithm will find a solution at ϵ very quickly. Then, we use the solution at $\lambda = \epsilon$ to compute the solution at $\lambda = 2\epsilon$ and so forth, thereby tracing out the MQRE path that converges to a Markov equilibrium of the game.[22]

8.4.2 Steady-State Equilibrium Dynamics

A proposal strategy associates to each status quo a vector of probabilities of proposing at each state. The voting strategies associate a probability of voting yes to each possible status quo–proposal pair. Because the equilibrium strategy space is so large, to describe the properties of equilibrium behavior it is convenient to use the stationary distribution over outcomes induced by equilibrium strategies. The equilibrium strategies generate a Markov process with a stationary transition matrix. This transition matrix associates each state $x' \in X$ to a probability distribution $\varphi(x \,|\, x')$ over states $x \in X$ in the following periods. For a given initial

---

[22] There are some numerically tricky issues when λ becomes very large, and the algorithm can take a long time to converge for high values of λ, but conceptually it is straightforward, and convergence is not difficult to achieve.

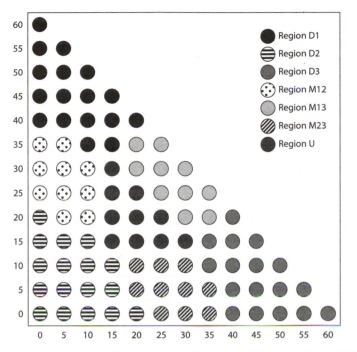

FIGURE 8.7. Allocation regions. The vertical coordinate represents agent 1's payoff, the horizontal represents agent 3's payoff. The payoff of agent 2 is the residual. (Source: Battaglini and Palfrey, 2012.)

distribution $\text{Prob}^0[x]$ over the status quos, we can therefore define the equilibrium distribution of states at t recursively as

$$\text{Prob}^t[x] = \sum_{x' \in X} \varphi(x \mid x') \text{Prob}^{t-1}[x'].$$

The probability function $\text{Prob}^t[x]$ converges to a stationary distribution $\text{Prob}^*[x]$ as $t \to \infty$. This distribution represents the frequency of the states that we would expect to observe in the long run, so it provides one of the fundamental properties of the Markov equilibrium.

The states can be clustered into coarser regions, (D1, D2, D3, U, M12, M13, M23), as marked in figure 8.7. The D regions correspond to *dictatorial* allocations where one player receives the lion's share of the pie. The M regions correspond to *majoritarian* allocations where a coalition of two players receives most of the pie, with nearly equal shares, while the third player receives only a small amount or nothing. The U region consists of *universal* allocations, where the pie is equally, or nearly equally, shared. Conditional on being, say, in D1, we can use the stationary distribution of the computed Markov equilibrium to

TABLE 8.4. Limiting MPE transition matrix of the seven regions with linear utility ($r = 0$) and $\delta = 0.83$. (Source: Battaglini and Palfrey, 2012.)

		SQ_{t+1}						
		D1	D2	D3	M12	M13	M23	U
	D1	0.34	0.33	0.33	0	0	0	0
	D2	0.33	0.34	0.33	0	0	0	0
	D3	0.33	0.33	0.34	0	0	0	0
SQ_t	M12	0.33	0.33	0.25	0.08	0	0	0
	M13	0.33	0.25	0.33	0	0.08	0	0
	M23	0.25	0.33	0.33	0	0	0.08	0
	U	0.01	0.01	0.01	0.31	0.31	0.31	0.03

derive the probability of transition to state M12 (the M region corresponding to the coalition of players 1 and 2). Doing this for all pairs of regions gives a representation of the steady-state equilibrium dynamics of the infinitely repeated game in a simple 7×7 matrix. This allows one to describe the dynamics in a concise way.

LINEAR UTILITIES

With linear utilities, the proposer has an incentive to form a minimal winning coalition and maximize his immediate payoff. This is because, on the one hand, there is discounting of future payoffs, while at the same time there is no incentive to smooth consumption, perhaps by giving up a little today for a higher allocation in the future. This is evident in the equilibrium transition matrix that is shown in table 8.4, using the states described in figure 8.7. Given the symmetry of the equilibrium this can be illustrated with three "representative" states: D1, M12, and U. The dynamics implied by table 8.4 are represented by graphically illustrating the transition probabilities from these three states.

Figure 8.8(a) makes clear that the short-run effect dominates. For example, suppose the initial state is D1, where agent 1 receives most of the pie. In this case the state will stay at D1 with 34% probability and move to D2 and D3 each with 33% probability; that is, with 100% probability the state will remain in the extreme regions. This occurs because in D1 each agent will propose almost all the payoff for himself, with a minimal share going to a single coalition partner.[23] The probability of remaining in D1 is higher because some of these proposals may be rejected with small positive probability.

It is interesting to note the dynamics evolving from a status quo in U. In this case the state does not jump directly to any of the D regions with high probability

---

[23] This differs slightly from the MPE with a continuous allocation space, which is discussed below.

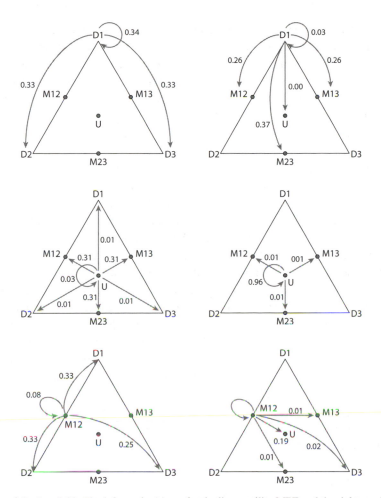

FIGURE 8.8. $\delta = 0.83$. The left graphs (a) are for the linear utility MPE and the right graphs (b) are for the concave utility ($r = 0.95$) MPE. The top graphs illustrate transitions from D regions, the middle graphs from U, and the bottom graphs from M. (Source: Battaglini and Palfrey, 2012.)

(in total only 3% of the time). Much more likely the state will transition to an M region. This is because it is very difficult for i to convince any other agent to vote for a Di proposal. This can happen only if the state in U is bordering a region Dj, $j \neq i$, by offering to $k \notin \{i, j\}$ (a currently disadvantaged agent) a more advantageous payoff in Di. From an M state, however, the system moves to a D state with very high probability, more than 90% of the time. From U, the system moves with high probability to an M region.

In the long run, therefore, the D states are absorbing, and the allocations rotate randomly around the three D regions.[24] Hence the stationary distribution of states is concentrated around the vertices, that is, on states in which a single agent receives a payoff between 0.9 and 1.0. When agents are risk neutral, therefore, they behave as if they were myopic, simply choosing allocations that maximize their current payoff.

This finding is consistent with the analysis in Kalandrakis (2004) who characterized an equilibrium of the bargaining game when the state space is the unit simplex (and so the unit of account is infinitesimal). There is, however, a slight difference. Kalandrakis (2004) shows that in the long run only the most extreme states are chosen (i.e., only states in which one agent receives the entire pie). In the equilibrium presented above, this does not occur; indeed with a strictly positive probability at least one of the other agents receives a positive payment. This difference is due to the fact that in the model studied here the proposer must divide the pie in discrete units of $\frac{1}{12}$ of the total size. With a continuum, the equilibrium must have voters voting in favor of the proposal when they are indifferent. (Otherwise, the proposer would have an incentive to sweeten the offer by an infinitesimal amount.) With a discrete pie, this is no longer the case.[25] In the Markov equilibrium selected as the limit in our computations, the proposer does not want to make offers that leave the other players just indifferent, because in the quantal response equilibrium, they would vote in favor of the proposal only 0.5 of the time even for large values of λ, while they would accept all better offers (including the cheapest one) with probability 1. The proposer therefore has an incentive to offer something to his coalition partner. Of course as the grid becomes arbitrarily fine, these equilibria become essentially identical.

CONCAVE UTILITIES

With strictly concave utility functions, agents are averse to sequences of outcomes in which the status quo—and hence their own share of the pie—changes every period. Hence the incentives for more symmetric distributions are greater because such distributions generate less variance across time. Among the least efficient outcomes would be allocation sequences where a single voter, the

---

[24] In the logit MQRE the D states are not completely absorbing, because all states are reachable with positive probability from all other states. The transition table for lower values of λ is smoothed out or "flattened" relative to table 8.4.

[25] Recall from the analysis in chapter 2 that a similar property arises in looking at subgame-perfect equilibrium of the ultimatum game. With a perfectly divisible pie, the only subgame-perfect equilibrium is for the proposer to offer zero and the responder to accept any offer. However, with a discrete grid, there is also a subgame-perfect equilibrium where the proposer makes the smallest positive offer, and the responder accepts only positive offers.

TABLE 8.5. Limiting MPE transition matrix of the seven regions with concave utility ($r = 0.95$) and $\delta = 0.83$. (Source: Battaglini and Palfrey, 2012.)

		SQ_{t+1}						
		D1	D2	D3	M12	M13	M23	U
	D1	0	0	0	0.27	0.27	0.27	0.19
	D2	0	0	0	0.27	0.27	0.27	0.19
	D3	0	0	0	0.27	0.27	0.27	0.19
SQ_t	M12	0	0	0.02	0.77	0.01	0.01	0.19
	M13	0	0.02	0	0.01	0.77	0.01	0.19
	M23	0.02	0	0	0.01	0.01	0.77	0.19
	U	0	0	0	0.01	0.01	0.01	0.98

proposer, appropriates the entire pie in each period. Though in this case an agent is receiving one-third of the pie *on average* (all the pie one-third of the time and none of the pie two-thirds of the time), this gives a lower discounted utility than receiving exactly one-third in *every* period. Proposers can avoid such "rotating dictator" outcomes by choosing a division that is closer to the centroid of the simplex. By allocating a higher share to an agent, the proposer exposes himself *less* to expropriation in the future, because it makes it harder for a future proposer to extract a larger share of the surplus by forming a coalition with an excluded agent. A proposal close to the centroid is harder to overturn and reduces the volatility of the proposer's future payoffs.

To see that rotation of outcomes on the vertices is no longer an equilibrium, consider table 8.5 and figure 8.8, which describe the seven-region transition matrix in the case with $r = 0.95$. The differences between the transition function and stationary distributions between concave and linear utilities are striking. Starting from a D state, when $r = 0.95$ the game *never* remains in a D state, but usually moves to a state M12, M13, or M23 (over 80% of the time) and occasionally to region U. When the state is in a majoritarian region it usually remains in the same region but again move to U with significant probability, 20% of the time. Once region U is reached, it is essentially an absorbing state, rarely moving to the M region and never to a D state.

The long-run incentives to move toward the center are also evident in the stationary distribution, where most of the mass is in the U region, with a small remainder in the M regions. With $r = 0.95$ the probability that agent i receives a payoff between $\frac{1}{3}$ and $\frac{2}{3}$ in the stationary distribution is almost 1, while the corresponding probability of those intermediate payoffs with linear utility is 0.

8.4.3 Using MQRE for Estimation

Just as we illustrated earlier how logit QRE can be used to estimate risk-aversion parameters in auctions and matrix games, MQRE can be used to simultaneously estimate λ and r in this dynamic bargaining game, using data from the Battaglini and Palfrey (2012) experiment. The estimation is done using standard maximum-likelihood methods. Using the path-following algorithm used earlier to approximate MPE, we trace out the logit solution to the game; that is, we trace out a unique connected family of logit MQREs, $\{\rho(\lambda, r), \sigma(\lambda, r)\}$ for increasing values of λ and for r ranging from 0 to 1, where ρ denotes the mixed proposal strategy as a function of the status quo, and σ denotes the voting strategies, which depend on the proposal and the status quo. This defines a likelihood function $L(\widehat{\rho}, \widehat{\sigma}; r, \lambda)$, where $(\widehat{\rho}, \widehat{\sigma})$ are the observed proposal and voting choice frequencies in the data, using the 91-state grid used to compute the MQRE.

The concavity parameter estimate, $\widehat{r} = 0.50$, is large in magnitude and highly significant. The estimate is also very close to concavity estimates from a variety of very different laboratory studies, including auction experiments (Goeree, Holt, and Palfrey, 2002), to abstract game experiments (Goeree, Holt, and Palfrey, 2003), lottery choice experiments (Holt and Laury, 2002), and field data from auctions (Campo et al., 2011). The fitted theoretical MQRE strategies match the pattern of dynamic behavior in the data quite well, particularly in terms of the similarity between the theoretical transitions and the empirical ones.

To summarize, first, even in very complex stochastic games, Markov QRE can be applied to computationally approximate Markov perfect equilibrium as limit points of logit MQRE using path-following algorithms. Second, even in these complex games, one can use MQRE as a structural model to estimate economic parameters that can strongly affect the qualitative and quantitative characteristics of MPE. In this application, doing so led to a convincingly strong rejection of the hypothesis that the utility of the agents is linear. In the bargaining problem here, concavity of the utility function appears to be an essential ingredient of the empirical behavior that was observed in the laboratory.

8.5 CRISIS BARGAINING

Crisis bargaining is a workhorse class of models used to study international conflict (e.g., Fearon, 1995). Negotiations between nation states are often carried out in the shadow of the power that each could bring to bear in the event of an impasse. A standard way of modeling the negotiation process is to allow parties to make demands and counteroffers in an alternating sequence. Crisis bargaining models add an important twist to the standard offer/counteroffer model of economic bargaining to split a pie between two parties. In the economic context, if the bargainers fail to reach agreement then the pie simply disappears and payoffs are zero for both players. The twist for the crisis bargaining context is that impasse results in war, which is costly to both parties, rather than simply a failure to trade. If war occurs, one side wins the entire pie, according to some random process, with both sides incurring a cost.[26]

Unequal ex ante win probabilities determine the extent of power asymmetries, and there can also be a cost of delay, with the size of the pie declining in each successive bargaining period with no agreement. Standard game-theoretic predictions in such a model are quite sharp: although demands are pushed to the precipice with nothing left on the table, there is no conflict in equilibrium regardless of the degree of power asymmetry. Indeed, there is no equilibrium delay in reaching an agreement. These sharp predictions (no extra concessions, no delays, and no conflict), which are also a property of the economic models of offer/counteroffer bargaining with complete information, provide a simple and interesting framework for comparisons with observed bargaining behavior in settings with significant costs of conflict and delay.

Sieberg et al. (2013) reports results of a laboratory experiment designed to investigate the effects of power asymmetries on conflict rates in a two-stage bargaining game with a random conflict outcome. If there is no agreement in the first stage, the amount to be divided shrinks from 10 to 9, with payoffs in dollar amounts. A failure to reach an agreement in the second stage results in the allocation of the remaining amount, 9, to one of the parties, with the win probability, p, for the first mover ("proposer") being either 0.2, 0.4, 0.6, or 0.8, depending on the treatment. The proposer's first stage demand, x_1, is implemented if it is accepted. A "responder" who rejects the initial demand may counter by demanding an amount, x_2, between 0 and 9, which is implemented if it is accepted by the original proposer, who would receive $9 - x_2$. If the proposer rejects, then the random conflict allocation of the full 9 amount is realized, with each player suffering a conflict cost, $c = 2$. Subjects in the experiment kept the same role, proposer or responder, and were randomly matched for 5 periods with the same proposer win probability, followed by a second treatment with 5 rounds of random matching, using a different proposer win probability. Twelve sessions with 12 subjects each were run, with all possible treatment pairs and orders involving one high proposer win probability (0.6 or 0.8) and one low win probability (0.2 or 0.4). All demands were constrained to be integer amounts.

The unique subgame-perfect Nash equilibrium, assuming risk neutrality, is determined via standard backward induction arguments. If negotiations reach the second stage, the responder's counteroffer is minimally greater than the proposer's expected payoff from conflict. This minimal second stage counteroffer corresponds to a maximal responder demand, which is what the proposer offers

---

[26] This section is based on Sieberg et al. (2013).

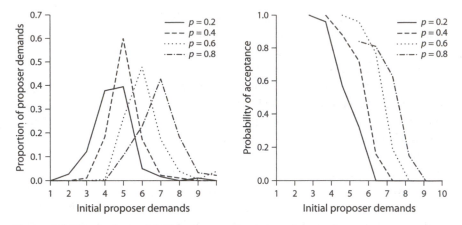

FIGURE 8.9. Distributions of initial proposer demands and responder acceptance rates as functions of proposer demands by treatment.

to the responder in the first stage by demanding the residual. For example, if the proposer win probability is 0.8, then the proposer's expected payoff from conflict is $0.8(9) - 2 = 5.2$, so a responder offer of 6 (demand of 3) would be accepted in the second stage. In this manner, responder second stage demands are 9, 7, 5, and 3 for proposer win probabilities of $p \in \{0.2, 0.4, 0.6, 0.8\}$, respectively. Since a higher proposer win probability results in a lower responder second stage demand, it follows that initial proposer demands are increasing in the proposer win probability, with initial demands of 1, 3, 5, and 7 as the win probabilities increase from 0.2 to 0.8. The observed distributions of proposer demands for each treatment are shown in the left panel of figure 8.9, with progressively higher demand distributions moving from left to right, corresponding to higher win probabilities. Notice that the demand distribution for $p = 0.8$ on the right-hand side peaks at 7, the Nash prediction under risk neutrality, but the other demand distributions are too high and too close together relative to the Nash predictions of 1, 3, and 5.

The observed proportions of initial demands that are accepted by responders, shown in the right panel figure 8.9, show a smooth, "inverse S" pattern which suggests quantal responses instead of sharp responses predicted by a Nash equilibrium. For the $p = 0.8$ treatment, for example, all proposer demands at or below 7 (offers of 3 or above) should be accepted and all higher initial proposer demands should be rejected, which is clearly not the case. In general, the responder accepts proportions that are too low and too close together relative to the spread of 2 in the Nash predictions with acceptance cutoffs of 1, 3, 5, and 7 for the four treatments.

The overall picture that arises from figure 8.9 is that proposers with power use it aggressively by making demands that generally exceed Nash predictions. Responder acceptance proportions, which exhibit a regular inverse S-shape, exhibit countervailing concessions, with some willingness to accept demands above Nash predictions, especially for lower proposer win probabilities. A significant proportion of the negotiations go into the second stage, with a conflict rate of about 0.25 across all treatments, well above the Nash prediction of no conflict.

To address these deviations from game-theoretic predictions, Sieberg et al. (2013) estimate a logit AQRE. In particular, the quantal continuation values for the second stage are calculated in a standard manner for AQRE, that is, with each player perceiving their own subsequent decisions as being played by agents with the same level of precision. For example, the probability that a proposer with a win probability of p will accept the division associated with a responder second stage demand of x_2 can be calculated as a ratio of exponentials of expected payoffs multiplied by a logit precision parameter λ:

$$\sigma_{A1}(x_2|p) = \frac{e^{\lambda(9-x_2)}}{e^{\lambda(9-x_2)} + e^{\lambda(9p-c)}},$$

where the payoffs for the two terms in the denominator correspond to accepting the offer of $9 - x_2$ or opting for the conflict expected payoff of $9p - c$. The equilibrium assumption is that the responder's beliefs match these proposer acceptance probabilities. These beliefs can be used to calculate the responder expected payoffs for each possible value of x_2 between 0 and 9, denoted by $U_2(x_2|p)$, taking account of the responder's expected conflict payoff in the event of a rejection. Then the logit equilibrium probability associated with a second stage responder demand of $x \in \{0, \dots, 9\}$ is given by a ratio of exponentials of expected payoffs associated with each demand, weighted by the logit precision parameter in the usual way:

$$\sigma_2(x|p) = \frac{e^{\lambda U_2(x|p)}}{\sum_{k=0}^{9} e^{\lambda U_2(k|p)}}.$$

The responder's continuation value for the second stage is then calculated by multiplying the responder expected payoffs $U_2(x|p)$ and corresponding probabilities associated with each second stage demand given above. The equilibrium probabilities of initial proposer demands, $\sigma_1(k|p)$, $k = 0, \dots, 9$, and responder acceptance probabilities for the first stage, are calculated analogously by computing a value for continuation for each player. In this manner, a single precision parameter affects all of the various acceptance and rejection probabilities and the demand probabilities for each possible demand that can be made by the proposer and the responder.

To summarize, the model generates acceptance and rejection probabilities for all possible first- and second-stage demands for a given proposer win probability p. Estimation involves finding the value of λ that maximizes the probability of seeing the data counts that were actually observed in the experiment. The likelihood function is a product of terms, each involving a probability of a particular outcome (e.g., initial proposer demand of 8, rejected), raised to a power representing the number of observations in that category. Thus the data for a given treatment consist of the numbers of accepted and rejected demands for each of the 10 possible initial demands and the 9 possible responder demands, for a total of $10 + 10 + 9 + 9 = 38$ data counts for accepted or rejected demands. The likelihood function with 4 treatments, therefore, has 4×38 terms, each consisting of an acceptance or rejection probability raised to a power that corresponds to the data count for that category.

The maximum-likelihood estimation was done separately for the first-half (rounds 1–5) and second-half (rounds 6–10) data. This estimated value of the precision parameter for the first half, $\lambda = 0.762$ (standard error 0.025), is lower (more noise) than the second-half estimate, $\lambda = 1.110$ (standard error 0.031). The profile of AQRE probabilities for initial proposer demands was used to calculate predictions for the average proposer demands for each treatment in the second half, which are shown in the dotted line in the top panel of figure 8.10. For comparison, the average observed proposer demands for the second half are shown by the solid line, which lies above the dashed-line locus of Nash predictions. It is apparent that the QRE model tracks the tendency for initial proposer demands to be too high (except at $p = 0.8$) and too flat relative to Nash predictions.

The middle panel of figure 8.10 shows the averages for second stage responder demands by treatment, which are too low and a little too flat relative to the dashed-line Nash predictions, and again, this pattern of deviations from Nash is captured by the AQRE line.

Even though proposer demands tend to be above Nash predictions for proposer win probabilities below 0.8, the concessionary behavior of responders did not result in more conflicts for low proposer win probabilities. The bottom panel of figure 8.10 shows observed conflict rates (solid line) in the second part (rounds 6–10), which are quite close to the QRE predictions and well above the Nash predictions of no conflict (dashed line at the base of the figure). In particular, QRE predicts a significant proportion of conflicts observed in the data, and the approximate invariance of conflict rates to proposer win probabilities.

An analysis of proposer continuation values provides some insight as to why initial proposer demands might tend to be too high and too flat relative to Nash predictions. The proposer continuation values implied by the fitted QRE choice

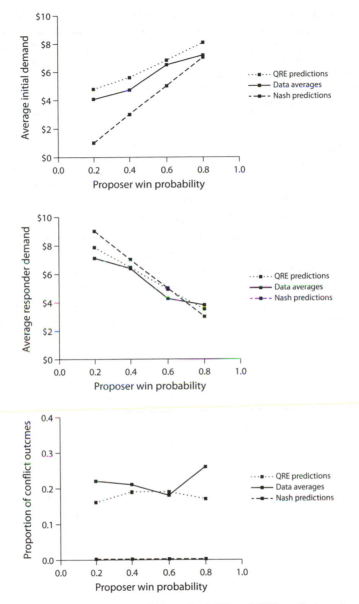

FIGURE 8.10. Average proposer demands (rounds 6–10) and QRE predictions (top panel), average responder demands (rounds 6–10) and QRE predictions (middle panel), and conflict rates (rounds 6–10) and QRE predictions.

TABLE 8.6. Continuation values inferred from QRE and Nash models.

Proposer win probabilities	$p = 0.2$	$p = 0.4$	$p = 0.6$	$p = 0.8$
Proposer continuation values (QRE)	0.99	2.54	4.25	5.89
Proposer continuation values (Nash)	0	2	4	6
Responder continuation values (QRE)	7.05	5.15	3.28	1.47
Responder continuation values (Nash)	9	7	5	3

probabilities are shown in the top row of table 8.6. For example, a proposer with a win probability of 0.2 (and who has an initial demand rejected) can expect to earn 0.99 in the final stage, largely because responders are less aggressive than predicted in this stage in anticipation of "irrational" rejections. The Nash continuation value after a rejected initial proposal would be 0 for this proposer, since the 0.2 chance of winning 9 in a conflict is less than the conflict cost of 2, so a responder second-stage counteroffer of 0 "should" be accepted by a perfectly rational proposer. The other proposer continuation values in the top row are also greater than the Nash continuation values in the second row, although the difference goes away for the highest proposer win probability of 0.8. The important point is that the higher the proposer's continuation value from rejection, the higher the optimal initial demand, which explains why proposer demands lie above Nash levels for win probabilities below 0.8.

The bottom part of table 8.6 shows the inferred continuation values for responders following their rejection of an initial proposer demand. For example, with a proposer win probability of 0.2, the responder continuation value is 9 since the proposer has no power when the conflict cost exceeds the proposer's expected gain from conflict. Note that the inferred responder QRE continuation values in the third row are below their respective Nash continuation values in the bottom row, because of the possibility that proposers will sometimes reject aggressive responder demands and force the responder into a costly conflict. This possibility reduces the optimal responder demand, which helps explain the qualitative direction of deviations from Nash predictions in figure 8.10.

To summarize, maximum-likelihood estimation of the two-stage conflict bargaining model that adds a "noise" dimension to the standard game-theoretic analysis provides a logit precision parameter that, in turn, generates predicted decision patterns, which are similar to the patterns observed in the experimental data:

1) Proposer initial demands are increasing in proposer win probabilities, as expected, but are above Nash levels and are "too flat."
2) Responder final demands are decreasing in proposer win probabilities, but are below Nash levels and are also too flat.

3) Conflict rates are significant and are roughly invariant to changes in proposer win probabilities.

Separate estimation for the part 1 and part 2 data suggests that noise declines in the final 5 rounds, which explains the reduced flatness in the proposer and responder demand patterns, and the reduction in conflict rates observed in the second part.

9

Applications to Economics

Many models of social interactions do not restrict the number of available actions
to be finite. This is especially true in economics where prices, efforts, and physical
inputs are typically assumed to be real-valued variables. As demonstrated by
the traveler's dilemma example of chapter 7, in continuous games, QRE does
not merely predict a (possibly asymmetric) spread around the Nash equilibrium.
Deviations by one player alter others' expected payoff profiles, which induce
further changes etc., creating a "snowball" effect that causes choices to be
substantially different from Nash predictions. This chapter explores whether the
equilibrium effects of noisy behavior can cause similar large deviations from
standard predictions in economically relevant situations.

The traveler's dilemma has the interesting property that payoffs are determined
by the *minimum* of players' decisions, a property that arises naturally in many
economic interactions. For example, in a market for a relatively homogeneous
product, the firm with the lowest price would be expected to obtain a larger
market share. To study such environments, a simple price-competition game is
considered in section 9.1, which is also partly motivated by the possibility of
changing a payoff parameter that has no effect on the unique Nash equilibrium,
but which may be expected to affect QRE. In the application here, intuition would
suggest that a reduction in the market share advantage enjoyed by the low-price
firm would raise the distribution of prices.

In many production settings, each player's contribution to a joint effort is
critical in the sense that a reduction in one player's effort causes a bottleneck that
reduces the payoffs to all. In this case, it is the minimum of all players' efforts
that determines payoffs, along with the cost of a player's own effort. This is a
"coordination game" when the payoff function is such that nobody would wish to
reduce the minimum effort that determines the joint production, but a unilateral
increase from a common effort level is individually costly and has no effect on
the minimum. In the minimum-effort coordination game studied in section 9.2,
any common effort in the range of feasible effort levels is a Nash equilibrium, but
one would expect that an increase in the cost of individual effort or an increase in

the number of players who are trying to coordinate would reduce the effort levels observed in an experiment.

Non-price resource allocations are often made in response to lobbying and other "rent-seeking" activities. There is experimental evidence that indicates the total social costs of such activities often consume an amount that exceeds the value of the prize being awarded, especially with large numbers of contestants. Such *overdissipation of rent* theoretically implies negative expected payoffs, which cannot occur in the Nash equilibrium. Section 9.3 presents an analysis of the logit equilibrium and rent dissipation for a rent-seeking contest that is modeled as an "all-pay auction."

QRE in games with continuous types *and* continuous actions are considerably more difficult to characterize analytically. The last two applications in this chapter are auctions with private information. Two applications are considered: one with independent private values (section 9.4) and the other with independent private signals and a common value (section 9.5). In both cases the type space for private information is discretized, which simplifies both the analysis and the estimation. Similar techniques could in principle be applied to games where the type space is very fine, although the computational issues would be more challenging.

9.1 IMPERFECT PRICE COMPETITION

As with the prisoner's dilemma, the traveler's dilemma example of chapter 7 provides a paradigm designed to make an important point about strategic interactions in an intentionally simplified setting with minimal institutional detail. The minimum-claim feature of the traveler's dilemma, however, arises naturally in many types of economic models, especially those involving price competition.[1] This section presents an analysis of price competition in an experimental setting designed to evaluate the "out-of-sample" predictive accuracy of the logit equilibrium model that was estimated for the traveler's dilemma data.

Capra et al. (2002) report a laboratory experiment based on duopoly with simultaneous price choices from the range [60,160]. The lower of the two prices will be denoted by P_{min}, and the firm selecting this price will, naturally, obtain a larger market share. Demand is perfectly inelastic at a quantity of $1 + \alpha$, with the sales for the low-price firm normalized to 1, and the sales for the high-price firm being α units, with $\alpha < 1$. In the event of a tie, the demand is shared equally, with each firm selling $(1 + \alpha)/2$ units. The product is homogeneous, so the high-price

---

[1] Indeed, Camerer (2003) discusses the traveler's dilemma in a market context.

firm has to meet the lower price to sell any units, but the process of meeting the lower price results in some loss of sales.[2] For simplicity, there are no costs, so the low-price firm earns P_{min} and the other firm, with sales of α, earns αP_{min}.

As long as the market share of the high-price firm is smaller ($\alpha < 1$), the usual Bertrand logic applies and either seller would have an incentive to undercut any common price. In addition, a unilateral price increase above a common level would be unprofitable because it would reduce market share and would not alter the minimum price. Therefore, the unique Nash equilibrium for this game is the lowest price of 60.[3] To summarize, the Nash price is the lowest price when α is below 1. When $\alpha = 1$, however, market shares are equal and each firm has a weakly dominant strategy of charging the maximum price.[4]

The starkly competitive nature of the assumed price competition in a Nash equilibrium is implausible if one expects the degree of buyer inertia to affect pricing behavior. When $\alpha = 0.9$, for example, there is little lost in the way of sales from having the high price, and a firm might be willing to risk a price increase if there is some small chance that the other firm will do the same. In contrast, lower prices might result from lower values of α. The motivation behind the design was the expectation that average price might be smoothly increasing in the degree of buyer inertia, parameterized by α, instead of staying at the Nash prediction and then jumping sharply to the upper limit when $\alpha \geq 1$.

The experiment was run at the University of Virginia with 6 groups of 10 subjects each. Participants were randomly paired for a sequence of 10 rounds, and in each round they would select prices simultaneously from the interval [60, 160] (in pennies). Instead of using a wide range of buyer inertia parameter values in different sessions, as in Capra et al. (1999), multiple sessions were conducted for each of two treatments: 3 sessions with $\alpha = 0.2$, and 3 sessions with $\alpha = 0.8$.

The average prices in penny amounts for the final 5 rounds of each session are shown in table 9.1. These average prices are higher for each of the 3 high-α

---

[2] For example, if some of the buyers were covered by "meet-or-release" clauses, then these buyers would not be able to switch without asking the high-price firm to meet the lower price, which it would do to avoid losing all sales. A large fraction of the other buyers (who do not have meet-or-release clauses in purchase contracts) would presumably go straight to the firm with the lower price, and for this reason, the firm with the lower initial price would have higher sales. See Holt and Scheffman (1987) for an analysis of the effects of meet-or-release and other best-price provisions on market outcomes.

[3] It can be shown that there is no mixed-strategy Nash equilibrium as long as there is a finite upper bound on prices. If demand were inelastic at any price, no matter how high, then a mixed-strategy Nash equilibrium exists, but it has nonintuitive comparative statics properties, i.e., an increase in α (the sales for the high-price firm) would result in *lower* prices on average.

[4] See Dufwenberg et al. (2007) for an experiment that implemented Bertrand competition ($\alpha = 0$) in the presence of price floors.

TABLE 9.1. Comparison of average prices in the low-α and high-α treatments.
(Standard errors in parentheses.)

Treatment	Sessions 1, 2	Sessions 3, 4	Sessions 5, 6	Pooled	Logit predictions
Low $\alpha = 0.2$	63 (14)	72 (20)	73 (32)	69 (13)	78 (7)
High $\alpha = 0.8$	102 (14)	126 (31)	134 (17)	121 (13)	128 (6)

sessions than is the case for each of the 3 low-α sessions, despite the fact that the Nash equilibrium price is 60 for all 6 sessions. A simple nonparametric test can be constructed by noting that there are 20 different ways that 6 numbers from 2 categories can be ranked, that is, "6 choose 3" $= \frac{6!}{3!3!} = 20$. Each of these 20 outcomes is equally likely under the null hypothesis of no effect, but the one shown in the table is the most extreme in the direction of the treatment effect. Therefore, the null can be rejected with a 5% ($\frac{1}{20}$) level of significance for a one-tailed test.

9.1.1 Learning and Convergence to Equilibrium

Figure 9.1 shows the average prices, by round, for each of the 6 sessions, with the high-buyer-inertia sessions shown by the dashed lines at the top. In addition, the thick line shows the average over all sessions in a treatment. Prices for the high-buyer-inertia sessions start high and stay high, with no tendency to approach the Nash prediction of 60. In contrast, the prices in the low-buyer-inertia sessions (with $\alpha = 0.2$) start at an intermediate level and tend to decline slowly toward the Nash prediction. Obviously, these dynamic patterns cannot be explained by a static equilibrium model, suggesting the value of considering models of learning and adjustment along the lines described in chapter 5.

To obtain a computationally tractable learning model, one defines a grid on the strategy space. Let the 101 penny amounts on the range [60, 160] be indexed by $j = 1, \ldots, 101$. Consider a specific person with index i, and let this person's beliefs about the other's price be represented by 101 probabilities, denoted p^i_j for $j = 1, \ldots, 101$, that sum to 1. In the empirical implementation of this model, the initial beliefs are uniform, so $p^i_j = \frac{1}{101}$. An easy way to generalize this setup is to give each of the 101 prices a "weight" of 1, and then let the belief corresponding to a particular price be its weight divided by the sum of the weights for all prices. Then if a particular price is observed, its weight can be increased, which will increase the belief probability associated with that price. One way to do this is to add 1 to the weight of a price that is observed. Specific prices that are observed recently may have a larger impact on beliefs than prices observed in prior rounds, and one way to deal with such "recency effects" is to discount all weights based on past observations by a factor, ρ, with $0 < \rho < 1$. Thus a price with a weight w

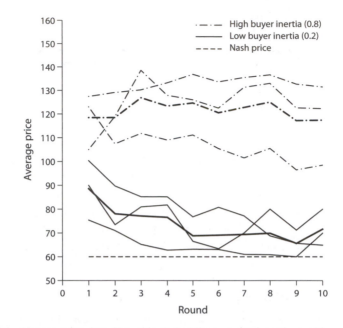

FIGURE 9.1. Average price data. Key: thin dashed lines are for 3 sessions with $\alpha = 0.8$, thin continuous lines are for 3 sessions with $\alpha = 0.2$, thick dashed and continuous lines are averages pooled across all 3 sessions for a given treatment. (Source: Capra et al., 2002.)

that is observed in a particular round would have its weight increased to $\rho w + 1$, and all other prices would have their weights decreased from w to ρw. And a price observed t rounds in the past would have the weight of that observation reduced by a factor of ρ^t, which yields a model of "geometric fictitious play."[5] In the limit when $\rho \to 1$, there is no discounting and each price observation has the same impact on beliefs, regardless of how long ago it was observed.

To summarize, the simplest geometric fictitious-play learning model begins with uniform initial belief probabilities for each price that are updated as the individual observes specific prices selected by others in a sequence of rounds. At a specific point in the experiment, the beliefs of person i are represented by probabilities p_j^i for the price with index j, and these beliefs about the other seller's price can be used to calculate the expected payoff of selecting the price with index j, which will be denoted by U_{ij}, as before. These expected payoffs, one for each possible price choice, in turn can be used to determine choice probabilities via the logit response function. In early rounds, the beliefs determined by the

---

[5] Cheung and Friedman (1997) provide a test of the geometric fictitious play model for 2×2 games, and significant generalizations of this model are discussed in Camerer (2003).

learning rule will typically not closely match the choice probabilities determined by the logit response function, that is, the equilibrium consistency of choice and belief probabilities is *not* imposed in a learning model, since it is not an equilibrium model.

Note that these calculations must be made for specific values of the precision parameter, λ, that is used in the logit response function, and for the recency parameter, ρ, used in the learning rule. Then specific parameter values can be used to *simulate* the learning process, by setting up a group of 10 simulated players, each with equal initial weights for each price (a uniform prior). The resulting beliefs and expected payoffs determine the choice probabilities, and actual choices can be randomly generated. Each simulated player is randomly matched with another, and the partner's simulated decision is used to update each player's beliefs before the next set of simulated prices is randomly generated.

Capra et al. (2002) used the traveler's dilemma data from section 7.1 to obtain an estimate of 0.75 for ρ, and then ran a set of 1000 simulated sessions for each treatment of the price competition game. A simulated session involved 10 players, who were randomly matched for 10 rounds. With $\alpha = 0.2$, the price averages start at about 93 and then decline before leveling off at about 75, as shown by the thick lower line in figure 9.2. For the high-α treatment, simulated prices start high, at about 130, and stay at that level for all rounds; see the thick dashed line in the upper part of the figure. The smooth dotted lines show the range of plus or minus 2 standard deviations for the simulated sessions for each treatment. The thin kinked lines (dashed for $\alpha = 0.8$ and solid for $\alpha = 0.2$) show the average prices for a typical result of a single simulated session for each treatment. These patterns in simulated session prices are similar to the data patterns from the experiment plotted in figure 9.1. In fact, this simulation exercise was done *before the experiment was run*, and the particular values of α (0.2 and 0.8) were selected to make it likely that the price paths would separate and lie on opposite sides of the midpoint of the range of feasible prices.

Notice that the simulated sessions shown in figure 9.2 have price sequences that do not completely settle down in later rounds, as is also the case for human subjects shown in figure 9.1. Continuing randomness is introduced by the logit response functions, but there may be a second factor at work in the simulations, since the recency parameter will magnify the effects of more recent observations. This would cause individuals with different recent observation "histories" to have different belief distributions, even after very long histories of play. The dispersion effect of heterogeneous histories could be part of the explanation for the fact that the logit standard errors for both treatments (in the far right column of table 9.1) are about half as large as the standard errors for the pooled data. Since all belief distributions are identical in a symmetric quantal response equilibrium, it will be

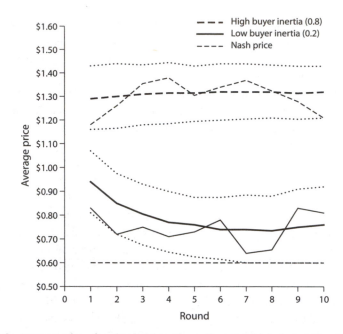

FIGURE 9.2. Average prices for simulated sessions. Key: the thick dashed line is for 1000 simulations with a high α, and the thin dashed line is for a single representative simulation. The thick continuous line is for 1000 simulations with a low α, and the thin continuous line is for a single representative simulation. The dotted lines show the range around the overall averages (plus or minus 2 standard deviations). (Source: Capra et al., 2002.)

the case that heterogeneity caused by different histories across simulated subjects will produce a steady-state choice distribution that is "flatter" than the logit quantal response equilibrium distribution. The steady-state distribution, for any ρ, corresponds to the *stochastic learning equilibrium* (SLE) that was introduced in chapter 5. The next section derives the SLE for the two extreme cases: $\rho = 1$, which implies perfect recall and results in a QRE, and $\rho = 0$, that is, Cournot beliefs.

9.1.2 Logit QRE and Logit SLE Price Distributions

The first step is to characterize the logit QRE choice probabilities for the pricing game. The payoff to a player who sets a price p when the rival sets a price q is given by

$$
u(p, q) = \begin{cases} p & \text{if } p < q, \\ \frac{1}{2}(1 + \alpha)p & \text{if } p = q, \\ \alpha q & \text{if } p > q. \end{cases} \tag{9.1}
$$

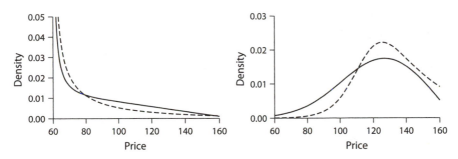

FIGURE 9.3. Logit equilibrium densities (dashed lines) and stochastic learning equilibrium densities (solid lines) for the pricing game with $\alpha = 0.2$ (left panel) and $\alpha = 0.8$ (right panel) using an "out-of-sample" precision parameter $\lambda = 0.12$ that was estimated for the traveler's dilemma game in section 7.1.

Assuming a symmetric outcome where both players use a common price distribution, $F(p)$, that has no mass points, the expected payoff is readily calculated as

$$U(p) = p(1 - F(p)) + \alpha \int_{60}^{p} q f(q) dq.$$

The first term is the product of the player's price and the probability that the other's price is higher (times the sales quantity, which has been normalized to 1). The second term on the right is an integral over prices below the player's own price, p, which are multiplied by α, the sales quantity when the player's price is high.

The symmetric logit equilibrium density is implicitly defined by a fixed-point equation,

$$f(p) = \frac{e^{\lambda U(p)}}{\int_{60}^{160} e^{\lambda U(p')} dp'}. \tag{9.2}$$

No closed-form solution to (9.2) exists, but a numerical solution for the logit choice density can be easily computed.[6] For $\lambda = 0.12$, which was the maximum-likelihood estimate for the traveler's dilemma experiment in section 7.1, the logit equilibrium densities are shown by the dashed lines in figure 9.3: the left panel

---

[6] As explained in section 2.7, the easiest way to solve for the equilibrium density is using the logit differential equation (2.21). For the pricing game, the logit differential equation is given by

$$f'(p) = \lambda f(p)(1 - F(p) - (1 - \alpha)p f(p)),$$

using the fact that the derivative of the expected payoff is given by $U'(p) = 1 - F(p) - (1 - \alpha)p f(p)$. Since $U'(p)$ is a decreasing function of the price choice, p, proposition 2.11 implies uniqueness. Furthermore, $U'(p)$ is an increasing function of α, so an increase in the market-share parameter will raise prices in the sense of first-degree stochastic dominance.

corresponds to $\alpha = 0.2$ and the right panel to $\alpha = 0.8$. The equilibrium densities can be used to calculate expected values and standard deviations (for a sample of size 10) of the equilibrium price choices. These logit predictions for the expected prices are shown in the last column of table 9.1. Notice that the predictions track the average prices "out-of-sample" for each treatment, shown in the adjacent column just to the left of the predictions.[7]

The logit QRE corresponds to the long-run outcome of the learning model in the previous section when $\rho = 1$. To see this, note that the expected payoffs in (9.2) are based on the "correct" distribution of choices. When beliefs are based on a finite sample drawn from this distribution, or on a sample where earlier observations are "geometrically weighted," then beliefs will not be entirely correct. In this sense, QRE arises as the outcome of a learning process only when players have had the opportunity to sample infinitely many times and perfectly recall the entire sample ($\rho = 1$).

Next consider the polar case where players use only a sample of size 1, which corresponds to $\rho = 0$. The steady-state outcome of the learning process with such "Cournot" beliefs corresponds to the stochastic learning equilibrium (SLE) introduced in chapter 5.[8] The SLE equilibrium condition is

$$f(p) = \int_{60}^{160} f(q) \frac{e^{\lambda u(p,q)}}{\int_{60}^{160} e^{\lambda u(p',q)} dp'} dq, \qquad (9.3)$$

with $u(p, q)$ defined in (9.1). On the right-hand side, a particular rival's price, q, is observed with probability $f(q)$, and, together with the logit rule, determines the better response to the belief that q will be selected again by the rival. The equilibrium condition is that the resulting density of choices (on the left) matches the density from which the rival's price was sampled (on the right).

There is no closed-form solution to (9.3) either, but it can be solved numerically. The solid lines in the left and right panels of figure 9.3 correspond to $\alpha = 0.2$ and $\alpha = 0.8$ respectively, using the same "out-of-sample" estimate $\lambda = 0.12$. Note the SLE densities are "flatter" than their QRE counterparts, which confirms the intuition spelled out at the end of the previous section.

---

[7] In general, the value of the precision parameter λ need not be the same for different groups of subjects in different strategic situations, since this parameter will depend on the degree of randomness in individual decisions, which may be influenced by the difficulty of the decision task, confusion about payoffs, the cognitive abilities of the subjects, and any heterogeneity in various nonmonetary motivations of the participants. In this case, however, some of the key aspects of the experiment (subject pool, matching protocol, payoff levels, and complexity of the game) were essentially the same as with the traveler's dilemma experiment discussed in section 7.1.

[8] Indeed, the steady-state outcome of the learning process for any value of ρ, including $\rho = 1$, or for any other learning model for that matter, corresponds to a stochastic learning equilibrium.

9.2 MINIMUM-EFFORT COORDINATION GAMES

Coordination problems arise naturally in many economic activities that involve joint production or other interactions with positive externalities. For example, if the assembly of a product involves two components that are made by different individuals, then excess production by one person can be costly in the sense that it does not increase the number of assembled units completed. In this case, the output of the joint production process is determined by the minimum of the two production efforts, which results in a need to coordinate. Economists have long been interested in coordination games, given the possibility that coordination failure may result in unfavorable outcomes for all. After the prisoner's dilemma, the coordination game is perhaps the most widely studied paradigm in game theory.

In theoretical models with multiple, Pareto-ranked Nash equilibria, it used to be common to *assume* that players could somehow coordinate on the equilibrium that is best for all. This assumption might be reasonable when there is a well-institutionalized coordination device or norm, but laboratory experiments have shown that such coordination is not to be expected uniformly, especially with large numbers of players. In particular, Van Huyck, Battalio, and Beil (1990) found convergence to the Nash equilibrium that is *worst* for all, at least in sessions with large numbers of players (about 15).

Example 2.9 introduces the minimum-effort game with a continuum of decisions. In that example, all nonnegative efforts were allowed, but instead assume that each player, $i = 1, \ldots, n$, chooses an effort, e_i, from a bounded interval $[\underline{e}, \overline{e}]$. Payoffs are determined by the minimum effort minus the cost of the person's own effort:

$$u_i(e_1, \ldots, e_n) = \min(e_1, \ldots, e_n) - ce_i, \qquad (9.4)$$

where $0 < c < 1$ is a common effort cost parameter. As explained in section 2.7, this cost assumption ensures that any common effort level is a Nash equilibrium.

This raises the question, however, of how such a level would be known if there are multiple equilibria. The payoff parameters, which affect the costs of deviation in each direction, would presumably have some impact on what players think others will do. Note that an increase in effort reduces one's payoff by c, and a downward deviation reduces the minimum by 1 and the cost by c, so the cost of a downward deviation from a known common effort is $1 - c$. The costs of deviating are the same in each direction if $c = 1 - c$, that is, $c = \frac{1}{2}$, and the payoff loss from an increase in effort is greater than the payoff loss from a decrease if $c > \frac{1}{2}$. These quick calculations at least provide some intuition for why effort levels might be negatively correlated with costs. Recall that the risk-dominant equilibrium for the 2×2 coordination game in table 2.5 (in section 2.3) involved low effort if the effort cost was less than $\frac{1}{2}$ and high effort otherwise. There is no widely accepted way of generalizing the notion of risk dominance to games with more than two feasible decisions, but there is a related and somewhat more general notion, the *maximization of potential*, which does correspond to risk dominance in the 2×2 case.

9.2.1 Stochastic Potential and Logit QRE

The general idea is to find a single function, the potential, which when maximized with respect to each player's decision variable, yields a Nash equilibrium. The class of potential games includes a well-known variety of interesting games.[9] The hope is that the maximization of potential will help select among multiple Nash equilibria in games where other selection devices fail. For the minimum-effort game the potential function is

$$\text{Prob}[e_1, \ldots, e_n] = \min(e_1, \ldots, e_n) - c \sum_{i=1}^{n} e_i. \tag{9.5}$$

This potential function is identical to player i's payoff in (9.4) except for the addition of a linear term, which, from player i's point of view, reflects *others'* effort costs: $\sum_{j \neq i} c e_j$. When optimizing the potential function with respect to e_i, this additional term makes no difference. In other words, maximization of potential yields the same optimal effort levels as when each player $i = 1, \ldots, n$ maximizes the individual payoff u_i with respect to e_i, resulting in a Nash equilibrium.

Maximization of the potential function in (9.5) requires that all efforts be equal (to avoid wasted excess effort), so the potential can be expressed as a function of a common effort, e: $P = e - nce$, which is maximized at the lowest possible effort $\underline{e}$ when $c > \frac{1}{n}$ and at the highest possible effort $\bar{e}$ when $c < \frac{1}{n}$. Thus the critical value of c is $\frac{1}{2}$ when there are two players, and an increase in the number of players makes it harder to coordinate on the highest effort outcome in the sense that the relevant range of costs is reduced.[10] Nevertheless, the prediction based on the maximization of potential is at one of the extreme ends of the set of feasible

---

[9] The class of games that allows for a potential includes all 2×2 games, and some versions of public goods and Cournot oligopoly games. See Monderer and Shapley (1996) and Anderson, Goeree, and Holt (2004) for more examples.

[10] It can be shown that the maximization of potential for the 2×2 coordination game (in table 2.5) is at low efforts when $c > \frac{1}{2}$ and at high efforts when $c < \frac{1}{2}$, so in this case, the predictions based on potential and risk-dominance coincide (Goeree and Holt, 2004). See Holt (2006, chapter 26) and Goeree and Holt (2003a) for an overview of theoretical and experimental work on coordination games.

efforts, unless c is exactly $\frac{1}{n}$. Intuitively, one might expect more of a smooth transition from low to high effort levels as the cost of effort is increased in the neighborhood of the critical point.

One way to accomplish this is to consider a stochastic generalization of the potential function, as in chapter 5:

$$P_S = \int_{\underline{e}}^{\bar{e}} \prod_{i=1}^{n} (1 - F_i(e)) de - c \sum_{i=1}^{n} \int_{\underline{e}}^{\bar{e}} (1 - F_i(e)) de$$

$$- \frac{1}{\lambda} \sum_{i=1}^{n} \int_{\underline{e}}^{\bar{e}} f_i(e) \log(f_i(e)) de. \tag{9.6}$$

The first two terms on the right are the expectation of the potential function in (9.5). Their sum is maximized by a degenerate distribution that corresponds to a Nash equilibrium. The final term on the right is the standard expression for entropy, which is maximized by a uniform density that corresponds to random choice behavior. Therefore, the precision parameter λ determines the relative weight of payoff incentives and noise.

> ***Proposition 9.1*** (Logit QRE and stochastic potential for the minimum-effort game): *The stochastic potential (9.6) is maximized by the logit QRE.*

The proof can be found in Anderson, Goeree, and Holt (2001). To glean some intuition, note that the derivative of the stochastic potential with respect to $F_i(e)$ yields the first-order condition

$$- \prod_{j \neq i} (1 - F_j(e)) + c + \frac{1}{\lambda} \frac{f_i'(e)}{f_i(e)} = 0,$$

which is the logit differential equation (2.21) for the minimum-effort coordination game. The stochastic potential (9.6) is strictly concave so it has a unique maximizer. Anderson, Goeree, and Holt (2001) further show that the solution is necessarily symmetric, that is, $F_i(e) = F(e)$ for $i = 1, \ldots, n$. To summarize, maximization of stochastic potential for the minimum-effort game is characterized by a single distribution $F(e)$ that solves equation (2.24). Finally, recall from example 2.9 that the logit QRE distribution is first-order stochastically decreasing in the number of players, n, and the effort cost, c, unlike the Nash equilibrium, which is unaffected by changes in these parameters.

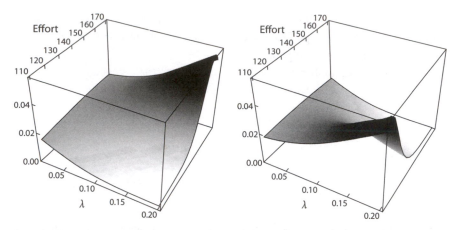

FIGURE 9.4. Logit equilibrium correspondence for the continuous minimum-effort game with two players and $c = \frac{1}{4}$ (left) and $c = \frac{3}{4}$ (right). The logit QRE density is uniform for $\lambda = 0$ and limits to one of the degenerate Nash equilibria as λ grows large. In line with potential maximization, the selected Nash equilibrium distribution puts all mass on $\bar{e} = 170$ when effort costs are low ($c < \frac{1}{2}$, left) and on $\underline{e} = 110$ when effort costs are high ($c > \frac{1}{2}$, right).

For the case of $n = 2$ players, the logit QRE can be solved analytically.[11] Figure 9.4 shows the logit QRE correspondence, which consists of a density for each value of λ, for low effort costs ($c = \frac{1}{4}$) on the left, and for high effort costs ($c = \frac{3}{4}$) on the right. When λ grows large, logit QRE puts all mass on the highest-possible effort when costs are low and on the lowest-possible effort when costs are high. In other words, the logit solution for the continuous minimum-effort game corresponds to the Nash equilibrium that maximizes potential. In contrast with finite games, the other symmetric Nash equilibria of the continuous minimum-effort game are not approachable.

9.2.2 A Coordination-Game Experiment

The invariance of Nash predictions to changes in the effort cost was the motivation for several coordination-game experiments reported in Goeree and Holt (2005a). They began with a simple two-person minimum-effort coordination game, with an effort range from 110 to 170. Subjects were recruited in groups of 10, and played a series of 10 rounds with random matching. Since the critical point (based on maximum potential) was $c = \frac{1}{2}$ for these two-person games, the

---

[11] The logit QRE distribution is

$$F(e) = \frac{\exp(\lambda K(e - \underline{e})) - 1}{\exp(\lambda K(e - \underline{e}))/(1 - c + K) - 1/(1 - c - K)},$$

where K is implicitly determined by the condition $F(\bar{e}) = 1$.

experimental design involved cost values, $c = \frac{1}{4}$ and $c = \frac{3}{4}$, that were not too close to $c = \frac{1}{2}$ and not too extreme either. Three sessions were run under either treatment condition.

The round-by-round average effort decisions for these two treatments are plotted in figure 9.5, where the thin lines represent individual sessions and the thick lines represent averages pooled for all sessions in a treatment. Notice that average efforts for both treatments begin at about the midpoint of the range of feasible effort decisions, and the null hypothesis that the effort distributions for the two treatments are equal cannot be rejected at the 5% level using a standard Kolmogorov–Smirnov test. The average effort series, however, show a clear separation by the fifth round, and for the final 5 rounds the null hypothesis of no treatment effect can be rejected at the 5% level (one-tailed test).[12]

Besides providing qualitative comparative statics predictions, the logit QRE analysis of the previous section can be used to generate quantitative predictions for average prices in the low-cost and high-cost treatments. Table 9.2 contains the average effort levels for each session, for rounds 6–10, with standard deviations in parentheses. The averages for the pooled data are 159 for the low-cost sessions and 126 for the high-cost sessions. The logit QRE predictions are based on a precision parameter of $\lambda = 0.135$, which is the maximum-likelihood estimate obtained from data of both the low-cost and high-cost sessions. As before, for given λ, the resulting equilibrium effort distribution can be calculated using the logit differential equation (2.24) and used to compute expected efforts for each treatment level. These predictions, shown in the far-right column of table 9.2, are quite close to the corresponding averages for the pooled data.

Finally, consider the interesting patterns of adjustment in figure 9.5, with a more or less symmetric spreading out of decisions from the midpoint. Goeree and Holt (1999) report estimates for the geometric learning model discussed in the previous section and show that simulations based on the estimated parameters reproduce the symmetric spreading pattern in the data.

9.3 ALL-PAY AUCTIONS

Many allocation mechanisms have the property that a prize is awarded on the basis of costly activities of potential recipients.[13] For example, exclusive licenses might

---

[12] The intuition behind the test, as explained previously, is that there are "6 choose 3" = 20 possible ways that the average efforts for the 6 sessions could have been ranked by treatment, and of these, the most extreme ranking was observed, i.e., with all 3 low-cost sessions producing the top three effort averages. Hence the probability of this outcome under the null hypothesis is $\frac{1}{20}$, or 5%.

[13] This section includes material from Anderson, Goeree, and Holt (1998a). Also see Goeree, Anderson, and Holt (1998) for a QRE analysis of a "war of attrition," which is essentially a second-price all-pay auction.

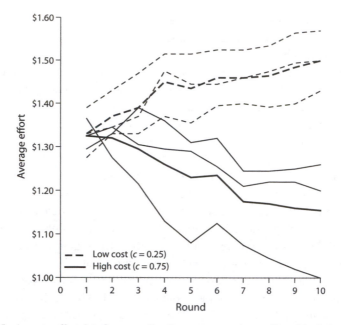

FIGURE 9.5. Average effort data for a coordination-game experiment. Key: thin dashed lines are for 3 sessions with $c = \frac{1}{4}$, thin continuous lines are for 3 sessions with $c = \frac{3}{4}$, and thick lines, dashed and continuous, are averages pooled across all 3 sessions for a given treatment. (Source: Goeree and Holt, 2004.)

be awarded to the person who mounts the most intensive lobbying effort (Tullock, 1967), or to those who wait in line the longest (Holt and Sherman, 1982). Tullock (1980) initiated a large literature in which it is assumed that the probability of obtaining the prize is an increasing function of the lobbying effort. The limiting case, in which the award is always made to the person with the maximum effort, is known as an "all-pay auction," since the losing contenders are not reimbursed for their efforts. An auction-like selection mechanism is appropriate for cases in which the efforts have some value to the person making the allocation or cases in which the efforts are observed, and rewarding high effort is considered to be fair, for example, standing in line. Since lobbying activities entail social costs, the main focus of this extensive literature is on the aggregate costs, which tend to "dissipate" the net social value of the prize being awarded.

Because of the similarity to auction games, in the theoretical literature efforts in rent-seeking games are typically referred to as bids: the prize goes to the highest bidder, but all must pay their bid amounts. The simplest case is one in which the prize has a known value of V that is the same for all. Each bidder, $i = 1, \ldots, n$, simultaneously makes a bid, $b_i \in [0, B]$, for some $B \geq V$.

TABLE 9.2. Data averages and logit QRE distribution averages for $\lambda = 0.135$. (Standard errors in parentheses. Source: Goeree and Holt, 2004.)

Cost	Session 1	Session 2	Session 3	Pooled	QRE predictions
Low ($c = \frac{1}{4}$)	151 (10)	166 (5)	159 (12)	159 (11)	154 (12)
High ($c = \frac{3}{4}$)	131 (11)	112 (5)	135 (11)	126 (14)	126 (12)

The Nash equilibrium bidding function serves as a benchmark for QRE comparisons, and is derived as follows. If others' bids are known and are less than V, then the best response would be to bid slightly above the others. Everyone bidding V, however, is not an equilibrium as it results in negative profits. Hence, the Nash equilibrium will involve randomization. In a symmetric mixed-strategy Nash equilibrium characterized by a common bid distribution $F_N(b)$, the probability of winning with a bid of b is $F_N(b)^{n-1}$, and the expected payoff is[14]

$$U(b) = V F_N(b)^{n-1} - b. \tag{9.7}$$

Since exit, with a zero payoff, is always an option, the expected payoff in a mixed-strategy Nash equilibrium must be nonnegative. This implies that the lowest bid over which randomization occurs must be 0, which generates an expected payoff of 0. All bids over that are selected with positive probability will, therefore, have zero expected payoffs, and the Nash distribution can be found by equating the expected payoff in (9.7) to 0, to obtain $F_N(b) = (b/V)^{1/(n-1)}$ for $b \in [0, V]$ with associated density

$$f_N(b) = \frac{1}{(n-1)V} \left(\frac{b}{V} \right)^{(2-n)/(n-1)}. \tag{9.8}$$

Since the distribution function, $F_N(b)$, is increasing in n and decreasing in V, it follows that bids will be increased in the sense of first-degree stochastic dominance when there is a reduction in the number of bidders, or when there is an increase in the prize value, both of which are intuitive properties. Notice that the Nash equilibrium density, $f_N(b)$, is constant (uniform) for $n = 2$ and decreasing in b for $n > 2$.[15]

Next, consider the symmetric logit QRE, which will be characterized by the logit differential equation as in earlier sections of this chapter. First, note that the expected payoff derivative is

$$U'(b) = (n-1)V F(b)^{n-2} f(b) - 1, \tag{9.9}$$

---

[14] Note that the N subscript here indicates a Nash equilibrium distribution.

[15] Baye, Kovenock, and de Vries (1996) show the Nash equilibrium is unique for this formulation of the all-pay auction.

which is substituted into the logit differential equation to obtain

$$f'(b) = \lambda f(b)\big((n-1)VF(b)^{n-2}f(b) - 1\big). \tag{9.10}$$

The expected payoff derivative (9.9), viewed as a function of F, f, and b, does not depend on b directly (but only indirectly through the distribution and density functions), so proposition 2.12 applies. It follows that an increase in V will raise the QRE bid distribution in the sense of first-degree stochastic dominance.

The effect of increasing n is more complicated. Consider first the case of two bidders, $n = 2$, for which the logit differential equation reduces to

$$f'(b) = \lambda f(b)(Vf(b) - 1),$$

which can be integrated to obtain

$$f(b) = \frac{1}{V}\frac{1}{1 - Ke^{\lambda b}}, \tag{9.11}$$

where K is a constant of integration obtained by equating the integral of the density over $[0, B]$ to 1:

$$K = \frac{1 - e^{\lambda(V-B)}}{1 - e^{\lambda V}}. \tag{9.12}$$

If the upper bound $B = V$, so that bidding for a sure loss is ruled out, then $K = 0$ and the density in (9.11) is constant on its support, independent of λ. The resulting uniform distribution is the same as the Nash equilibrium distribution (with $n = 2$) computed previously. This is intuitive as logit QRE interpolates between random behavior ($\lambda = 0$) and Nash behavior ($\lambda = \infty$), but if the two coincide then so does logit QRE. If overbidding errors are allowed, that is, $B > V$, then the constant, K, determined by (9.12) is negative and the logit equilibrium density is everywhere decreasing on its support.[16] The associated distribution function is

$$F(b) = F_N(b) - \left(\frac{B}{V} - 1\right)\log\left(\frac{f(b)}{f(B)}\right),$$

which shows that the logit bid distribution first-order stochastically dominates the Nash bid distribution. Since rents are "fully dissipated" in the Nash equilibrium, that is, bidders' expected payoffs are 0, this implies that rents are "overdissipated" in the logit QRE. This result may appear somewhat obvious when trembles that allow for sure losses are built into the model. However, the next section shows that with three or more bidders, overdissipation occurs even when such trembles are precluded by assuming $B = V$.

---

[16] When $\lambda \to \infty$, the constant of integration $K \to 0$ since $B > V$, so $f(b) \to f_N(b)$. This is the familiar result that logit QRE limits to Nash when the precision parameter grows large.

9.3.1 Rent Dissipation

In order to characterize the extent to which the rent associated with the prize value is dissipated, recall that ex ante expected payoffs are 0 for all bidders in a Nash equilibrium, so the rent is fully dissipated. In laboratory experiments, Davis and Reilly (1998) report negative profits and rents that are more than fully dissipated, so the obvious question is whether such excess dissipation can occur in a quantal response equilibrium. It turns out that overdissipation depends on the number of bidders.

> **Proposition 9.2** (Rent dissipation in an all-pay auction): *In the logit QRE, there is overdissipation of rents in the symmetric-value all-pay auction with $B = V$ when there are 3 or more bidders.*

Proof: The logit equilibrium density for the symmetric n-person all-pay auction can be written as

$$f(b) = f(0)e^{\lambda(VF(b)^{n-1}-b)},$$

where $f(0)$ is determined by the requirement that the density integrates to 1. Recall that the Nash equilibrium density for the symmetric game, $f_N(b)$ is a decreasing function of b when $n > 2$. The ratio of the logit and Nash densities is

$$\frac{f(b)}{f_N(b)} = \frac{f(0)}{f_N(b)}e^{\lambda(VF(b)^{n-1}-b)}. \tag{9.13}$$

The mixed-strategy Nash equilibrium satisfies $V F_N(b)^{n-1} - b = 0$ so the term in the exponent on the right-hand side of (9.13) is 0 whenever the logit distribution crosses the Nash distribution, $F(b) = F_N(b)$, and $f(b)/f_N(b) = f(0)/f_N(b)$ at any such crossing. Since $f_N(b)$ is decreasing when $n > 2$, the ratio of densities is increasing at successive crossings. It follows that there can be at most two crossings, that is, at the lower and upper bound of the support, and that the density ratio is less than 1 at the lower bound and greater than 1 at the upper bound. Hence, $F_N(b) > F(b)$ at all interior points. Since the logit distribution lies below the Nash distribution that fully dissipates rents, logit QRE produces higher expected bids, which in turn implies that rents are overdissipated. ∎

Since overdissipation occurs even if trembles that result in sure losses are ruled out, that is, $B = V$, it is intuitive that overdissipation is even more likely when $B \geq V$. This can be verified from the explicit solution for the logit QRE

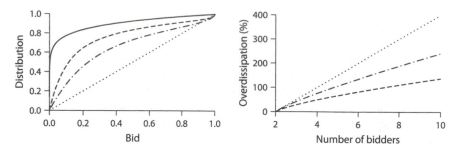

FIGURE 9.6. The left panel shows logit equilibrium distributions for an all-pay auction with $B = V = 1$ and $n = 10$ bidders: the dotted line corresponds to $\lambda = 0$, the dashed-dotted line to $\lambda = 4$, the dashed line to $\lambda = 10$, and the solid line to $\lambda = \infty$ (i.e., Nash equilibrium bid distribution). The right panel shows the degree of overdissipation, as a percentage of the prize $V = 1$, in the logit QRE.

distribution

$$F(b) = \left(\frac{1}{\lambda V} \right)^{1/(n-1)} \Gamma^{(-1)} \left(\frac{1 - e^{-\lambda b}}{1 - e^{-\lambda B}} \Gamma \left(\lambda V, \frac{1}{n-1} \right), \frac{1}{n-1} \right)^{1/(n-1)}, \quad (9.14)$$

with $\Gamma(z, a) = \int_0^z t^{a-1} e^{-t} dt$ the incomplete gamma function that is strictly increasing for positive z. It is straightforward to verify that $F(b)$ is decreasing in the upper bound on the bids, B, and strictly increasing in the number of bidders, n. Moreover, using standard properties of the incomplete gamma function one can show that $\lim_{\lambda \to \infty} F(b) = F_N(b)$.

The left panel of figure 9.6 shows logit equilibrium distributions for various values of the precision parameter ($\lambda = 0$, $\lambda = 4$, and $\lambda = 10$) in an all-pay auction with $n = 10$ bidders and $B = V = 1$. The solid line shows the Nash distribution, which results in exact dissipation of the rent ($V = 1$). The right panel shows the degree of overdissipation under logit QRE for different numbers of bidders (using the same three precision parameters).

Davis and Reilly (1998) report experimental results for an all-pay auction with 4 bidders, $c = 1$, and *no upper bound on bids*. The implication of proposition 9.2 is that rents will be overdissipated, since there would be overdissipation with $B = V$ and the logit distribution (9.14) is decreasing in B.[17] The social costs of rent seeking in the experiment consistently exceeded the prize value, so subjects lost money on average. The losses were more prevalent in early periods. This propensity for losses was dealt with by providing subjects with a relatively large initial cash balance.

---

[17] Note that for λ sufficiently low enough, there will be overdissipation even when $B < V$ as long as $B > V/n$.

9.3.2 Own-Payoff Effects in Asymmetric All-Pay Auctions

Consider next the case of value asymmetries, which can lead to nonintuitive comparative statics results for a Nash equilibrium. In particular, an increase in one bidder's value will not alter the Nash density for that bidder. Suppose that $V_1 > V_2$. With asymmetries, the probability that one person has the high bid is determined by the other person's equilibrium distribution. Therefore, the expected payoff for bidder 2 is $V_2 F_1(b) - b$. In a mixed-strategy Nash equilibrium, this expected payoff is constant over the range of randomization, and therefore F_1 is a function of V_2, but not of V_1, and vice versa. In contrast, the logit equilibrium exhibits "own-payoff" effects, which is demonstrated with an asymmetric, two-bidder all-pay auction.

> **Proposition 9.3** (Own-payoff effects in an asymmetric all-pay auction): *In a two-bidder all-pay auction, an increase in a bidder's value raises that bidder's logit equilibrium bids (in the sense of first-degree stochastic dominance).*

Proof: The differential equations characterizing logit QREs are given by

$$f_1'(b) = \lambda f_1(b)(V_1 f_2(b) - 1), \qquad f_2'(b) = \lambda f_2(b)(V_2 f_1(b) - 1).$$

Multiplying the left equation by V_2 and the right one by V_1 and subtracting yields

$$V_2 f_1'(b) - V_1 f_2'(b) = -\lambda(V_2 f_1(b) - V_1 f_2(b)).$$

This equation can be integrated to obtain[18]

$$V_1 f_2(b) - V_2 f_1(b) = Ae^{-\lambda b},$$

with A a constant of integration, which can be determined by integrating one more time to obtain

$$V_1 F_2(b) = V_2 F_1(b) + (V_1 - V_2)\frac{1 - e^{-\lambda b}}{1 - e^{-\lambda B}}, \tag{9.15}$$

using the fact that $F_1(B) = F_2(B)$, which implies $A = \lambda(V_2 - V_1)/$ $(1 - e^{-\lambda B})$. Bidder 1's logit density is given by $f_1(b) = f_1(0)e^{\lambda U_1(b)}$ with $f_1(0)$ a constant that ensures the density integrates to 1. Using (9.15), bidder

---

[18] This equation expresses $f_2(b)$ as a function of $f_1(b)$ (and vice versa), and these can be substituted back into the appropriate logit differential equations to obtain a single differential equation for each of the logit densities, which can be solved explicitly; see Anderson, Goeree, and Holt (1998b, note 15). Our main interest here, however, is on the effects of a change in a bidder's value on that bidder's own bid distribution.

1's expected payoff, $U_1(b) = V_1 F_2(b) - b$, can be expressed in terms of $F_1(b)$:

$$f_1(b) = f_1(0) \exp\left(\lambda\left(V_2 F_1(b) + (V_1 - V_2)\frac{1 - e^{-\lambda b}}{1 - e^{-\lambda B}} - b\right)\right).$$

If bidder 1's value is increased to $\tilde{V}_1 > V_1$, the new equilibrium condition is

$$\tilde{f}_1(b) = \tilde{f}_1(0) \exp\left(\lambda\left(V_2 \tilde{F}_1(b) + (\tilde{V}_1 - V_2)\frac{1 - e^{-\lambda b}}{1 - e^{-\lambda B}} - b\right)\right).$$

Note that at any crossing of the distribution functions, $F_1(b) = \tilde{F}_1(b)$, these expressions imply that the ratio of slopes is given by

$$\frac{\tilde{f}_1(b)}{f_1(b)} = \frac{\tilde{f}_1(0)}{f_1(0)} \exp\left(\lambda(\tilde{V}_1 - V_1)\frac{1 - e^{-\lambda b}}{1 - e^{-\lambda B}}\right). \tag{9.16}$$

The right-hand side of (9.16) is strictly increasing in b since $\tilde{V}_1 > V_1$ by assumption, and therefore, the ratio of the densities is increasing at successive crossings. There can thus be only two crossings, since if there were more, the ratio of densities would either decrease and then increase or the reverse, a contradiction. Since there are only two crossings and the ratio in (9.16) is increasing, it must be less than 1 at $b = 0$ and greater than 1 at $b = B$. It follows that $F_1(b) > \tilde{F}_1(b)$ for all interior values of b.   ∎

This section assumed that prize values are common knowledge, which contrasts with most of the auction literature where models bidders' values are privately known. The next section looks at the application of QRE to auctions with private information, which are modeled as Bayesian games. In the economic theory literature, these auction games nearly always consider a continuum of types (private information), and a continuum of possible bids.[19] To make the QRE analysis of such games tractable, the auctions are modeled here as games with a finite number of types and a finite number of possible bids. In principle, the number of possible types and bids could be arbitrarily large, so that it approximates a game with continuous types and bids, but to obtain computational results, the assumption of finiteness (and a relatively small number of bids) is maintained here.

---

[19] In experimental studies of such auctions, the number of types and the number of possible bids is restricted (either implicitly or explicitly) to be finite, e.g., pennies.

9.4 PRIVATE-VALUE AUCTIONS

In 1989, Glenn Harrison published an article in the *American Economic Review* (Harrison, 1989) that in some ways foreshadowed the development of quantal response equilibrium as a stochastic theory of games. The main point of his article was that in the private-value auctions economists conducted in the laboratory, the payoffs are (locally) flat at the optimal bidding function, and this has since come to be known as the "flat maximum critique." Most experiments indicated that subjects in these experiments bid higher than the equilibrium bids, which were derived under the assumption of risk neutrality. A proposed explanation was that bidding in auctions is risky, and this apparent overbidding was consistent with risk-averse bidders. The main point of Harrison (1989) was that the expected payoff losses from bidding above the equilibrium bid were in fact very low. He then argued that the observed overbidding could be an artifact of the flat payoff structure, and may have nothing whatsoever to do with risk aversion, or any other preference-based explanation that one might dream up. This article was met with a flurry of comments from several economists who were early entrants to the growing field of experimental economics, and the comments and replies were published in the *Review*.

One of the more relevant comments for this discussion of error effects was provided by Friedman (1992), who pointed out that the flat maximum critique by itself does not explain the observed systematic overbidding, because expected payoff losses from underbidding relative to equilibrium are just as small as the expected payoffs from overbidding. There would have to be some kind of asymmetry with respect to under- versus overbidding to account for the data. Harrison's idea of the flat maximum, together with Friedman's insight that departures from equilibrium should be affected by asymmetries from under- or overbidding, inspired a laboratory experiment where two very similar first-price private-value auction games are compared. Both games have the exact same Bayes–Nash equilibrium bidding function, but in one game the loss in expected payoffs from overbidding are much greater than the expected payoff loss from underbidding, while the opposite is the case for the other auction game.

Quantal response equilibrium takes account of the flatness of payoffs, in a way that incorporates not only the weak incentives to play a fully optimal strategy, but also recognizes that this can have equilibrium ripple effects that feed back and lead players to adjust their bidding behavior in response to the knowledge that the other bidders are quantal responding instead of best responding. Consistent with Friedman's conjecture, the logit quantal response equilibrium predicts lower bidding when the payoff losses from underbidding relative to Nash are reduced. The design uses a very coarse value and bidding structure, which is simple enough to perform a structural estimation where one can simultaneously measure the effect of payoff flatness (by estimating the logit precision parameter) and at the same time estimate risk-aversion parameters (or other preference parameters). Thus, not only does the design allow for a test of the Friedman conjecture based on the flat maximum critique, but it also allows a formal statistical test for the presence of risk aversion that takes account of the flat payoff structure.

9.4.1 Two Simple Auction Games

There are two bidders competing in a first-price sealed-bid private-value auction.[20] Values are independent draws from a uniform distribution over six possible values. In the Low game, the set of values is $V_L = \{0, 2, 4, 6, 8, 11\}$, and the set of allowable (integer) bids is $B = \{0, 1, 2, \ldots, 11\}$. In the High game the set of values is $V_H = \{0, 3, 5, 7, 9, 12\}$ and the set of allowable (integer) bids is $B = \{0, 1, 2, \ldots, 12\}$. In both cases the unique symmetric pure-strategy Bayes–Nash equilibrium is the same: bid 0 with the lowest value, 1 with the second lowest value, and so forth. Because the positive values in the High game are shifted up by 1, this implies different expected payoffs from deviating from an equilibrium than in the Low game. Figure 9.7 compares the payoff losses in the two games from deviating one or two bidding increments above and below the equilibrium (e.g., the payoff losses from bidding 1, 2, 3, 4, or 5 if you have the third highest value and the Bayes–Nash prediction is to bid 3). These expected payoff losses are calculated under the assumption that the bidder faces a uniform distribution of bids over the set $\{0, 1, 2, 3, 4, 5\}$, which is true in the unique Bayes–Nash equilibrium.

An experiment to compare bidding behavior in these two games was reported in Goeree, Holt, and Palfrey (2002), using subjects from Caltech and the University of Virginia. There were 4 sessions in each location, all conducted with 10 subjects, using pencil and paper rather than computers. In each round, each subject rolled a fair six-sided die in private to determine their value for that round (this roll was observed also by the experimenter), and then wrote down an allowable bid. At the end of a round, each subject was told the bid of the subject they were matched with, and their resulting payoff. There were 15 rounds in each session and subjects were randomly and anonymously rematched each round. The values and allowable bids, described above, were denominated in US dollars and subjects were paid the sum of their earnings in all 15 rounds.

Figure 9.7 clearly shows that the expected losses from overbidding by one dollar are about three times as much in the Low treatment than the High treatment. Similarly, the expected losses from underbidding by one dollar are about three

---

[20] This section includes material from Goeree, Holt, and Palfrey (2002).

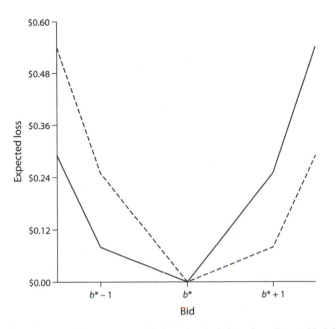

FIGURE 9.7. Expected payoff losses in equilibrium (as deviations from Bayes–Nash bids). The solid line is for the Low game and the dashed line is for the High game.

times as much in the High treatment than the Low treatment. As a consequence, QRE predicts lower bidding in the Low game than in the High game. The data show exactly this relative pattern of bidding. Figure 9.8 shows the entire time series of average bids, for each value, aggregated over all the data in the experiment.

The QRE response parameter, λ, and constant relative risk-aversion parameter, $r \in [0, 1)$, are jointly estimated using the power Luce specification of regular QRE. In equilibrium, the probability of bidding b with a value of v is then given by

$$\sigma(b|v) = \frac{U(b|v;r)^\lambda}{\sum\limits_{b' \in B} U(b'|v;r)^\lambda},$$

where $U(b|v;r)$ is the equilibrium expected payoff from bidding b with value v if the risk-aversion parameter is r:

$$U(b|v;\rho) = \frac{(v-b)^{1-r}}{1-r} P_{\text{win}}(b),$$

where the probability of winning is given by

$$P_{\text{win}}(b) = \frac{1}{6} \sum_{v \in V} \sum_{b' < b} \sigma(b'|v) + \frac{1}{2} \cdot \frac{1}{6} \sum_{v \in V} \sigma(b|v).$$

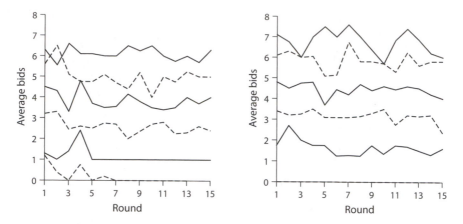

FIGURE 9.8. Time series of bids as a function of value. Low game in left panel; High game in right panel. The lowest curve in each graph shows the bids for the lowest value $v = 0$, and so forth.

The first term is the probability that b is strictly higher than the other bid, and the second term corresponds to the case where b is tied for the highest bid.

The estimated values are $\widehat{\lambda} = 10$ and $\widehat{r} = 0.52$.[21] The risk neutral model ($r = 0$) is strongly rejected by a likelihood ratio test, and the estimated values of r when estimated separately for the two games are not significantly different from each other. Similar estimates of r are obtained if one uses a logit QRE specification instead.[22]

The accuracy of the fit to the aggregate data is illustrated in the next two figures. Figure 9.9 shows, for both treatments, the Bayes–Nash, empirical, and fitted average bids as a function of value. The fitted bids were done using out-of-sample estimates. That is, the fitted values for the Low game are based on the estimates of $\widehat{\lambda}$ and $\widehat{r}$ obtained from using only the High game data, and vice

[21] The payoff numbers used in the estimation were expressed in integer dollar amounts, which is why the precision estimate of 10 is so much higher than the precision estimates of about 0.1 for some of the other games discussed earlier in the chapter.

[22] An alternative hypothesis about overbidding is that subjects simply enjoy winning, so there is a discontinuity in the payoff function, with subjects strictly preferring to win with a bid equal to value than to lose. This can be represented by incorporating a win bonus, w, into the utility function, and implies an expected utility given by

$$U(b|v; \rho, w) = \left(\frac{(v - b)^{1-r}}{1 - r} + w \right) P_{\text{win}}(b).$$

When one estimates this more complicated three-parameter QRE model, the estimated value of the risk-aversion parameter r is virtually unchanged and there is no significant improvement in fit by allowing the additional "joy of winning" parameter.

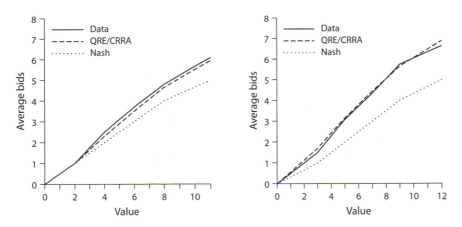

FIGURE 9.9. Average bids by value. Low game in left panel; High game in right panel. Comparison of the risk neutral Nash equilibrium, the data, and the fitted model.

versa. Even though the Nash equilibria are the same for each prize value, the observed bid averages in figures 9.8 and 9.9 are generally lower for the Low game (left panels) as compared with the comparable averages for the High game (right panels).

The estimates go beyond making predictions about the *average* bids for each value, and in fact make predictions about the entire joint distribution of bids and values. Figure 9.10 compares the empirical (left panels) and fitted joint distribution of bids and values for the Low and High auctions. The fitted distributions are based on out-of-sample estimates, in the same way as figure 9.9. The distributions of the empirical and predicted bids shown in these figures are almost indistinguishable.[23]

9.5 COMMON-VALUE AUCTIONS

In common-value auctions there are a number of interesting effects of quantal response equilibrium on the distribution of bids. The first is that bids are now stochastic functions of signals, so bidders may overbid or underbid relative to the Bayesian Nash equilibrium. A second and more subtle effect is that, because quantal response adds noise to the bidding functions, this decreases the linkage

---

[23] When these paired figures were shown to a colleague at Caltech, he remarked confidently that he could see which one was based on theoretical calculations and which one was data, but then guessed incorrectly.

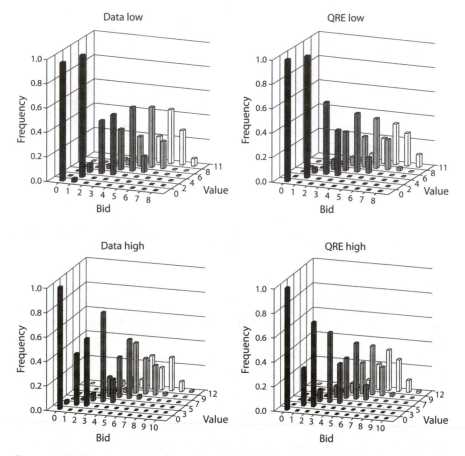

FIGURE 9.10. Top: empirical (left panel) and fitted (right panel) joint distribution of bids and values for the Low auction game. Bottom: empirical (left panel) and fitted (right panel) joint distribution of bids and values for the High auction game.

between bids and signals, and hence weakens the winner's curse effect. If one thinks of the extreme case, where $\lambda = 0$, bids are completely uninformative about signals. The winner's curse arises precisely because the information contained in submitting the highest bid is that everyone else's signal estimate of the value was lower, so the conditional expected value of the object when you submit the winning bid is lower than the expected value conditional only on your signal. The diminishing of the winner's curse then decreases the expected losses from overbidding, relative to the equilibrium. Thus, since the cost of overbidding is less, an indirect effect of QRE is to bid higher. Intuitively, adding extra noise to the bidding function of all bidders has a qualitatively similar effect to having partially cursed beliefs, as it reduces the correlation between bids and signals. It is also similar to the effect of garbling all the signals.

9.5.1 The Maximum-Value Auction Game

Consider a very simple two-person *second-price auction* for a single valuable object, called the *maximum-value auction*, where the joint distribution of signals and value is very extreme.[24] Each bidder draws a signal, x_i, from a uniform distribution on the set $\{0, 1, 2, 3, 4, 5, 6, 7, 8, 9, 10\}$, and the common value is equal to $V = \max\{x_1, x_2\}$.[25] It is easy to show that the game is solvable in two stages of iteration of weakly dominated strategies. In the first iteration, bidding one's own signal weakly dominates bidding any lower amount. Given that nobody is underbidding, it follows that bidding one's own signal weakly dominates bidding any higher amount. Thus, the unique Bayesian Nash equilibrium (BNE) that survives iterated elimination of weakly dominated strategies is to submit a bid equal to one's private signal. That is, $B^*(x_i) = x_i$.

Ivanov, Levin, and Niederle (2010) reported results of an experiment to study bidding behavior in this auction game. The experiment was conducted at Ohio State University and used a variant on the strategy method to elicit bidding functions. Over a sequence of 11 rounds each subject was assigned a signal, drawn from the distribution without replacement, and asked to submit a bid. Bids could be any experimental dollar amount between $0.00 and $1,000,000.00. Each experimental dollar was worth US$0.50. They were randomly and anonymously matched with another bidder in each round, to determine payoffs, and there was no feedback at all until final payment at the end of the experiment. These 11 rounds of bidding constituted phase I of the experiment. Because a subject was assigned a different signal in each of the 11 rounds, a complete bid function could be constructed for each subject. Phase II also consisted of a sequence of 11 rounds of bidding, using the same procedures but with one exception. Rather than being matched against another subject in the room, they were matched against a computer that was programmed to bid exactly according to the same subject's bidding function that was elicited in phase I.[26]

The data exhibited extreme overbidding, which had some interesting patterns that will be described shortly. To give an idea of how much overbidding there

---

[24] This section includes material from Camerer, Nunnari, and Palfrey (2014).

[25] This information structure has been used in other common-value games. A buyer–seller bargaining game with this maximum-value structure was studied by Carrillo and Palfrey (2009), and an asset market where the asset's value was equal to the maximum signal among the traders was studied in Angrisani et al. (2011).

[26] Subjects were fully informed about this. Ivanov, Levin, and Niederle (2010) ran several variations on this game, but the analysis here is limited to this primary treatment.

was, 2.27% of bids were equal to $1,000,000.00 and 7.66% of bids were at least $100.00. The question is why. Are these bidders (undergraduates at Ohio State University) crazy? Do they understand the rules? A partial answer is provided by the logit QRE model and the answers to the last two questions are "probably not" and "probably," respectively.

9.5.2 Logit QRE for the Maximum-Value Auction Game

To develop some intuition about how logit QRE behavior might differ from Nash equilibrium it is instructive to consider the two extremes of QRE. At one extreme, if players are perfectly rational, then beliefs will correspond to the symmetric Nash equilibrium where every player believes that the opponent will always bid his signal. However, this means that equilibrium payoffs are extremely flat. In fact, it is completely flat above one's signal: that is, every bid above one's signal is a best reply to a signal-bidding opponent in this second-price auction. Thus, in a high-rationality quantal response equilibrium when λ is high, payoff functions are nearly flat on the entire bidding space bounded below by $b = x$. In contrast, bidding below one's signal leads to relatively large expected payoff losses, implying an asymmetry between over- and underbidding. This suggests that models based on QRE might help explain overbidding in the maximum-value auction game.

At the opposite extreme, $\lambda = 0$, your opponent is bidding completely randomly. But this implies zero correlation between your opponent's signal and his bid, so there is in fact no possibility of a winner's curse. In fact, simple calculations show that the optimal bid response to a random opponent when $x_i = 0$ is 5 and the optimal bid increases monotonically to 10 when your value is 10. In fact, for all values of x_i except 10, the optimal bid is strictly above x_i.

Hence, the equilibrium effect of quantal responses pushes in the direction of overbidding at both extremes of beliefs about the opponent's quantal response behavior, $\lambda = 0$ or $\lambda = \infty$. Furthermore, the intuition from the optimal reply to a totally random opponent implies that the incentives to bid above signal are highest for relatively low signals, and decline for higher signals. This is shown to be the case in Camerer, Nunnari, and Palfrey (2014). Average bids in a quantal response equilibrium exhibit a *"hockey-stick" pattern*: relatively flat overbidding for low signals converging to near-signal bidding for higher signals.

9.5.3 Estimating HQRE for the Maximum-Value Auction Game

Chapter 5 developed the heterogeneous quantal response equilibrium (HQRE) model, where all players have common rational expectations about the joint probability distribution of actions and signals, but each player may have a different response parameter, λ_i. In the maximum-value auction, a player i's expected utility from bid b, conditional on receiving private signal x, is

$$U(b|x) = \big((E(V)|b > b_{-i}) - E(b_{-i}|b > b_{-i})\big)\, \text{Prob}[b > b_{-i}]$$
$$+ \tfrac{1}{2}\big((E(V)|b = b_{-i}) - b\big)\, \text{Prob}[b = b_{-i}], \qquad (9.17)$$

where V is the maximum of x and x_{-i}. The first term is the expected difference between the value and the opponent's bid multiplied by the probability of winning, conditional on submitting the winning bid. The second term corresponds to the event of a tie, which is broken randomly: (the expected net benefit conditional on tying) $\times$ (the probability of a tie) $\times \tfrac{1}{2}$. Thus the logit equilibrium choice probabilities of i's bids are given by

$$\sigma_i(b|x) = \frac{e^{\lambda_i U(b|x)}}{\sum_{b'} e^{\lambda_i U(b'|x)}}. \qquad (9.18)$$

Because the game has so many strategies, the empirical payoff approach is used to estimate the λ_i's. That is, one uses the empirical joint distributions of signals and bids (i.e., the distributions observed in the experiments) to compute the empirical expected utility of each bid given each private signal, $\widehat{U}(b_i|x_i)$. Then estimate, for each subject, using maximum-likelihood estimation, the $\widehat{\lambda}_i$ that best fits that subject's observed joint frequency distribution of bids and signals. From $\widehat{\lambda}_i$ and $\widehat{U}$ one can back out the predicted distribution over bids (conditional on the signal) for subject i. Aggregating over all subjects, this procedure generates a predicted aggregate joint distribution of bids for each signal.[27]

COMBINING HQRE WITH BEHAVIORAL MODELS

Using the same basic approach, one can also estimate hybrid models that combine HQRE with alternative models that have been proposed to explain overbidding in common-value auction games. Camerer, Nunnari, and Palfrey (2014) analyze two such hybrid models: one that combines cursed equilibrium with HQRE (CE-HQRE) and a second that combines HQRE with a level-k model of strategic sophistication (LK-HQRE).

As explained earlier in the analysis of the compromise game, in a cursed equilibrium players fail to account for the correlation between other players' bids

---

[27] For tractability, bids are binned for the estimation. See Camerer, Nunnari, and Palfrey (2014) for details.

and signals, but have correct expectations about the marginal distribution of bids. Intuitively, cursed beliefs exacerbate the overbidding behavior that is predicted by QRE in this common-value auction game because players ignore, to the extent captured by a "cursedness" parameter χ, the connection between the bid and signal of the other player. Therefore, bidders with low signals do not fully take into account the fact that a low bid indicates the other player's signal is likely to be low, and hence they overbid to take advantage of the perceived bargain. Fully cursed expectations ($\chi = 1$) lead to a change to the formula for the expected utility of a player, compared with (9.17), which is now given by

$$U^{\chi=1}(b|x) = \big(E(V) - E(b_{-i}|b > b_{-i})\big)\,\text{Prob}[b > b_{-i}]$$
$$+ \tfrac{1}{2}\big(E(V) - b\big)\,\text{Prob}[b = b_{-i}]. \tag{9.19}$$

For the estimation, the degree to which players are cursed is a free parameter that is allowed to vary between 0 and 1. This is called a χ-*cursed equilibrium*, where $\chi \in [0, 1]$ equals the probability a player assigns to other players playing their average distribution of actions irrespective of type rather than their type-contingent strategy (to which he assigns probability $1 - \chi$). That is, the expected utility of a χ-cursed agent i from bidding b_i when observing signal x_i, is given by

$$U^{\chi}(b|x) = (1 - \chi)U(b|x) + \chi U^{\chi=1}(b|x). \tag{9.20}$$

As in the previous section, agents have idiosyncratic logit response parameters, resulting in the modified logit choice equations

$$\sigma^{\chi}(b|x) = \frac{e^{\lambda_i U^{\chi}(b|x)}}{\sum_{b'} e^{\lambda_i E U^{\chi}(b'|x)}}. \tag{9.21}$$

For estimation, the empirical joint distribution of signals and bids in phase I is plugged into (9.20) to compute the conditional (cursed) expected utilities, $\widehat{U}^{\chi}_{\text{cursed}}(b|x)$. Then, estimated values of the individual $\widehat{\lambda}_i$'s and the cursed parameter $\widehat{\chi}$ are fitted using maximum likelihood.

The hybrid model combining HQRE with a level-k model is somewhat different. The predicted behavior of a level-0 player is purely random, that is, for every signal, every bid is equally probable. Thus, level-0 players are exactly like $\lambda_i = 0$ players in the HQRE model. A level-1 player quantal responds to random play. Level-2 players quantal respond to level-1 players and so forth. To keep the model simple only the first four level types are considered, and to keep the model parsimonious, a one-parameter distribution of types (truncated Poisson with parameter τ) is used in the estimation.

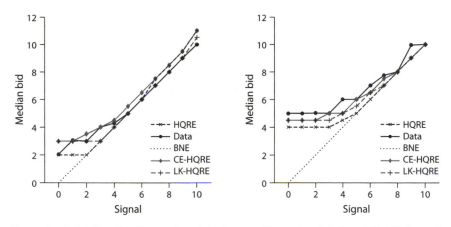

FIGURE 9.11. Median bids. Data and models. Camerer, Nunnari, and Palfrey (2014) (left panel) and Ivanov, Levin, and Niederle (2010) (right panel).

The models were estimated using the phase I data.[28] The estimated models recover the basic pattern of overbidding in the data, with significant overbidding for low signals, relative to the symmetric Nash equilibrium, with the amount of overbidding diminishing for higher signals. The right-hand panel of figure 9.11 graphs the median observed bids as a function of signal, and also graphs the medians of the fitted distribution of bids for each of the three models. This panel shows that all three models track the data from Ivanov, Levin, and Niederle (2010) in very similar ways.

Camerer, Nunnari, and Palfrey (2014) also report a similar analysis from their own replication study using Caltech students as subjects. Far less overbidding is observed by the Caltech subjects, so at first blush the data look very different from Ivanov, Levin, and Niederle (2010).[29] The observed and estimated median bidding functions for the Camerer, Nunnari, and Palfrey (2014) data are shown in the left panel of figure 9.11. In spite of the large difference in magnitude of the overbidding between the two data sets, they have a strikingly similar pattern, which is picked up by all three models. Both data sets exhibit significant overbidding, with substantially more overbidding for low signals than high signals, resulting in the same pattern of bids as a function of signals that was evident in the Ivanov, Levin, and Niederle (2010) data, but somewhat shifted

[28] Camerer, Nunnari, and Palfrey (2014) also report the results of out-of-sample fits to the phase II bidding behavior, using the phase I estimated values of λ_i, χ, and τ. Those out-of-sample estimates also yield median bids that track the data for all three models.

[29] The average (across all signals) of phase I bids in the Camerer, Nunnari, and Palfrey (2014) data is 6, but in the Ivanov, Levin, and Niederle (2010) data the average bid is 36,679.

down. The models closely track these patterns. As a result, the qualitative conclusions reached using the same methodology of estimating hybrid models of HQREs are nearly identical for both data sets. The main differences are that (1) the distribution of λ_i's in the Caltech sample is higher than the distribution in the Ohio State sample, and (2) the variance of bids conditional on signals is lower in the Caltech sample.

10

Epilogue: Some Thoughts about Future Research

The central theme of this book is that the quantal response equilibrium framework provides a rigorous and usable statistical generalization of the workhorse Nash equilibrium model that is so widely used in economics and other social sciences. The generalization involves introducing, in a natural and relatively simple mathematical way, behavioral noise that pervades human decision making. This framework is developed theoretically for both extensive- and normal-form games in initial chapters of the book. These early chapters provide a unified treatment of existence and equilibrium properties, including uniqueness (where applicable), along with comparisons to the properties of Nash equilibria. In many settings, it is possible to derive useful comparative statics results, or even closed-form solutions for the QRE distributions of decisions. Subsequent chapters considered the formation of (nonequilibrium) beliefs from a process of either introspection in novel settings or learning from observation over time in other settings. Such learning processes, combined with quantal response behavior, can converge to QRE in some limiting cases, and the convergence properties of both evolutionary and steady-state quantal learning models are developed. These various theoretical results produce comparative statics predictions that are consistent with data from laboratory experiments. In addition to qualitative comparisons, the estimation routines presented in chapter 6 can be used to estimate the effects of behavioral forces and specific parameters such as inequity aversion, cursedness, altruism, risk preferences, and so forth. Finally, chapters 7–9 present a more detailed development of theory and data analysis for specific applications of interest to pure game theorists, political scientists, and economists. In a nutshell, QRE provides a unified and general treatment of equilibrium, learning, and estimation for games of strategy.

Before offering our thoughts about interesting directions for future research, it is worth underscoring the basic philosophy behind the QRE framework. First, and perhaps foremost, is that QRE is an *equilibrium theory*. Just as the theory of general competitive equilibrium does not aim to explain the first trade and the first price observed in a market, QRE does not aim to explain single actions by individuals or the profile of actions by a single group. In a sense, this insistence on rational expectations and internal consistency of beliefs and actions makes it even more surprising that QRE can organize aggregate data reasonably from so many different laboratory data sets, where subjects face unfamiliar tasks and have only a few minutes to make decisions. The final application of chapter 9, the maximum-value auction game, illustrates this point very well. While the fit to the data is improved somewhat by injecting behavioral limitations and biases (cursed equilibrium and strategic unsophistication), the main patterns in the maximum-value auction data are captured by QR *Equilibrium*; see figure 9.11. An important insight gained from this analysis of equilibrium interactions is that the effect of including noise in models of decision making in a game can be much more pronounced than just adding some dispersion around a standard Nash prediction. In the traveler's dilemma game of chapter 7, for example, observed choice data for some parameterizations are far from the unique Nash prediction, and are clustered at the opposite end of the set of feasible decisions. QRE models can explain why data distributions conform to Nash predictions in some treatments and diverge in others, in a manner that is consistent with intuition about the asymmetric risks of adjusting decisions in one direction or another (see figures 7.1–7.3).

The second key aspect of the QRE framework is that it is stochastic, and does not rule out any action a priori. Strictly dominated actions may be low probability events in a quantal response equilibrium, but the probability is never zero. This is important for analyzing data using maximum-likelihood approaches, in order to avoid the zero-likelihood problem. We would even go a step further and argue that modeling choice in games as stochastic is also the right approach in a more fundamental sense. From the standpoint of an observer who cannot observe the many latent factors influencing a specific choice by a specific subject in a cognitive task, that subject's choice contains a stochastic element. This view is well documented in the psychology literature, and supported by neuro-scientific measurement as well.

The third aspect of the QRE framework that we would like to underscore is the idea that choice probabilities are monotone in payoffs, but this is a restriction that is *not* imposed on the original general formulation of structural QRE. In that original formulation, QRE is in fact a fully general theory in the sense that there exist error structures to generate any profile of choice probabilities in any game—even nonmonotone choice probabilities (Goeree, Holt, and Palfrey, 2005; Haile, Hortaçsu, and Kosenok, 2008). However, to obtain interesting empirical restrictions it is assumed in applications that QRE has a "regular" structure, which imposes a requirement of monotonicity. And the data nearly always exhibit choice probabilities that are monotone in expected payoffs. To our knowledge, there has been no published research to date that applies non-i.i.d. error structures that can exhibit nonmonotonicities. This might be an interesting avenue of research to explore, but such cases would seem likely to arise only in anomalous or unusual situations, much like the possibility of upward-sloping demand curves in neoclassical general equilibrium theory.

The first nine chapters of this book provide a broad survey of research in QRE theory over the first two decades since its introduction. It is natural at this point to ask where it is going in the next two decades. This chapter offers some concluding thoughts in response to that question. The discussion is organized in three sections: "Theory," "Applications," and "Methodology." Each section briefly describes a few examples of areas that offer interesting opportunities for future research in the QRE framework, or bringing QRE analysis to bear on specific problems of interest to social scientists and game theorists.

10.1 THEORY

10.1.1 Repeated Games

Chapter 5 proposed a way to extend the QRE approach to study stationary equilibria in simple repeated games with perfect monitoring. First, while that small section of the book develops the basic framework for studying stationary QRE in repeated games, this has not been pursued further to establish theoretical results for specific games. This seems like a potentially valuable direction to go. Much is known about the properties of repeated game equilibria for this simplest class of games, and the techniques developed to analyze the subgame-perfect equilibria might be adaptable to studying stationary QRE. An initial foray in this direction can be found in Nayyar (2009), where some initial results are presented for an approximate folk theorem for the logit QRE model of the prisoner's dilemma as λ approaches ∞ and the discount factor approaches 1. General results for repeated games with more strategies or more players have not been obtained, or at least have not been published at this point in time.

Second, specific repeated games, such as the battle of the sexes game, the stag hunt game, or oligopolistic competition, would also be natural candidates to investigate. Some of these games have been run in the laboratory with strikingly replicable findings: rapid convergence to alternation equilibria in the battle of the sexes game and rapid convergence to the payoff dominant equilibrium in the stag hunt game. It would be interesting to see whether QRE could help explain these phenomena.

Third, repeated game theory considers a much broader set of games than the very simple class addressed in chapter 5, with this broader set allowing for imperfect monitoring, private types, limited foresight, limited memory, renegotiation of strategies, and many other variations. It would seem useful to extend QRE theory to incorporate these more complicated games. There might even be interesting theoretical links between QRE and the limitations that are studied in the more complex environments, such as imperfect monitoring.

10.1.2 Correlated Equilibrium, Preplay Communication, and Mechanism Design

All of the versions of QRE developed to date have looked at games where there are no explicit opportunities for players to coordinate their actions. There are at least three approaches to the noncooperative study of coordination in traditional game theory: correlated strategies, preplay communication, and contracting. Extending the QRE framework to allow for correlated strategies leads to an equilibrium concept that is a noisy version of correlated equilibrium, called *correlated quantal response equilibrium* (CQRE). Palfrey and Pogorelskiy (2014) have some initial results about CQRE, including the relationship between (regular) CQRE and a version of the revelation principle. They find, for example, that the implementation of a subset of CQRE is not possible through mediated communication if players quantal respond to the mediator's instructions. That is, there is not an equivalence of CQRE across various formulations of CQRE, in contrast to the more familiar correlated equilibrium (Fudenberg and Tirole, 1991, pp. 55–58).

Allowing for correlated strategies also opens up the possibility of developing QRE versions of cooperative equilibrium concepts based on correlated strategies such as the coalition-proof equilibrium of Moreno and Wooders (1996, 1998).[1] It also provides a formal method for modeling unmediated as well as mediated preplay communication within the QRE framework.

As shown by Myerson (1986, 2008), there is also a close connection between mechanism design theory and preplay communication. Mechanism design theory traditionally assumes that players are perfect best responders, and proceeds by developing the implications of (exact) incentive compatibility to characterize the set of mechanisms that can implement different allocation rules. A QRE approach to mechanism design theory would inject stochastic choice into the strategies of the players, and it is not difficult to see that the standard revelation principle may

---

[1] Relatedly, a QRE version of coalition-proof Nash equilibrium (CPNE) could be easily developed within the existing QRE framework, which allows only for independent mixing, as CPNE allows only independent mixing by coalitions.

TABLE 10.1. Endogenous λ example.

	L	R
U	40, 2	0, 1
D	40, 1	0, 40

no longer hold.[2] For example, it might make a difference to offer a sequence of prices in a seller–buyer bargaining game, rather than a simple one-shot version, even if the Nash equilibrium outcomes of the two games are equivalent. This harks back to the example in chapter 3 that illustrates the violations of strategic invariance that can arise in extensive-form games.

10.1.3 Endogenizing the Quantal Response Parameter

A question we often hear is, where does λ come from? The cheap answer is that it is an exogenous "skill" parameter that each player is endowed with, as developed in the HQRE section of chapter 4. That is, we are agnostic and treat it as a reduced-form summary of a player's understanding of the rules, ability to predict expected payoffs from different strategies, or past experience in similar environments. An alternative approach is to think of λ as a *choice* variable for each player. This would make sense, for example, if the incentives to play accurately are different for different players.

Some initial unpublished results were developed independently by McKelvey, Palfrey, and Weber (1997) and by Rogers and Goeree (2001). In those two models, player i has an opportunity to invest in λ_i, where the cost of investment is given by a positive, increasing, and convex function $C_i(\lambda_i)$, which is subtracted from the game payoffs $U_i = \sum_j \sigma_{ij} U_{ij}$. For instance, assuming quadratic costs and using the logit version for the choice probabilities, σ_{ij}, yields a first-order condition that equates the optimal λ_i to the variance of the expected payoffs, $\lambda_i^* = \mathrm{Var}(U_i)$. This result is intuitive in that investing in precision, that is, higher λ, makes sense only if making better choices pays off.

Consider, for instance, the 2×2 game in table 10.1. In this game, Row has absolutely no incentive to "invest" in λ, because the expected payoffs for the two choices are equal and the quantal response strategy will be $(0.5, 0.5)$ for any λ. Thus, for any positive cost function, Row's optimal λ is 0. The situation is far different for the column player, who has a strong incentive to invest in λ in order to play optimally (R) with higher probability when Row is totally random. This will result in an equilibrium pair of investments $(0, \lambda^*)$, where the exact value of $\lambda^* > 0$ would depend on the cost function.

---

[2] There are some existing non-QRE models of mechanism design with error-prone behavior. See, for example, Eliaz (2002).

This defines a two-stage game, and there are two parts of the equilibrium: first the endogenous choice of λ, and second the actual play of the game. This is called an *endogenous rationality equilibrium (ERE)*.

10.2 APPLICATIONS

10.2.1 Bargaining Games: Legislative Bargaining, Offer/Counteroffer

Bargaining games, particularly those with a potentially infinite horizon (e.g., Rubinstein, 1982; Baron and Ferejohn, 1989) pose a number of complex issues for characterizing the properties of (stationary) quantal response equilibrium. The section in chapter 8 that considers MQRE in a dynamic political bargaining game illustrates some of the difficulties. On the other hand, it seems feasible to apply similar numerical techniques to better understand QRE in these models. These games seem well suited for using QRE because the subgame-perfect Nash equilibria involve making offers that responders are indifferent between accepting and rejecting. In a QRE, such offers would be rejected with probability $\frac{1}{2}$, and even offers that leave a small amount of surplus to the responders would be rejected with high probability. This could lead to substantially different properties of QRE, compared with subgame-perfect Nash equilibrium. Intuition suggests, for example, that the proposer advantage will be diminished in QRE relative to Nash equilibrium.

Virtually all experiments on Baron–Ferejohn and Rubinstein bargaining games find that proposer power is much less than standard theory predicts. Using the empirical payoff approach, it should be possible to do QRE estimation using the laboratory data from such experiments. The estimation could also incorporate parameters from social preference models, which could plausibly play an important role in these purely distributive games. Data from double auction bargaining games (e.g., Radner and Schotter, 1989) could also be usefully analyzed using the empirical payoff approach.

10.2.2 The Winner's Curse

The last application in chapter 9, the maximum-value auction game, illustrates that QRE can at least partially explain the winner's curse in that auction. The intuition is quite clear, and is related to an intuition behind the finding that winner's-curse behavior can arise in level-k models of bidding behavior (Crawford and Iriberri, 2007b). If opposing bidders bid randomly, then there is no correlation between bids and signals, which completely eliminates the winner's curse. Random bidders appear in level-k theory as level-0 players. In QRE, *all* bidders' strategies are subject to payoff-monotone randomness, not just one type of player. This randomness dampens the correlation between bids and private information, which results in (loosely speaking) something like partially cursed bidding behavior by all bidders. With HQRE or TQRE, bids can be driven up by the existence of $\lambda = 0$ bidders (equivalent to level-0 bidders) in the population, as appeared to be the case in the analysis of the maximum-value auction game data of chapter 9.

A natural question then is whether a similar specification of HQRE as in Camerer, Nunnari, and Palfrey (2014) can explain the data from more standard common-value auctions (e.g., Kagel and Levin, 1986). One question is simply whether HQRE captures the main features of overbidding in those auctions, as was the case in the maximum-value auction game. But deeper issues could also be explored. As in Camerer, Nunnari, and Palfrey, one could also include parameters that allow for either partial cursedness (χ) or a distribution of level-k types (τ) and by doing so measure the improvement of fit from the additional parameters. The technique for estimation would require the empirical payoff approach, but this would seem to be feasible.

The winner's curse also arises in common-value buyer–seller trading games, such as those analyzed in Carrillo and Palfrey (2011). The signal structure in some of the games they study is identical to the maximum-value auction game, but in a bilateral-bargaining setting rather than a second-price auction. It would be interesting to apply similar techniques to that setting, and, as in Camerer, Nunnari, and Palfrey (2014), also structurally estimate models that include parameters from level-k theory and cursed equilibrium.

10.3 METHODOLOGY

Last but not least, there are many open methodological questions about QRE that remain to be explored. A major limiting factor for the theoretical analysis of QRE is the characterization of equilibrium in games with large strategy spaces or many players. In most cases, it is not feasible to derive these properties analytically, so one needs to fall back on numerical methods to compute approximations of equilibrium.[3] This is also related to recent research on the computation of Nash equilibrium in large games, and path-following algorithms exploiting the generically smooth properties of logit QREs have been used to create fast algorithms for Nash computations. As is the case with natural sciences and

---

[3] There are a few important exceptions where theoretical properties have been established analytically, e.g., when games satisfy the local payoff property; see chapter 2.

engineering, computation is fast becoming an integral tool for social scientists. The formal, behaviorally relevant structure of QRE is ideally suited for further development on this front.

A second set of methodological questions involves the estimation methods used in the application of QRE to analysis of experimental data. The most common approach, and the one used in most of the applications here, involves a two-step procedure. In the first step the logit QRE correspondence (or some other parametric family of regular QRE) is computed numerically. Then the data is fitted to this correspondence using maximum-likelihood estimation to obtain a value of λ and any other parameters of interest. There are of course alternative measures of fit that could be used for the estimation, such as choosing the parameter values that minimize the sum of squared deviations of the data from the theoretical choice probabilities. There has not been a careful study of which method might be better, or which parametric specifications might be better or worse.[4] For complicated games, the alternative method is the empirical payoff approach which does not require the computation of the equilibrium correspondence, and hence is relatively easy to implement. Chapter 6 compares the estimates from the empirical payoff approach with the estimates from the equilibrium computation approach in an example with asymmetric matching pennies data. The estimates were very close, but it would be nice to know whether the two methods will generally produce similar estimates in medium-sized data sets.

A third set of methodological questions concern the application of QRE to field data. In experimental data sets, it is typically the case that the experimenter controls (and hence observes) the payoff information, including information that according to the model is privately known to a player. With field data, this information must be estimated simultaneously with other parameters of the QRE model, which can lead to identification issues. For example, Bajari and Hortaçsu (2005) use the empirical payoff approach to estimate field auction data, where the distribution of valuations has to be estimated, rather than directly measured as in an experiment. This limits the richness of the models they can estimate. For example, there could be identification issues if they tried to simultaneously estimate the logit equilibrium parameter, a risk-aversion parameter, and the distribution of valuations. Thus, there are unique challenges to applying QRE estimation to field data, but some of these problems can likely be overcome through clever estimation strategies. Applying other game-theoretic structural

---

[4] In large samples it shouldn't make much difference, but experimental data sets are typically rather small. There has been some discussion about the use of maximum likelihood rather than mean squared deviation (MSD). See, e.g., Morgan and Sefton (2002), who present some arguments in favor of maximum likelihood.

estimation methods to analyze field data in industrial organization applications has proved to be very successful and one hopes that QRE estimation could usefully be added to the applied econometrician's toolbox.

10.4 CLOSING REMARKS

These various applications and extensions are just a few of the possibilities that come to mind. Indeed, it has been difficult for the authors to bring this project to a close, as the process of collecting and integrating results has stimulated major extensions and exciting insights, only a sample of which is incorporated into the preceding chapters. The expanding experimental literature in economics, political science, and other social sciences will continue to provide new patterns and paradoxes that guide and discipline further theoretical developments. QRE theories are well situated in the sense that they generalize and extend the elegant structures of classical game theory, but in a behaviorally relevant manner that is motivated by laboratory data and that provides a general theoretical foundation for estimation. The QRE framework helps us to think outside the box in terms of devising new experiments, and can provide novel intuitions about strategic interactions for real-world applications. At the same time, as a statistical theory of games, it allows us to look inside the box in an internally consistent and rigorous way, to better understand the underlying processes and forces that drive strategic behavior in interactive games.

References

Ali, N., J. K. Goeree, N. Kartik, and T. R. Palfrey (2008) "Information Aggregation in Standing and Ad Hoc Committees," *American Economic Review*, **98**, 181–186.

Anderson, L. R. (1994) *Information Cascades*, unpublished doctoral dissertation, University of Virginia.

Anderson, L. R. (2001) "Payoff Effects in Information Cascade Experiments," *Economic Inquiry*, **39**, 609–615.

Anderson, L. R., and C. A. Holt (1997) "Information Cascades in the Laboratory," *American Economic Review*, **87**, 847–862.

Anderson, S. P., J. K. Goeree, and C. A. Holt (1998a) "Rent Seeking with Bounded Rationality: An Analysis of the All-Pay Auction," *Journal of Political Economy*, **106**, 828–853.

Anderson, S. P., J. K. Goeree, and C. A. Holt (1998b) "A Theoretical Analysis of Altruism and Decision Error in Public Goods Games," *Journal of Public Economics*, **70**, 297–323.

Anderson, S. P., J. K. Goeree, and C. A. Holt (2001) "Minimum-Effort Coordination Games: Stochastic Potential and Logit Equilibrium," *Games and Economic Behavior*, **34**, 177–199.

Anderson, S. P., J. K. Goeree, and C. A. Holt (2002) "The Logit Equilibrium: A Perspective on Intuitive Behavioral Anomalies," *Southern Economic Journal*, **69**, 21–47.

Anderson, S. P., J. K. Goeree, and C. A. Holt (2004) "Noisy Directional Learning and the Logit Equilibrium," *Scandinavian Journal of Economics*, **106**, 81–602.

Angrisani, M., A. Guarino, S. Huck, and N. C. Larson (2011) "No-Trade in the Laboratory," *B.E. Journal of Theoretical Economics (Advances)*, **11**, 1935–1704.

Arad, A., and A. Rubinstein (2012) "The 11–20 Money Request Game: A Level-k Reasoning Study," *American Economic Review*, **102**, 3561–3573.

Aumann, R. (1995) "Backward Induction and Common Knowledge of Rationality," *Games and Economic Behavior*, **8**, 6–19.

Austen-Smith, D., and J. S. Banks (1996) "Information Aggregation, Rationality, and the Condorcet Jury Theorem," *American Political Science Review*, **90**, 34–45.

Bajari, P., and A. Hortaçsu (2005) "Are Structural Estimates of Auction Models Reasonable? Evidence from Experimental Data," *Journal of Political Economy*, **113**, 703–741.

Banks, J., C. F. Camerer, and D. Porter (1994) "Experimental Tests of Nash Refinements in Signaling Games," *Games and Economic Behavior*, **6**, 1–31.

Baron, D., and J. Ferejohn (1989) "Bargaining in Legislatures," *American Political Science Review*, **83**, 1181–1206.

Bas, M., C. Signorino, and R. Walker (2008) "Statistical Backwards Induction: A Simple Method for Estimating Recursive Strategic Models," *Political Analysis*, **16**, 21–40.

Basu, K. (1994) "The Traveler's Dilemma: Paradoxes of Rationality in Game Theory," *American Economic Review*, **84**, 391–395.

Battaglini, M., and T. R. Palfrey (2012) "The Dynamics of Distributive Politics," *Economic Theory*, **49**, 739–777.

Battaglini, M., R. Morton, and T. R. Palfrey (2008) "Information Aggregation and Strategic Abstention in Large Laboratory Elections," *American Economic Review*, **98**, 194–200.

Battaglini, M., R. Morton, and T. R. Palfrey (2010) "The Swing Voter's Curse in the Laboratory," *Review of Economic Studies*, **77**, 61–89.

Baye, M. R., D. Kovenock, and C. G. de Vries (1996) "The All-Pay Auction with Complete Information," *Economic Theory*, **8**, 291–305.

Baye, M. R., and J. Morgan (2004) "Price Dispersion in the Lab and on the Internet: Theory and Evidence," *RAND Journal of Economics*, **35**, 449–465.

Beja, A. (1992) "Imperfect Equilibrium," *Games and Economic Behavior*, **4**, 18–36.

Bernheim, D. (1984) "Rationalizable Strategic Behavior," *Econometrica*, **52**, 1007–1028.

Bernheim, D. and D. Ray (1989) "Collective Dynamic Consistency in Repeated Games," *Games and Economic Behavior*, **1**, 295–326.

Bikhchandani, S., D. Hirshleifer, and I. Welch (1992) "A Theory of Fads, Fashion, Custom, and Cultural Change as Informational Cascades," *Journal of Political Economy*, **100**, 992–1026.

Binmore, K. (1987) "Modelling Rational Players (Part I)," *Economics and Philosophy*, **3**, 179–214.

Block, H., and J. Marschak (1960) "Random Orderings and Stochastic Theories of Responses," in *Contributions to Probability and Statistics*, Olkin, Ghurye, Hoeffding, Madow, and Mann (eds.), Stanford University Press, 97–132.

Blume, L. E. (1993) "The Statistical Mechanics of Strategic Interaction," *Games and Economic Behavior*, **5**, 387–424.

Blume, L. E. (1995) "The Statistical Mechanics of Best-Response Strategy Revision," *Games and Economic Behavior*, **11**, 111–145.

Brandts, J., and C. A. Holt (1992) "An Experimental Test of Equilibrium Dominance in Signaling Games," *American Economic Review*, **82**, 1350–1365.

Brandts, J., and C. A. Holt (1993) "Adjustment Patterns and Equilibrium Selection in Experimental Signaling Games," *International Journal of Game Theory*, **22**, 279–302.

Breitmoser, Y., J. H. W. Tan, and D. J. Zizzo (2010) "Understanding Perpetual R&D Races," *Economic Theory*, **44**, 445–467.

Brown, A., C. F. Camerer, and D. Lovallo (2007) "To Review or Not to Review? Limited Strategic Thinking at the Box Office," *Working Paper*, California Institute of Technology.

Brown, J., and R. Rosenthal (1990) "Testing the Minimax Hypothesis: A Re-examination of O'Neill's Game Experiment," *Econometrica*, **58**, 1065–1081.

Camerer, C. F. (2003) *Behavioral Game Theory*, Princeton, NJ: Princeton University Press.

Camerer, C. F., T. Ho, and J. Chong (2004) "A Cognitive Hierarchy Model of Behavior in Games," *Quarterly Journal of Economics*, **119**, 861–898.

Camerer, C. F., and D. Lovallo (1999) "Overconfidence and Excess Entry: An Experimental Approach," *American Economic Review*, **89**, 306–318.

Camerer, C. F., S. Nunnari, and T. R. Palfrey (2014) "Quantal Response and Nonequilibrium Beliefs Explain Overbidding in Maximum-Value Auctions," *Social Science Working Paper No. 1349*, California Institute of Technology.

Campo, S., E. Guerre, I. Perrigne, and Q. Vuong (2011) "Semiparametric Estimation of First-Price Auctions with Risk Averse Bidders," *Review of Economic Studies*, **78**, 112–147.

Capra, C. M., J. K. Goeree, R. Gomez, and C. A. Holt (1999) "Anomalous Behavior in a Traveler's Dilemma," *American Economic Review*, **89**, 678–690.

Capra, C. M., J. K. Goeree, R. Gomez, and C. A. Holt (2002) "Learning and Noisy Equilibrium Behavior in an Experimental Study of Imperfect Price Competition," *International Economic Review*, **43**, 613–636.

Carrillo, J., and T. R. Palfrey (2009) "The Compromise Game: Two-sided Adverse Selection in the Laboratory," *American Economic Journal: Microeconomics*, **1**, 151–181.

Carrillo, J., and T. R. Palfrey (2011) "No Trade," *Games and Economic Behavior*, **71**, 66–87.

Cason, T. N., and V. L. Mui (2005) "Uncertainty and Resistance to Reform in Laboratory Participation Games," *European Journal of Political Economy*, **21**, 708–737.

Chen, H., J. Friedman, and J. Thisse (1997) "Boundedly Rational Nash Equilibrium: A Probabilistic Choice Approach," *Games and Economic Behavior*, **18**, 32–54.

Cheung, Y., and D. Friedman (1997) "Individual Learning in Normal Form Games: Some Laboratory Results," *Games and Economic Behavior*, **19**, 46–76.

Condorcet, Marie Jean Antoine Nicolas de Caritat marquis de (1785) *Essai sur l'application de l'analyse à la probabilité des decisions rendues à la pluralité des voix*, Paris, France: Imprimerie Royale.

Costa-Gomes, M., and V. P. Crawford (2006) "Cognition and Behavior in Two-Person Guessing Games: An Experimental Study, *American Economics Review*, **96**, 1737–1768.

Coughlan, P., R. D. McKelvey, and T. R. Palfrey (1999) "Experiments on Two-Person Games with Incredible Threats," *Manuscript*, California Institute of Technology.

Crawford, V. P., and N. Iriberri (2007a) "Fatal Attraction: Salience, Naïveté, and Sophistication in Experimental 'Hide-and-Seek' Games," *American Economics Review*, **97**, 1731–1750.

Crawford, V. P., and N. Iriberri (2007b) "Level-k Auctions: Can a Non-Equilibrium Model of Strategic Thinking Explain the Winner's Curse and Overbidding in Private-Value Auctions?" *Econometrica*, **75**, 1721–1770.

Davis, D. D., and R. J. Reilly (1998) "Do Too Many Cooks Spoil the Stew? An Experimental Analysis of Rent-Seeking and a Strategic Buyer," *Public Choice*, **95**, 89–115.

Diekmann, A. (1986) "Volunteer's Dilemma: A Social Trap Without a Dominant Strategy and Some Empirical Results," in *Procedural Effects of Social Behavior: Essays in Honor of Anatol Rapoport*, A. Diekmann and P. Mitter (eds.), Heidelberg: Physica.

Dufwenberg, M., and U. Gneezy (2000) "Price Competition and Market Concentration: An Experimental Study," *International Journal of Industrial Organization*, **18**, 7–22.

Dufwenberg, M., U. Gneezy, J. K. Goeree, and R. Nagel (2007) "Price Floors and Competition," *Economic Theory*, **33**, 211–224.

Dvoretsky, A., A. Wald, and J. Wolfowitz (1951) "Elimination of Randomization in Certain Statistical Decision Procedures and Zero-Sum Two-Person Games," *Annals of Mathematical Statistics*, **22**, 1–21.

El-Gamal, M., R. D. McKelvey, and T. R. Palfrey (1993a) "A Bayesian Sequential Experimental Study of Learning in Games," *Journal of the American Statistical Association*, **88**, 428–435.

El-Gamal, M., R. D. McKelvey, and T. R. Palfrey (1993b) "Computational Issues in the Statistical Design and Analysis of Experimental Games," *International Journal of Supercomputer Applications*, **7**, 189–200.

El-Gamal, M., R. D. McKelvey, and T. R. Palfrey (1994) "Learning in Experimental Games," *Economic Theory*, **4**, 901–922.

El-Gamal, M., and T. R. Palfrey (1995) "Vertigo: Comparing Structural Models of Imperfect Behavior in Experimental Games," *Games and Economic Behavior*, **8**, 322–348.

El-Gamal, M., and T. R. Palfrey (1996) "Economical Experiments: Bayesian Efficient Experimental Design," *International Journal of Game Theory*, **25**, 495–517.

Eliaz, K. (2002) "Fault Tolerant Implementation," *Review of Economic Studies*, **69**, 589–610.

Erev, I., and A. E. Roth (1998) "Predicting How People Play Games: Reinforcement Learning in Experimental Games with Unique, Mixed Strategy Equilibria," *American Economic Review*, **88**, 848–881.

Eyster, E., and M. Rabin (2005) "Cursed Equilibrium," *Econometrica*, **73**, 1623–1672.

Farrell, J., and E. Maskin (1989) "Renegotiation in Repeated Games," *Games and Economic Behavior*, **1**, 361–369.

Fearon, J. (1995) "Rationalist Explanations for War," *International Organization*, **49**, 379–414.

Feddersen, T., and W. Pesendorfer (1996) "The Swing Voter's Curse," *American Economic Review*, **86**, 408–424.

Feddersen, T., and W. Pesendorfer (1998) "Convicting the Innocent: The Inferiority of Unanimous Jury Verdicts under Strategic Voting," *American Political Science Review*, **92**, 12–35.

Feddersen, T., and W. Pesendorfer (1999) "Elections, Information Aggregation, and Strategic Voting," *Proceedings of the National Academy of Sciences of the USA*, **96**, 10572–10574.

Fehr, E., and K. M. Schmidt (1999) "A Theory of Fairness, Competition, and Cooperation," *Quarterly Journal of Economics*, **114**, 817–868.

Fey, M., R. D. McKelvey, and T. R. Palfrey (1996) "Experiments on the Constant Sum Centipede Game," *International Journal of Game Theory*, **25**, 269–287.

Foster, D., and P. Young (1990) "Stochastic Evolutionary Game Dynamics," *Theoretical Population Biology*, **38**, 219–232.

Franzen, A. (1995) "Group Size and One Shot Collective Action," *Rationality and Society*, **7**, 183–200.

Friedman, D. (1992) "Theory and Misbehavior: A Comment," *American Economic Review*, **82**, 1374–1378.

Friedman, D., and D. Ostrov (2010) "Gradient Dynamics in Population Games: Some Basic Results," *Journal of Mathematical Economics*, **46**, 691–707.

Friedman, D., and D. Ostrov (2013) "Evolutionary Dynamics over Continuous Action Spaces for Population Games That Arise from Symmetric Two-Player Games," *Journal of Economic Theory*, **148**, 743–777.

Friedman, J., and C. Mezzetti (2005) "Random Belief Equilibrium in Normal Form Games," *Games and Economic Behavior*, **51**, 296–323.

Fudenberg D., and D. Kreps (1993) "Learning Mixed Equilibria," *Games and Economic Behavior*, **5**, 320–367.

Fudenberg D., and J. Tirole (1991) "Game Theory," *Games and Economic Behavior*, Cambridge, MA: MIT Press.

Gale, J., K. Binmore, and L. Samuelson (1995) "Learning to Be Imperfect: The Ultimatum Game," *Games and Economic Behavior*, **8**, 56–90.

Geanakoplos, J. (1994) "Common Knowledge," in *Handbook of Game Theory, Volume 2*, R. Aumann and S. Hart (eds.), Amsterdam: North Holland, 1437–1496.

Goeree, J. K., S. P. Anderson, and C. A. Holt (1998) "The War of Attrition with Noisy Players," in *Advances in Applied Microeconomics, 7*, M. Baye (ed.), Greenwich, CT: Jai Press, 15–29.

Goeree, J. K., and C. A. Holt (1999) "Stochastic Game Theory: For Playing Games, Not Just for Doing Theory," *Proceedings of the National Academy of Sciences of the USA*, **96**, 10564–10567.

Goeree, J. K., and C. A. Holt (2000) "Asymmetric Inequality Aversion and Noisy Behavior in Alternating-Offer Bargaining Games," *European Economic Review*, **44**, 1057–1068.

Goeree, J. K., and C. A. Holt (2001) "Ten Little Treasures of Game Theory and Ten Intuitive Contradictions," *American Economic Review*, **91**, 1402–1422.

Goeree, J. K., and C. A. Holt (2003a) "Coordination Games" in *Encyclopedia of Cognitive Science, Vol. 2*, L. Nadel (ed.), London: Nature Publishing Group, McMillan, 204–208.

Goeree, J. K., and C. A. Holt (2003b) "Learning in Economics Experiments," in *Encyclopedia of Cognitive Science, Vol. 2*, L. Nadel (ed.), London: Nature Publishing Group, McMillan, 1060–1069.

Goeree, J. K., and C. A. Holt (2004) "A Model of Noisy Introspection," *Games and Economic Behavior*, **46**, 365–382.

Goeree, J. K., and C. A. Holt (2005a) "An Experimental Study of Costly Coordination," *Games and Economic Behavior*, **51**, 349–364.

Goeree, J. K., and C. A. Holt (2005b) "An Explanation of Anomalous Behavior in Models of Political Participation," *American Political Science Review*, **99**, 201–213.

Goeree, J. K., C. A. Holt, and S. K. Laury (2002) "Altruism and Noisy Behavior in One-Shot Public Goods Experiments," *Journal of Public Economics*, **83**, 257–278.

Goeree, J. K., C. A. Holt, and T. R. Palfrey (2002) "Quantal Response Equilibrium and Overbidding in First-Price Auctions," *Journal of Economic Theory*, **104**, 247–272.

Goeree, J. K., C. A. Holt, and T. R. Palfrey (2003) "Risk Averse Behavior in Generalized Matching Pennies Games," *Games and Economic Behavior*, **45**, 97–113.

Goeree, J. K., C. A. Holt, and T. R. Palfrey (2005) "Regular Quantal Response Equilibrium," *Experimental Economics*, **8**, 347–367.

Goeree, J. K., C. A. Holt, and T. R. Palfrey (2008) "Quantal Response Equilibrium," in *The New Palgrave: A Dictionary of Economics*, second edition, 783–787.

Goeree, J. K., C. A. Holt, and A. M. Smith (2015) "An Experimental Examination of the Volunteer's Dilemma," *Working Paper*, Department of Economics, University of Technology Sydney.

Goeree, J. K., P. Louis, and J. Zhang (2015) "Noisy Introspection in the 11–20 Game," *Working Paper*, University of Technology Sydney.

Goeree, J. K., T. R. Palfrey, and B. Rogers (2006) "Social Learning with Private and Common Values," *Economic Theory*, **28**, 254–264.

Goeree, J. K., T. R. Palfrey, B. Rogers, and R. McKelvey (2007) "Self-Correcting Information Cascades," *Review of Economic Studies*, **74**, 733–762.

Guarnaschelli, S., R. D. McKelvey, and T. R. Palfrey (2000) "An Experimental Study of Jury Decision Rules," *American Political Science Review*, **94**, 407–423.

Güth, W., R. Schmittberger, and B. Schwarze (1982) "An Experimental Study of Ultimatum Bargaining," *Journal of Economic Behavior and Organization*, **3**, 376–388.

Guyer, M., and A. Rapoport (1972) "2 × 2 Games Played Once," *Journal of Conflict Resolution*, **16**, 409–431.

Haile, P. A., A. Hortaçsu, and G. Kosenok (2008) "On the Empirical Content of Quantal Response Equilibrium," *American Economic Review*, **98**, 180–200.

Harrison, G. (1989) "Theory and Misbehavior in First Price Auctions," *American Economic Review*, **79**, 749–762.

Harsanyi, J. C. (1967) "Games with Incomplete Information Played by 'Bayesian' Players," *Management Science*, **14**, 159–182.

Harsanyi, J. C. (1973) "Games with Randomly Distributed Payoffs: A New Rationale for Mixed-Strategy Equilibrium Points," *International Journal of Game Theory*, **2**, 1–23.

Harsanyi, J. C. (1975) "The Tracing Procedure: A Bayesian Approach to Defining a Solution for *n*-Person Noncooperative Games," *International Journal of Game Theory*, **4**, 61–94.

Harsanyi, J. C., and R. Selten (1988) *A General Theory of Equilibrium Selection in Games*, Cambridge: MIT Press.

Haruvy, E., Y. Lahav, and C. N. Noussair (2007) "Traders' Expectations in Asset Markets: Experimental Evidence," *American Economic Review*, **97**, 1901–1920.

Herrera, H., M. Morelli, and T. R. Palfrey (2014) "Turnout and Power Sharing," *Economic Journal*, **124**, 131–162.

Holt, C. A. (2006) *Markets, Games, and Strategic Behavior*, Boston: Addison-Wesley.

Holt, C. A., and S. Laury (2002) "Risk Aversion and Incentive Effects," *American Economic Review*, **92**, 1644–1655.

Holt, C. A., M. Porzio, and Y. Song (2015) "Price Bubbles, Expectations, and Gender in Asset Markets: An Experiment," University of Virginia working paper, June 2015.

Holt, C. A., and D. Scheffman (1987) "Facilitating Practices: The Effects of Advance Notice and Best-Price Policies," *Rand Journal of Economics*, **18**, 187–197.

Holt, C. A., and R. Sherman (1982) "Waiting-Line Auctions," *Journal of Political Economy*, **88**, 433–445.

Hofbauer, J., and W. H. Sandholm (2002) "On the Global Convergence of Stochastic Fictitious Play," *Econometrica*, **70**, 2265–2294.

Hörner, J. (2004) "A Perpetual Race to Stay Ahead," *Review of Economics Studies*, **71**, 1065–1088.

Ivanov, A., D. Levin, and M. Niederle (2010) "Can Relaxation of Beliefs Rationalize the Winner's Curse?: An Experimental Study," *Econometrica*, **78**, 1435–1452.

Jehiel, P. (2005) "Analogy-Based Expectation Equilibrium," *Journal of Economic Theory*, **123**, 81–104.

Kagel, J., and D. Levin. (1986) "The Winner's Curse and Public Information in Common Value Auctions," *American Economic Review*, **76**, 894–920.

Kahneman, D. (1988) "Experimental Economics: A Psychological Perspective," in *Bounded Rational Behavior in Experimental Games and Markets*, R. Tietz, W. Albers, and R. Selten (eds.), New York: Springer, 11–18.

Kahneman, D., and A. Tversky (1973) "On the Psychology of Prediction," *Psychological Review*, **80**, 237–251.

Kalandrakis, T. (2004) "A Three Player Dynamic Majoritarian Bargaining Game," *Journal of Economic Theory*, **16**, 294–322.

Kandori, M. (1997) "Evolutionary Game Theory in Economics," in *Advances in Economics and Econometrics: Theory and Applications, Seventh World Congress, Vol. 1*, D. Kreps and K. Wallis (eds.), Cambridge, UK: Cambridge University Press, 243–277.

Karlin, S. (1966) *A First Course in Stochastic Processes*, **80**, 237–251.

Kartal, M. (2015a) "Laboratory Elections with Endogenous Turnout: Proportional Representation versus Majoritarian Rule," *Experimental Economics*, **18**, 366–384.

Kartal, M. (2015b) "A Comparative Welfare Analysis of Electoral Systems with Endogenous Turnout," *Economic Journal*, **125**, 1369–1392.

Kohlberg, E., and J.-F. Mertens (1986) "On the Strategic Stability of Equilibria," *Econometrica*, **54**, 1003–1037.

Kreps, D. (1990) *A Course in Microeconomic Theory*, Princeton University Press: Princeton.

Kreps, D., P. Milgrom, J. Roberts, and R. Wilson (1982) "Rational Cooperation in the Finitely Repeated Prisoner's Dilemma," *Journal of Economic Theory*, **27**, 245–252.

Kreps, D., and R. Wilson (1982) "Sequential Equilibria," *Econometrica*, **50**, 863–894.

Kübler, D., and G. Weizsäcker (2004) "Limited Depth of Reasoning and Failure of Cascade Formation in the Laboratory," *Review of Economic Studies*, **71**, 425–441.

Ledyard, J. (1984) "The Pure Theory of Large Two Candidate Elections," *Public Choice*, **44**, 7–41.

Levine, D., and T. R. Palfrey (2007) "The Paradox of Voter Participation: A Laboratory Study," *American Political Science Review*, **101**, 143–158.

Lieberman, B. (1960) "Human Behavior in a Strictly Determined 3×3 Matrix Game," *Behavioral Science*, **5**, 317–322.

Lopez-Acevedo, G. (1997) "Quantal Response Equilibria for Posted Offer Markets," *Estudios Económicos*, **12**, 95–131.

Luce, R. D. (1959) *Individual Choice Behavior*, New York: Wiley.

Luce, R. D., and P. Suppes (1965) "Preference, Utility, and Subjective Probability," in *Handbook of Mathematical Psychology, Vol. III*, R. D. Luce, R. R. Bush, and E. Galanter (eds.), New York: Wiley, 252–410.

Ma, C. A., and M. Manove (1993) "Bargaining with Deadlines and Imperfect Player Control," *Econometrica*, **61**, 1313–1339.

Maskin, E., and J. Tirole (2001) "Markov Perfect Equilibrium," *Journal of Economic Theory*, **100**, 191–219.

McFadden, D. (1976) "Quantal Choice Analysis: A Survey," *Annals of Economic and Social Measurement*, **5**, 363–390.

McKelvey, R. D., A. M. McLennan, and T. L. Turocy (2014) "Gambit: Software Tools for Game Theory," version 14.1.0. http://www.gambit-project.org.

McKelvey, R. D., and T. R. Palfrey (1992) "An Experimental Study of the Centipede Game," *Econometrica*, **6**, 803–836.

McKelvey, R. D., and T. R. Palfrey (1994) "Quantal Response Equilibrium for Normal Form Games," *Social Science Working Paper #883*, March 1994, California Institute of Technology.

McKelvey, R. D., and T. R. Palfrey (1995) "Quantal Response Equilibria for Normal Form Games," *Games and Economic Behavior*, **10**, 6–38.

McKelvey, R. D., and T. R. Palfrey (1996) "A Statistical Theory of Equilibrium in Games," *Japanese Economic Review*, **47**, 186–209.

McKelvey, R. D., and T. R. Palfrey (1998) "Quantal Response Equilibria for Extensive Form Games," *Experimental Economics*, **1**, 9–41.

McKelvey, R. D., T. R. Palfrey, and R. A. Weber (1997) "Endogenous Rationality Equilibrium," *Manuscript*, California Institute of Technology.

McKelvey, R. D., T. R. Palfrey, and R. A. Weber (2000) "The Effects of Payoff Magnitude and Heterogeneity on Behavior in 2×2 Games with Unique Mixed Strategy Equilibria," *Journal of Economic Behavior and Organization*, **42**, 523–548.

McKelvey, R. D., and J. W. Patty (2006) "A Theory of Voting in Large Elections," *Games and Economic Behavior*, **57**, 155–180.

Meyer, D. J., J. B. Van Huyck, R. C. Battalio, and T. R. Saving (1992) "History's Role in Coordinating Decentralized Allocation Decisions: Laboratory Evidence on Repeated Binary Allocation Games," *Journal of Political Economy*, **100**, 292–316.

Monderer, D., and L. S. Shapley (1996) "Potential Games," *Games and Economic Behavior*, **14**, 124–143.

Moreno, D., and J. Wooders (1996) "Coalition-Proof Equilibrium," *Games and Economic Behavior*, **17**, 80–112.

Moreno, D., and J. Wooders (1998) "An Experimental Study of Communication and Coordination in Noncooperative Games," *Games and Economic Behavior*, **24**, 47–76.

Morgan, J., and M. Sefton (2002) "An Experimental Investigation of Unprofitable Games," *Games and Economic Behavior*, **40**, 123–146.

Myerson, R. B. (1986) "Multistage Games with Communication," *Econometrica*, **54**, 323–358.

Myerson, R. B. (2008) "Perspectives on Mechanism Design in Economic Theory," *American Economic Review*, **98**(3), 586–603.

Nagel R. (1995) "Unraveling in Guessing Games: An Experimental Study," *American Economic Review*, **85**, 1313–1326.

Nayyar, S. (2009) *Essays on Repeated Games*, PhD Dissertation, Princeton University.

Ochs, J. (1990) "The Coordination Problem in Decentralized Markets: An Experiment," *Quarterly Journal of Economics*, **105**, 545–559.

Ochs, J. (1995) "Games with Unique Mixed Strategy Equilibria: An Experimental Study," *Games and Economic Behavior*, **10**, 202–217.

O'Neill, B. (1987) "Nonmetric Test of the Minimax Theory of Two-Person Zerosum Games," *Proceedings of the National Academy of Sciences of the USA*, **84**, 2106–2109.

Östling, R., J. T. Wang, E. Chou, and C. F. Camerer (2011) "Testing Game Theory in the Field: Swedish LUPI Lottery Games," *American Economic Journal: Microeconomics*, **3**(3), 1–33.

Palfrey, T. R. (2007) "McKelvey and Quantal Response Equilibrium," in *A Positive Change in Political Science: The Legacy of Richard D. McKelvey's Most Influential Writings*, J. Aldrich, J. Alt, and A. Lupia (eds.), University of Michigan Press: Ann Arbor, 425–440.

Palfrey, T. R., and K. Pogorelskiy (2014) "Correlated Quantal Response Equilibrium," *Working Paper*, California Institute of Technology, Pasadena, CA.

Palfrey, T. R., and K. Poole (1987) "The Relationship between Information, Ideology, and Voting Behavior," *American Journal of Political Science*, **31**, 511–530.

Palfrey, T. R., and H. Rosenthal (1983) "A Strategic Calculus of Voting," *Public Choice*, **41**, 7–53.

Palfrey, T. R., and H. Rosenthal (1985) "Voter Participation and Strategic Uncertainty," *American Political Science Review*, **79**, 62–78.

Pearce, D. G. (1984) "Rationalizable Strategic Behavior and the Problem of Perfection," *Econometrica*, **52**, 1029–1050.

Popova, U. (2006) "Equilibrium Analysis of Signaling with Asymmetric Information in Poker Game," *Mimeo*, University of Virginia.

Radner, R., and A. Schotter (1989) "The Sealed-Bid Mechanism: An Experimental Study," *Journal of Economic Theory*, **48**, 179–220.

Rhode, P., and M. Stegeman (2001) "Non-Nash Equilibria of Darwinian Dynamics with Applications to Duopoly," *International Journal of Industrial Organization*, **19**, 415–453.

Rogers, B., and J. K. Goeree (2001) "Games That Make You Think," *Manuscript*, University of Virginia.

Rogers, B., T. R. Palfrey, and C. F. Camerer (2009), "Heterogeneous Quantal Response Equilibrium," *Journal of Economic Theory*, **144**, 1440–1467.

Rosenthal, R. (1981) "Games of Perfect Information, Predatory Pricing, and the Chain Store Paradox," *Journal of Economic Theory*, **25**, 92–100.

Rosenthal, R. (1989) "A Bounded-Rationality Approach to the Study of Noncooperative Games," *International Journal of Game Theory*, **18**, 273–292.

Roth, A. E., and I. Erev (1995) "Learning in Extensive-Form Games: Experimental Data and Simple Dynamic Models in the Intermediate Term," *Games and Economic Behavior*, **8**, 164–212.

Rubinstein, A. (1982) "Perfect Equilibrium in a Bargaining Model," *Econometrica*, **50**, 97–110.

Santos-Pinto, L., and J. Sobel (2005) "A Model of Positive Self-Image in Subjective Assessments," *American Economics Review*, **95**, 1386–1402.

Schmidt, D. (1992) "Reputation Building by Error-Prone Agents," *Mimeo*, California Institute of Technology.

Schotter, A., K. Weigelt, and C. Wilson (1994) "A Laboratory Investigation of Multiperson Rationality and Presentation Effects," *Games and Economic Behavior*, **6**, 445–468.

Schram, A., and J. Sonnemans (1996) "Why People Vote: Experimental Evidence," *Journal of Economic Psychology*, **17**, 417–442.

Sieberg, K., D. Clark, C. A. Holt, T. Nordstrom, and W. Reed (2013) "Experimental Analysis of Asymmetric Power in Conflict Bargaining," *Games and Economic Behavior*, **4**, 375–397.

Signorino, C. (1999) "Strategic Interaction and the Statistical Analysis of International Conflict," *American Political Science Review*, **93**, 279–298.

Stahl, D. O., and P. R. Wilson (1994) "Experimental Evidence on Players' Models of Other Players," *Journal of Economic Behavior and Organization*, **25**, 309–327.

Stahl, D. O., and P. R. Wilson (1995) "On Players' Models of Other Players: Theory and Experimental Evidence," *Games and Economic Behavior*, **10**, 218–254.

Sundali, J. A., A. Rapoport, and D. A. Seale (1995) "Coordination in Market Entry Games with Symmetric Players," *Organizational Behavior and Human Decision Processes*, **64**, 203–218.

Thaler, R. (1988) "Anomalies: The Ultimatum Game," *Journal of Economic Perspectives*, **2**, 195–206.

Thurstone, L. L. (1927) "A Law of Comparative Judgement," *Psychological Review*, **34**, 278–286.

Tullock, G. (1967) "The Welfare Costs of Tariffs, Monopolies, and Thefts," *Western Economic Journal*, **5**, 224–232.

Tullock, G. (1980) "Efficient Rent Seeking," in *Toward a Theory of the Rent Seeking Society*, J. Buchanan, R. Tollison, and G. Tullock (eds.), College Station: Texas A&M Press.

Tulman, S. (2015) "Altruism, Noise, and the Paradox of Voter Turnout: An Experimental Study," *Journal of Applied Mathematics*, **2015**(2), 1–22.

Turocy, T. L. (2005) "A Dynamic Homotopy Interpretation of the Logistic Quantal Response Equilibrium Correspondence," *Games and Economic Behavior*, **51**, 243–263.

Turocy, T. L. (2007) "Computation in Finite Games: Using Gambit for Quantitative Analysis," in *A Positive Change in Political Science: The Legacy of Richard D. McKelvey's Most Influential Writings*, J. Aldrich, J. Alt, and A. Lupia (eds.), University of Michigan Press: Ann Arbor, 475–487.

Van Damme, E. (1987) *Stability and Perfection of Nash Equilibria*, Berlin: Springer.

Van Huyck, J. B., R. C. Battalio, and R. O. Beil (1990) "Tacit Coordination Games, Strategic Uncertainty, and Coordination Failure," *American Economic Review*, **80**, 234–248.

Vega-Redondo, F. (1997) "The Evolution of Walrasian Behavior," *Econometrica*, **65**, 375–384.

von Neumann, J., and O. Morgenstern (1944) *Theory of Games and Economic Behavior*, Princeton: Princeton University Press.

Weizsäcker, G. (2003) "Ignoring the Rationality of Other Players: Theory and Experimental Evidence," *Games and Economic Behavior*, **44**, 145–171.

Yi, K. (2005) "Quantal-Response Equilibrium Models of the Ultimatum Bargaining Game," *Games and Economic Behavior*, **51**, 324–348.

Young, P. (1998) *Individual Strategy and Social Structure: An Evolutionary Theory of Institutions*, Princeton University Press.

Zauner, K. G. (1999) "A Payoff Uncertainty Explanation of Results in Experimental Centipede Games," *Games and Economic Behavior*, **26**, 157–185.

Zhang, B. (2013) "Quantal Response Methods for Equilibrium Selection in Games," *Working Paper*, Beijing Normal University.

Zhang, B., and J. Hofbauer (2013) "Quantal Response Methods for Equilibrium Selection in Bimatrix Games," *Working Paper*, Beijing Normal University.

Index